Phytohormones and Abiotic Stress Tolerance in Plants

Nafees A. Khan • Rahat Nazar • Noushina Iqbal
Naser A. Anjum
Editors

Phytohormones and Abiotic Stress Tolerance in Plants

Editors
Nafees A. Khan
Rahat Nazar
Noushina Iqbal
Aligarh Muslim University
Department of Botany
Aligarh
India
naf9@lycos.com
khan_rahatnazar@rediffmail.com
noushina.iqbal@gmail.com

Naser A. Anjum
Centre for Environmental and
Marine Stud
Department of Chemistry
Aveiro
Portugal
anjum@ua.pt

ISBN 978-3-642-25828-2 e-ISBN 978-3-642-25829-9
DOI 10.1007/978-3-642-25829-9
Springer Heidelberg Dordrecht London New York

Library of Congress Control Number: 2012933369

Printed on acid-free paper

Springer is part of Springer Science+Business Media (www.springer.com)

Preface

Plants are exposed to rapid and various unpredicted disturbances in the environment resulting in stressful conditions. Abiotic stress is the negative impact of nonliving factors on the living organisms in a specific environment and constitutes a major limitation to agricultural production. The adverse environmental conditions that plants encounter during their life cycle disturb metabolic reactions and adversely affect growth and development at cellular and whole plant level. Under abiotic stress, plants integrate multiple external stress cues to bring about a coordinated response and establish mechanism to mitigate the stress by triggering a cascade of events leading to enhanced tolerance. Responses to stress are complicated integrated circuits involving multiple pathways and specific cellular compartments, and the interaction of additional cofactors and/or signaling molecules coordinates a specified response to a given stimulus. Stress signal is first perceived by the receptors present on the membrane of the plant cells. The signal information is then transduced downstream resulting in the activation of various stress-responsive genes. The products of these stress genes ultimately lead to stress tolerance response or plant adaptation and help the plant to survive and surpass the unfavorable conditions. Abiotic stress conditions lead to production of signaling molecule(s) that induce the synthesis of several metabolites, including phytohormones for stress tolerance. Phytohormones are chemical compounds produced in one part and exert effect in another part and influence physiological and biochemical processes. Phytohormones are critical for plant growth and development and play an important role in integrating various stress signals and controlling downstream stress responses and interact in coordination with each other for defense signal networking to fine-tune defense. The adaptive process of plants response imposed by abiotic stresses such as salt, cold, drought, and wounding is mainly controlled by the phytohormones. Stress conditions activate phytohormones signaling pathways that are thought to mediate adaptive responses at extremely low concentration. Thus, an understanding of the phytohormones homeostasis and signaling is essential for improving plant performance under optimal and stressful environments.

Traditionally five major classes of plant hormones have been recognized: auxins, cytokinins, gibberellins, abscisic acid, and ethylene. Recently, other signaling molecules that play roles in plant metabolism and abiotic stress tolerance have also been identified, including brassinosteroids, jasmonic acid, salicylic acid, and nitric oxide. Besides, more active molecules are being found and new families of regulators are emerging such as polyamines, plant peptides, and karrikins. Several biological effects of phytohormones are induced by cooperation of more than one phytohormone. Substantial progress has been made in understanding individual aspects of phytohormones perception, signal transduction, homeostasis, or influence on gene expression. However, the physiological, biochemical, and molecular mechanisms induced by phytohormones through which plants integrate adaptive responses under abiotic stress are largely unknown. This book updates the current knowledge on the role of phytohormones in the control of plant growth and development, explores the mechanism responsible for the perception and signal transduction of phytohormones, and also provides a further understanding of the complexity of signal crosstalk and controlling downstream stress responses. There is next to none any book that provides update information on the phytohormones significance in tolerance to abiotic stress in plants.

We extend our gratitude to all those who have contributed in making this book possible. Simultaneously, we would like to apologize unreservedly for any mistakes or failure to acknowledge fully.

Aligarh, India Nafees A. Khan, Rahat Nazar, Noushina Iqbal
Aveiro, Portugal Naser A. Anjum

Contents

Chapter 1
Signal Transduction of Phytohormones Under Abiotic Stresses

F. Eyidogan, M.T. Oz, M. Yucel, and H.A. Oktem

Abstract Growth and productivity of higher plants are adversely affected by various environmental stresses which are of two main types, biotic and abiotic, depending on the source of stress. Broad range of abiotic stresses includes osmotic stress caused by drought, salinity, high or low temperatures, freezing, or flooding, as well as ionic, nutrient, or metal stresses, and others caused by mechanical factors, light, or radiation. Plants contrary to animals cannot escape from these environmental constraints, and over the course of evolution, they have developed some physiological, biochemical, or molecular mechanisms to overcome effects of stress. Phytohormones such as auxin, cytokinin, abscisic acid, jasmonic acid, ethylene, salicylic acid, gibberellic acid, and few others, besides their functions during germination, growth, development, and flowering, play key roles and coordinate various signal transduction pathways in plants during responses to environmental stresses. Complex networks of gene regulation by these phytohormones under abiotic stresses involve various *cis*- or *trans*-acting elements. Some of the transcription factors regulated by phytohormones include ARF, AREB/ABF, DREB, MYC/MYB, NAC, and others. Changes in gene expression, protein synthesis, modification, or degradation initiated by or coupled to these transcription factors and their corresponding *cis*-acting elements are briefly summarized in this work. Moreover, crosstalk between signal transduction pathways involving phytohormones is explained in regard to transcriptional or translational regulation under abiotic stresses.

F. Eyidogan (✉)
Baskent University, Ankara, Turkey
e-mail: fusunie@baskent.edu.tr

M.T. Oz • M. Yucel • H.A. Oktem
Department of Biological Sciences, Middle East Technical University, Ankara, Turkey

N.A. Khan et al. (eds.), *Phytohormones and Abiotic Stress Tolerance in Plants*,
DOI 10.1007/978-3-642-25829-9_1,

1.1 Introduction

Plants have successfully evolved to integrate diverse environmental cues into their developmental programs. Since they cannot escape from adverse constraints, they have been forced to counteract by eliciting various physiological, biochemical, and molecular responses. These responses include or lead to changes in gene expression, regulation of protein amount or activity, alteration of cellular metabolite levels, and changes in homeostasis of ions. Gene regulation at the level of transcription is one of the major control points in biological processes, and transcription factors and regulators play key roles in this process. Phytohormones are a collection of trace amount growth regulators, comprising auxin, cytokinin, gibberellic acid (GA), abscisic acid (ABA), jasmonic acid (JA), ethylene, salicylic acid (SA), and few others (Tuteja and Sopory 2008). Hormone responses are fundamental to the development and plastic growth of plants. Besides their regulatory functions during development, they play key roles and coordinate various signal transduction pathways during responses to environmental stresses (Wolters and Jürgens 2009).

A range of stress signaling pathways have been elucidated through molecular genetic studies. Research on mutants, particularly of *Arabidopsis*, with defects in these and other processes have contributed substantially to the current understanding of hormone perception and signal transduction. Plant hormones, such as ABA, JA, ethylene, and SA, mediate various abiotic and biotic stress responses. Although auxins, GAs, and cytokinins have been implicated primarily in developmental processes in plants, they regulate responses to stress or coordinate growth under stress conditions. The list of phytohormones is growing and now includes brassinosteroids (BR), nitric oxide (NO), polyamines, and the recently identified branching hormone strigolactone (Gray 2004).

Treatment of plants with exogenous hormones rapidly and transiently alters genome-wide transcript profiles (Chapman and Estelle 2009). In *Arabidopsis*, hormone treatment for short periods (<1 h) alters expression of 10–300 genes, with roughly equal numbers of genes repressed and activated (Goda et al. 2008; Nemhauser et al. 2006; Paponov et al. 2008). Not surprisingly, longer exposure to most hormones (≥ 1 h) alters expression of larger numbers of genes. Complex networks of gene regulation by phytohormones under abiotic stresses involve various *cis*- or *trans*-acting elements. Some of the transcription factors, regulators, and key components functioning in signaling pathways of phytohormones under abiotic stresses are described in this work. Moreover, changes in gene expression, protein synthesis, modification, or degradation initiated by or coupled to plant hormones are briefly summarized.

1.2 Auxins

Application of auxin to plant tissues brings out various responses including electrophysiological and transcriptional responses, and changes in cell division, expansion, and differentiation. Rapid accumulation of transcripts of a large number of genes which are known as primary auxin response genes occurs with auxin. Auxin gene families include the regulator of auxin response genes, auxin response factors (ARFs), and the early response genes, auxin/indole-3-acetic acid (Aux/IAA), GH3, small auxin-up RNAs (SAURs), and LBD (Abel et al. 1994; Abel and Theologis 1996; Guilfoyle and Hagen 2007; Hagen and Guilfoyle 2002; Iwakawa et al. 2002; Yang et al. 2006). Although the roles of these factors in specific developmental processes are not fully understood yet, it was suggested that many members of these gene families are also involved in stress or defense responses (Jain and Khurana 2009).

When auxin-treated cells were examined, it was proposed that part of the auxin response is mediated by modification of gene expression and that it does not require de novo protein synthesis. It was identified that three main families (Aux/IAA, GH3, and SAUR) of early auxin response genes were expressed within 5–60 min after auxin treatment (Tromas and Perrot-Rechenmann 2010).

With the tight cooperation of these genes, plants can properly respond to auxin signals and environmental stresses, as well as maintain natural growth and development. The DNA-binding domains of ARFs bind to auxin response elements (AuxREs) (TGTCTC) of auxin-responsive genes and regulate their expression (Fig. 1.1). ARFs bind with specificity to AuxRE in promoters of auxin response genes and function in combination with Aux/IAA repressors, which dimerize with ARF activators in an auxin-regulated manner. It was suggested that differences in AuxRE sequences and abundance may serve as the first level of complexity in the transcriptional regulation of auxin-responsive genes (Szemenyei et al. 2008).

Northern and reverse transcriptase PCR (RT-PCR) analyses suggested that ARF genes are transcribed in different tissues and organs in *Arabidopsis* and rice plants (Okushima et al. 2005; Wang et al. 2007a). Most ARFs have a DNA-binding domain at the N-terminal. ARFs are transcription factors involved in the regulation of early auxin response genes. It was proposed that ARFs act as activators if they contain a glutamine/serine/leucine-rich (QSL-rich) middle region or as repressors if they contain a serine or serine/proline/glycine-rich middle domain (Tromas and Perrot-Rechenmann 2010).

In the literature, it was shown that the expression of ARF genes responds to environmental or hormonal signals. ARF2, 7, and 19 transcripts increased to some level, and ARF1 transcripts decreased slightly in response to dark-induced senescence in leaves (Ellis et al. 2005). Responses of ARF genes to environmental factors were indicated to be small or negligible; therefore, it was suggested that unidentified factors should play a key role in regulating expression of these genes or regulation by environmental factors is highly specific to selected tissue type (Guilfoyle and Hagen 2007).

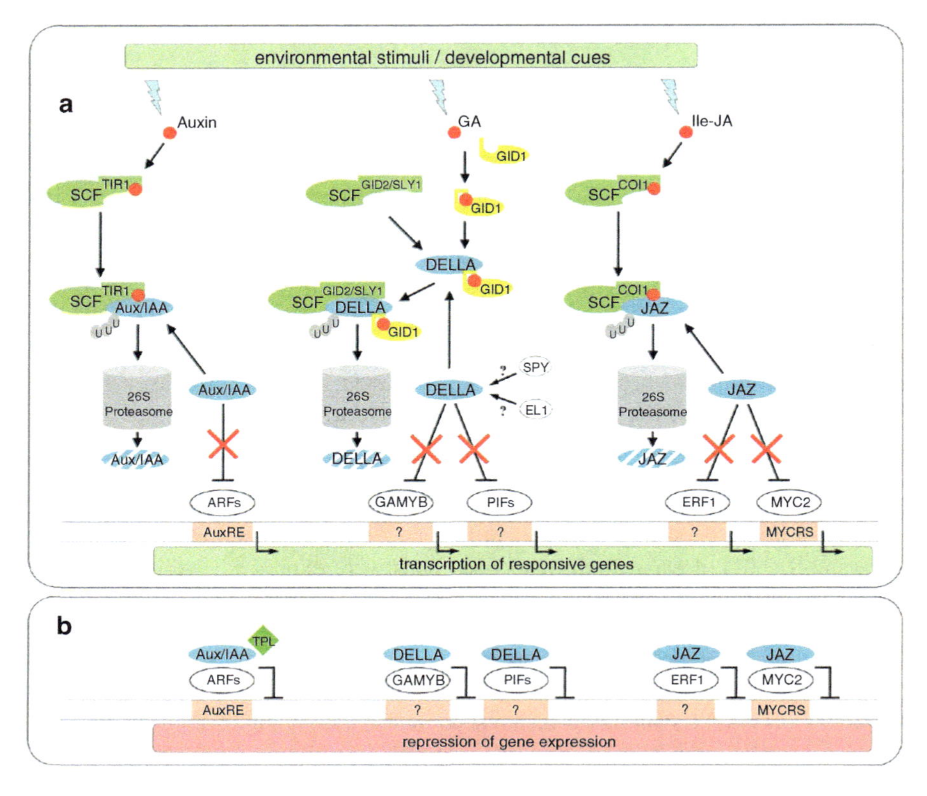

Fig. 1.1 Models for signal transduction pathways of auxin, gibberellic acid (GA), and jasmonoyl isoleucine (Ile–JA). (**a**) Upon phytohormone accumulation in a plant cell, repression on expression of responsive genes is relieved by degradation of transcriptional regulator. (**b**) In the absence or low levels of phytohormones, transcriptional regulators bind to certain transcription factors and repress gene expression. *Arrows* and *T-bars* indicate activation and inhibition, respectively

The Aux/IAA genes comprise a large class of auxin-inducible transcripts and have been identified in many plants. They encode short-lived nuclear proteins and act as repressors of auxin-regulated transcriptional activation (Berleth et al. 2004). Genetic and molecular studies showed that these proteins function as negatively acting transcription regulators that repress auxin response (Fig. 1.1). Aux/IAA proteins do not bind to AuxREs directly, but they regulate auxin-mediated gene expression by controlling the activity of ARFs. Aux/IAA proteins negatively regulate auxin-mediated transcription activity by binding ARFs through conserved domains (domains III and IV) found in both types of proteins (Ulmasov et al. 1997; Tiwari et al. 2003; Kim et al. 1997).

The Aux/IAA transcription factor has no DNA-binding domain, but together with ARF, it coregulates the transcription of auxin-responsive genes (Gray et al. 2001). With interactions between ARF and Aux/IAA proteins, the specific response to auxin is generated. Yeast two-hybrid and other physical assays in vivo have confirmed a number of interactions, such as the ARF–Aux/IAA interactions and the AtIAA1, 6, 12, 13, and 14 interactions with ARF5 or ARF7 (Hamann et al. 2004; Fukaki et al. 2005; Weijers et al. 2005; Wang et al. 2010). It was also reported that the domain I of Aux/IAA recruits topless (TPL), which acts as a transcriptional corepressor for ARF–Aux/IAA-mediated gene regulation during the auxin response (Szemenyei et al. 2008).

Derepression of auxin responses occurs after an increase in the intracellular auxin level. When auxin levels increase in nucleus, the targeted degradation of the Aux/IAA repressors by the 26 S proteasome is promoted (Fig. 1.1). Auxin increases the interaction of the domain II of Aux/IAAs with transport inhibitor response 1/auxin-related F-Box (TIR1/AFBs), F-box proteins of the E3 ubiquitin ligase complex Skp1/Cullin1/F-box-TIR1/AFBs ($SCF^{TIR1/AFBs}$). There is limited information about relative affinity of interaction between various Aux/IAAs and the different TIR1/AFBs F-box proteins. With the presence of Aux/IAA peptides, auxin binds to TIR1, but the mechanism is not clear.

The $SCF^{TIR1/AFBs}$ auxin signaling pathway is short and controls the auxin-induced changes of gene expression by targeting the degradation of transcriptional repressors. It was shown that multiple signaling components such as MAP kinases (Kovtun et al. 1998), IBR5 protein phosphatase (Strader et al. 2008), or RAC GTPases (Tao et al. 2002) participate in the regulation of early auxin response genes. Therefore, it is not clear whether the $SCF^{TIR1/AFBs}$ pathway is sufficient to tightly regulate auxin-regulated gene expression.

It was also shown that two additional proteins were involved in the regulation of auxin-responsive gene expression. First is the long-standing auxin-binding protein 1 (ABP1) receptor involved in very early auxin-mediated responses at the plasma membrane in *Arabidopsis* (Braun et al. 2008). Since TIR1/AFBs and Aux/IAAs are mainly located in the nucleus, physical interaction with ABP1 is highly unlikely. Second is the indole-3-butyric acid response 5 (IBR5) phosphatase which promotes auxin responses through a pathway different from TIR1-mediated repressor degradation (Strader et al. 2008).

The transcription of LBD genes is enhanced in response to exogenous auxin, indicating that the LBD gene family may act as a target of ARF (Lee et al. 2009). The LBD genes encode proteins harboring a conserved lateral organ boundaries (LOB) domain, which constitute a novel plant-specific class of DNA-binding transcription factors, indicative of its function in plant-specific processes (Husbands et al. 2007; Iwakawa et al. 2002).

It was reported that the transcription of GH3 genes is also related to ARF proteins. AtGH3-6/DFL1, AtGH3a, and At1g28130 expression was reduced in a T-DNA insertion line (*arf8-1*) and increased in overexpression lines of AtARF8. This indicates that the three GH3 genes are targets of AtARF8 transcriptional control. The control of free IAA level by AtARF8 in a negative feedback fashion might occur by regulating GH3 gene expression (Tian et al. 2004). In the *atarf7* or *atarf7/atarf19* mutants, downregulation of AtGH3-6/DFL1 and in rice, downregulation of OsGH3-9 and OsGH3-11 levels under IAA treatment was observed (Okushima et al. 2005; Terol et al. 2006). It was shown that multiple auxin-inducible elements were found in promoters of the GH3 gene family. This result confers auxin inducibility to the GH3 genes (Liu et al. 1994). GH3 genes were not only regulated by ARFs but also modulated by plant hormones, biotic and abiotic stresses, and other transcriptional regulators. Auxin-induced transcription is also modulated by tobacco bZIP transcription factor, BZI-1, which binds to the GH3 promoter (Heinekamp et al. 2004). A GH3-like gene, CcGH3, is regulated by both auxin and ethylene in *Capsicum chinense* L. (Liu et al. 2005). The upregulation of the GH3 genes in response to Cd was shown in *Brassica juncea* L. (Minglin et al. 2005). A GH3-5 gene in *Arabidopsis*, WES1, was shown to be induced by various stress conditions like cold, heat, high salt, or drought and by SA and ABA (Park et al. 2007). Auxin metabolism was induced by GH3 genes via R2R3-type MYB transcription factor, MYB96, and optimization of root growth was observed under drought conditions in *Arabidopsis* (Seo and Park 2009). Therefore, GH3-mediated auxin homeostasis is important in auxin actions which regulate stress adaptation responses (Park et al. 2007).

Accumulation of small auxin-up RNAs (SAURs) occurs rapidly and transiently with auxin in many plants (Woodward and Bartel 2005). The short half-lives of SAUR mRNAs appear to be conferred by downstream elements in the 3′ untranslated region of the messages (Sullivan and Green 1996). *Arabidopsis* mutants that stabilize downstream element-containing RNAs, and thus stabilize SAUR transcripts, have no reported morphological phenotype (Johnson et al. 2000), and although their function is not clearly established, they have been proposed to act as calmodulin-binding proteins. As in GH3 and Aux/IAA genes, most SAUR genes share a common sequence in their upstream regulatory regions, TGTCTC or variants, which was first identified from the promoter region of the pea PS-IAA4/5 gene (Ballas et al. 1993).

A wide variety of abiotic stresses have an impact on various aspects of auxin homeostasis, including altered auxin distribution and metabolism. Two possible molecular mechanisms have been suggested for altered distribution of auxin: first, altered expression of PIN genes, which mediate polar auxin transport; and second, inhibition of polar auxin transport by phenolic compounds accumulated in response to stress exposure (Potters et al. 2009). On the other hand, auxin metabolism is modulated by oxidative degradation of IAA catalyzed by peroxidases (Gazarian et al. 1998), which, in turn, are induced by different stress conditions. Furthermore, it has been shown that reactive oxygen species generated in response to various environmental stresses may influence the auxin response (Kovtun et al. 2000; Schopfer et al. 2002). Although these observations provide some clues, the exact mechanism of auxin-mediated stress responses still remains to be elucidated.

To address whether auxin-responsive genes were also involved in stress response in rice plants, their expression profile was investigated by microarray analysis under desiccation, cold, and salt stress. It was indicated that at least 154 auxin-induced and 50 auxin-repressed probe sets were identified that were differentially expressed, under one or more of the stress conditions analyzed. Among the 154 auxin-induced genes, 116 and 27 genes were upregulated and downregulated, respectively, under abiotic stress conditions. Similarly, among the 50 auxin-repressed genes, 6 and 41 genes were upregulated and downregulated, respectively. Moreover, 41 members of auxin-related gene families were found to be differentially expressed under at least one abiotic stress condition. Among these, 18 (two GH3, seven Aux/IAA, seven SAUR, and two ARF) were upregulated and 18 (one GH3, five Aux/IAA, eight SAUR, and four ARF) were downregulated under one or more abiotic stress conditions. However, another five genes (OsGH3-2, OsIAA4, OsSAUR22, OsSAUR48, and OsSAUR54) were upregulated under one or more abiotic stress conditions and downregulated under other stress conditions. Interestingly, among the 206 auxin-responsive (154 auxin-induced and 50 auxin-repressed) genes and 41 members of auxin-related gene families that were differentially expressed under at least one abiotic stress condition, only 51 and 3 genes, respectively, were differentially expressed under all three stress conditions (Jain and Khurana 2009).

It was indicated that the expression of Aux/IAA and ARF gene family members was altered during cold acclimation in *Arabidopsis* (Hannah et al. 2005). Molecular genetic analysis of the auxin and ABA response pathways provided evidence for auxin–ABA interaction (Suzuki et al. 2001; Brady et al. 2003). The role of IBR5, a dual-specificity phosphatase-like protein, supported the link between auxin and ABA signaling pathways (Monroe-Augustus et al. 2003).

Promoters of the auxin-responsive genes and members of auxin-related gene families differentially expressed under various abiotic stress conditions were analyzed to identify *cis*-acting regulatory elements linked to specific abiotic stress conditions. Although no specific *cis*-acting regulatory elements could be linked to a specific stress condition analyzed, several ABA and other stress-responsive elements were identified. The presence of these elements further confirms the stress responsiveness of auxin-responsive genes. The results indicated the existence of a complex system, including several auxin-responsive genes, that is operative during stress signaling in rice. The results of study suggested that auxin could also act as a stress hormone, directly or indirectly, that alters the expression of several stress-responsive genes (Jain and Khurana 2009).

It was shown that genes belonging to auxin-responsive SAUR and Aux/IAA family, ARFs and auxin transporter-like proteins are downregulated in the grapevine leaves exposed to low UV-B (Pontin et al. 2010). Similar results were also found in the study of pathogen resistance responses, where a number of auxin-responsive genes (including genes encoding SAUR, Aux/IAA, auxin importer AUX1, auxin exporter PIN7) were significantly repressed (Wang et al. 2007b), supporting the idea that downregulation of auxin signaling contributes to induction of immune responses in plants (Bari and Jones 2009).

Some of the plant glutathione *S*-transferases (GSTs) are induced by plant hormones auxins and cytokinins. The transcript level of GST genes was induced very rapidly in the presence of auxin. OsGSTU5 and OsGSTU37 were preferentially expressed in root and were also upregulated by auxin and various stress conditions (Jain et al. 2010).

1.3 Gibberellins

Gibberellins (GAs) are a large family of tetracyclic, diterpenoid phytohormones, which regulate plant growth. Bioactive GAs influence various developmental processes such as seed germination, stem elongation, pollen maturation, and transition from vegetative growth to flowering (Olszewski et al. 2002). Growth and stress are often opposed, and a retardation of development is generally observed under environmental stress conditions. Therefore, components of GA signaling are candidates for putative integrator of growth and stress signals. Moreover, crosstalk of GA signaling with various phytohormone signaling events, which function in response to stress, bestows an important role on GA under stress conditions. Crosstalk could potentially occur by altering expression levels of GA-signaling components or modulating their protein activity or stability (Fu and Harberd 2003; Achard et al. 2003, 2006).

Mutants of rice (*Oryza sativa*) and *Arabidopsis* deficient in GA biosynthesis or signaling were utilized to identify proteins that are essential for GA perception and signaling. The current model of GA signaling suggests binding of GA to a soluble GA-insensitive dwarf 1 (GID1) receptor (Ueguchi-Tanaka et al. 2005) (Fig. 1.1). The GID1–GA complex then interacts with DELLA repressor proteins, resulting in degradation of DELLA protein through a ubiquitin–proteasome pathway initiated by SCF (Skip/Cullin/F-box) complex (Sun 2011). The GA-specific F-box proteins, GID2 in rice (Sasaki et al. 2003), and sleepy1 (SLY1) and sneezy (SNE) in *Arabidopsis* (McGinnis et al. 2003; Strader et al. 2004) confer specificity to the SCF-type E3 ubiquitin ligase, $SCF^{GID2/SLY1}$, toward the DELLA–GID1–GA complex. $SCF^{GID2/SLY1}$ adds a polyubiquitin chain to the DELLA protein and hence induces its degradation by the 26 S proteasome complex (Fig. 1.1). The degradation of DELLA repressors by the 26 S proteasome activates GA action (Ueguchi-Tanaka et al. 2007).

The GID1 receptor, which encodes a soluble protein with similarity to hormone-sensitive lipases, was first identified in rice (Ueguchi-Tanaka et al. 2005). Its homologs GID1a, GID1b, and GID1c were identified and characterized as the major GA receptors in *Arabidopsis* (Nakajima et al. 2006; Griffiths et al. 2006). Subsequently, GA receptors in various plants such as cotton, barley, and fern have been identified (Aleman et al. 2008; Chandler et al. 2008; Yasumura et al. 2007). GID1 is a soluble nuclear-enriched receptor which interacts with DELLA proteins in a GA-dependent manner (Willige et al. 2007). Structural analysis of GID1 has revealed basis for GID1–GA and DELLA–GID1–GA interactions as well as evolutionary aspects of the GA receptor (Shimada et al. 2008; Murase et al. 2008; Ueguchi-Tanaka and Matsuoka 2010).

DELLA repressors are the key regulators of GA signaling (Schwechheimer 2008). Five DELLA proteins, namely, GA-insensitive (GAI), repressor of *ga1*-3 (RGA), RGA-like 1 (RGL1), RGL2, and RGL3, have been identified in *Arabidopsis* (Bolle 2004). On the other hand, single DELLA protein genes present in rice and barley genomes are slender rice1 (SLR1) (Ogawa et al. 2000; Ikeda et al. 2001) and slender 1 (SLN1) (Chandler et al. 2002; Fu et al. 2002), respectively. DELLA repressor loss-of-function mutants are taller than the wild-type plants and flower early, whereas transgenic plants overexpressing a DELLA protein are dwarf and flower late (Fu et al. 2002; Peng et al. 1997). The N-terminal domains of these repressors containing the DELLA motif play a regulatory role in GA-signal perception and GA-induced degradation (Dill et al. 2001). The absence of a typical basic DNA-binding domain suggests that DELLA proteins are more likely to function as transcriptional regulators instead of as transcription factors (Hussain and Peng 2003) (Fig. 1.1). Molecular studies showed that dwarf wheat varieties adopted during "green revolution" are affected in components of GA-signaling pathways, specifically orthologs of GAI (Peng et al. 1999).

Repressor activity of DELLA proteins might be controlled by mechanisms such as posttranslational modifications. Though initial studies had indicated phosphorylation of DELLA repressors as a prerequisite for GA-dependent degradation (Sasaki et al. 2003; Gomi et al. 2004), later studies have shown that DELLA proteins are phosphorylated in a GA-independent manner and phosphorylated as well as nonphosphorylated DELLA proteins are degraded in response to GA (Itoh et al. 2005). Requirement of DELLA dephosphorylation for subsequent degradation has been suggested in an *Arabidopsis* cell-free assay system and in tobacco BY2 cells (Wang et al. 2009; Hussain et al. 2005). Moreover, it was reported that phosphorylation of SLR1 by early flowering1 (EL1), encoding a serine/threonine protein kinase, might be critical for DELLA protein activity (Dai and Xue 2010). The *Arabidopsis* spindly (SPY) protein, which is an *O*-linked *N*-acetylglucosamine (GlcNAc) transferase, may function as a negative regulator of GA response. Though evidence of direct modification is lacking, it was suggested that SPY increases the activity of DELLA proteins, by adding a GlcNAc monosaccharide to serine/threonine residues (Silverstone et al. 2007). Thus, posttranslational modifications are clearly important for proper functioning or stability of the DELLA proteins, although the identities of the factors responsible for these modifications and modes of regulation remain to be determined.

Several putative direct targets of DELLA in *Arabidopsis* were identified by expression microarrays (Zentella et al. 2007; Hou et al. 2008). DELLA has induced expression of upstream GA biosynthetic genes and GA receptor genes, suggesting direct involvement of DELLA in maintaining GA homeostasis via a feedback mechanism. Other DELLA-induced target genes encode transcription factors/ regulators like basic helix-loop-helix (bHLH), MYB-like, and WRKY family proteins. Among DELLA targets were RING-type E3 ubiquitin ligases including XERICO which is important for ABA accumulation. Thus, DELLA inhibits GA-mediated responses in part by upregulating ABA levels through XERICO. This revealed a role of DELLA in mediating interaction between GA and ABA signaling pathways (Zentella et al. 2007). Recently, it was reported that in *Arabidopsis* scarecrow-like3 (SCL3) and DELLA antagonize each other in controlling both downstream GA responses and upstream GA biosynthetic genes (Zhang et al. 2011).

DELLA stability is indirectly affected by other phytohormone pathways or environmental cues through alteration of GA metabolism and bioactive GA levels.

Auxin induces root and stem elongation, at least in part, by upregulating GA biosynthetic genes and downregulating GA catabolism genes (Sun 2010). During cold and salt stresses, AP2 transcription factors such as CBF1 and dwarf delayed-flowering 1 (DDF1) induce expression of GA catabolism genes (Magome et al. 2004). Similarly, stabilization of DELLA by ABA treatment is achieved by reduction of GA accumulation (Sun 2010). Integrative role of DELLA repressors in salt stress, ABA, and ethylene responses was described, and it was stated that salinity activates ABA and ethylene signaling, two independent pathways whose effects are integrated at the level of DELLA function (Achard et al. 2006). Growth restraint conferred by DELLA proteins extends the duration of the vegetative phase and promotes survival under adverse conditions.

DELLA proteins play critical roles in protein–protein interactions within various environmental and phytohormone signaling pathways. They are involved in many aspects of plant growth, development, and adaptation to stresses (Feng et al. 2008; Harberd et al. 2009; Arnaud et al. 2010; Hou et al. 2010). It was hypothesized that GA signaling or DELLA proteins enable flowering plants to maintain transient growth arrest, giving them the flexibility to survive periods of adversity (Harberd et al. 2009). The binding of DELLA proteins to the phytochrome-interacting factor (PIF) proteins integrates light and GA-signaling pathways (Fig. 1.1). This binding prevents PIFs from functioning as positive transcriptional regulators of growth in the dark. Since PIFs are degraded in light, they can only function in the combined absence of light and presence of GA (Hartweck 2008). DELLA inhibits hypocotyl elongation by binding directly to PIF3 and PIF4 and preventing expression of PIF3/PIF4 target genes (Feng et al. 2008). The transcription factor PIF3-like5 (PIL5) directly promotes the transcription of the GAI and RGA DELLA protein genes before germination and thereby controls repressor protein abundance. In response to light, PIL5 is degraded, and the transcription of GAI and RGA is reduced, relieving the restraint on germination (Oh et al. 2007). In barley, activation of α-amylase expression is induced by GAMYB (Gubler et al. 1999). It has been demonstrated that GA response mediated through GAMYB is dependent on the DELLA proteins SLN1 and SLR1, in barley and rice, respectively (Gubler et al. 2002), in which the DELLA proteins act as negative regulators of GAMYB-mediated gene expression.

Recently, two homologous GATA-type transcription factors from *Arabidopsis*, namely, GNC (GATA, nitrate-inducible, carbon-metabolism involved) and GNL/CGA1 (GNC-Like/cytokinin-responsive GATA factor 1), were identified as GA-regulated genes. It was indicated that GNC and GNL/CGA1 are important downstream targets of DELLA proteins and PIF transcription factors and that they might be direct PIF targets (Richter et al. 2010). In another recent study, role of DELLA as a transcriptional activator has been revealed. It was shown that the jasmonic acid (JA) ZIM-domain 1 (JAZ1) protein, a key repressor of JA signaling, interacts in vivo with DELLA proteins. JAZ proteins inhibit the activity of MYC2, which regulate target genes including some of JA-responsive genes. Binding of DELLA to JAZ removes the repression on MYC2 and JA-responsive genes (Hou et al. 2010). In *Arabidopsis*, DELLA proteins were implicated in JA signaling or perception, and a role of DELLA in the regulation of plant–pathogen interactions was suggested (Navarro et al. 2008). Consequently, function of DELLA proteins as transcriptional repressors or activators grants these regulatory proteins a critical role at the crossroads of phytohormone signaling pathways during development or under various environmental conditions.

It is essential to identify the genes that are the final targets of GA-signaling pathway. GA function and GA-induced gene transcription in cereal aleurone cells have been reviewed (Olszewski et al. 2002; Sun and Gubler 2004). DNA microarrays have been utilized to dissect the transcriptional changes that promote GA-induced seed germination in *Arabidopsis*. Identified GA-responsive genes included the ones encoding for expansins, xyloglucan endotransglycosylase/hydrolases (XETs), aquaporins, a D-type cyclin, and a replication protein A, which are implicated in cell elongation and cell division (Ogawa et al. 2003). A cDNA microarray was employed to understand the molecular mechanisms by which GA and BRs regulate the growth and development in rice seedlings. Increased expression of XETs and downregulation of stress-related genes were observed after exogenous application of GA (Yang et al. 2004). In citrus, effects of GAs on internode transcriptome were investigated using a cDNA microarray. An overall upregulation of genes encoding proteins of the photosystems and chlorophyll-binding proteins, as well as of genes of the carbon fixation pathway, was observed (Huerta et al. 2008). In maize, transcriptional profiles of immature ears and tassels were investigated with microarrays at early stage of water stress. Transcripts upregulated in both organs included those involved in protective functions, detoxification of reactive oxygen species, nitrogen metabolism, and GA metabolism (Zhuang et al. 2008).

1.4 Cytokinins

Cytokinin signaling is similar to the two-component signal transduction pathways present in most bacteria and fungi. Hybrid histidine kinase (HK) receptors bind to cytokinin and then are autophosphorylated. Then phosphate group is transferred to histidine phosphotransfer proteins (HPs) (Fig. 1.2). The *Arabidopsis* HPs (AHPs) are a small family of proteins that act as intermediates in cytokinin signaling. The AHPs interact directly with various sensor HKs and type A and type B response regulators (RRs) in yeast two-hybrid assay. It was found that there were 23 *Arabidopsis* response regulators (ARRs) and nine related proteins (APRRs) in *Arabidopsis* (Schaller et al. 2002). The type B or transcription factor-type class also has 11 members. Each type B protein is composed of an N-terminal receiver domain and a long C-terminal part containing a single-repeat MYB-type DNA-binding domain (Sakai et al. 1998) called a GARP domain (Riechmann et al. 2000) and the proline- and glutamine-rich region frequently observed in eukaryotic transactivating domains (Tjian and Maniatis 1994). The ARRs are classified according to their C-terminal domains. Type A and type C have short C-termini, while type B ARRs have longer C-termini.

Transcription of type A ARRs is rapidly elevated by exogenous cytokinin (Brandstatter and Kieber 1998; Jain et al. 2006). In addition to transcriptional regulation, cytokinin treatment also results in an increase in the half-life of a subset of type A ARR proteins (To et al. 2007). Type A ARRs which are direct targets of the type B ARR transcription factors are negative regulators of cytokinin signaling. Consistent with their role as transcription factors, type B ARRs localize to the nucleus (Hwang and Sheen 2001; Asakura et al. 2003; Mason et al. 2005). Genetic and molecular analyses indicate that the type B ARRs are redundant positive elements in cytokinin signaling and are the immediate upstream activators of type A ARR gene expression (Hwang and Sheen 2001; Mason et al. 2005; Argyros et al. 2008). It was shown that type B ARRs are positive elements in cytokinin signaling (Ishida et al. 2008; Mason et al. 2005; Argyros et al. 2008) (Fig. 1.2).

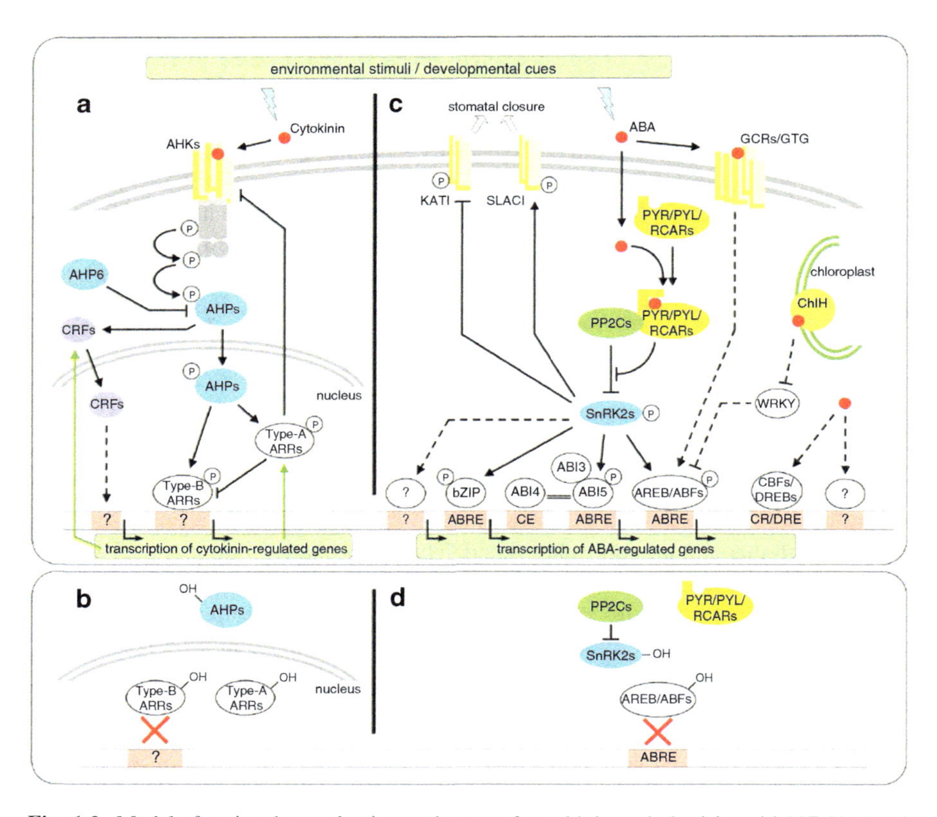

Fig. 1.2 Models for signal transduction pathways of cytokinin and abscisic acid (ABA). (**a**, **c**) Accumulation of phytohormones triggered by environmental stimuli or developmental cues initiates cascades of events involving phosphatases or kinases to induce expression of responsive genes. (**b**, **d**) In the absence or low levels of phytohormones, inactive transcription factors cannot induce gene expression. *Arrows* and *T-bars* indicate activation and inhibition, respectively. *Dashed arrows* or *T-bars* indicate possible interactions

To determine the target genes of the cytokinin-regulated transcriptional network, microarray analyses have been performed by different groups (Brenner et al. 2005; Rashotte et al. 2003; Rashotte et al. 2006). In addition to the type B ARRs, there are several other transcription factors that have been implicated by microarray analyses in the response to cytokinin. The cytokinin response factors (CRFs) act, along with the type B ARRs, to mediate the transcriptional response to cytokinin (Fig. 1.2). The CRFs have six family members, which are a subset of the AP2-like superfamily. Three of CRFs are transcriptionally upregulated by cytokinin in a type B ARR-dependent manner (Rashotte et al. 2006). Microarray analysis of cytokinin-regulated genes in a multiple *crf* mutant revealed that many genes regulated by type B ARRs are also regulated by CRFs.

It was indicated that the functions of the cytokinin-regulated genes reflect processes known to be targets of cytokinin signaling, including genes involved in cell expansion, other phytohormone pathways (auxin, ethylene, and GA), responses to pathogens, and regulation by light. Other, more directed approaches have identified individual genes regulated by cytokinin, including cyclinD3 (Riou-Khamlichi et al. 1999), which provides a mechanistic link between cytokinin and the regulation of the cell cycle. Additionally, other clusters of genes suggest unsuspected targets of cytokinin, including genes involved in trehalose-6-phosphate metabolism and potential effects in the redox state of the cell. Undoubtedly, there are many additional targets that remain to be identified. Moreover, the transcription factors responsible for the regulation of these targets and how they interact remain to be determined (Argueso et al. 2010).

It was also known that cytokinin function has been linked to a variety of abiotic stresses (Hare et al. 1997). When public microarray expression data was examined, it was revealed that the genes encoding proteins in the cytokinin signaling pathway were differentially affected by various abiotic stresses. For example, it was shown that cold stress appears to rapidly upregulate the expression of multiple type A ARRs and conversely to downregulate the expression of all three cytokinin receptors. Although there are no reports linking cytokinin to a rapid response to cold stress, these results can suggest a role for cytokinin in the response to cold stress (Argueso et al. 2009). After dehydration, the expression of the AHK2 and AHK3 genes was found to be induced (Tran et al. 2007), which was shown in the analysis of public microarray data (Argueso et al. 2009). Exposure of plants to drought results in a decrease in the level of cytokinins in the xylem sap (Bano et al. 1994; Shashidhar et al. 1996). A recent study has confirmed that isoprene-type cytokinins (zeatin and zeatin riboside) are decreased in the xylem in response to drought stress. Surprisingly, in the same study, it was found that the level of the aromatic cytokinin 6-benzylaminopurine (BAP) was elevated (Alvarez et al. 2008).

It was found that the expression of *Agrobacterium* isopentenyl transferase (IPT), rate-limiting enzyme in cytokinin biosynthesis, downstream of a drought/maturation-induced promoter resulted in a remarkable tolerance to extreme drought conditions in tobacco (Rivero et al. 2007). While wild-type plants died, transgenic plants had complete recovery after drought conditions. In addition to this, under water restriction, there was no yield loss (Rivero et al. 2007). This result was consistent with the notion that elevated cytokinin levels may promote survival in drought conditions. Similar results were obtained in another study, which suggested that endogenous cytokinin may play a role in conferring drought tolerance (Alvarez et al. 2008).

Especially in roots, the expressions of several of the CRF genes were down-regulated in response to salt stress. It was suggested that these genes may play an important role in mediating the input of cytokinin into the salt stress response pathway (Rashotte et al. 2006). In another study, one out of ten recently described rice RR genes had shown to be upregulated in seedlings exposed to a high concentration of salt (Jain et al. 2006). In developing kernels where the cytokinin role in response to water stress was previously studied (Brugiere et al. 2003), only specific genes for de novo biosynthesis (e.g., IPT2), degradation (e.g., CKX1, CKX4), and signal response (e.g., RR3) were active.

Cytokinins control many aspects of development and responses to the environment. Recent research highlighted the importance of cytokinin-regulated transcriptional networks in the regulation of these processes. As well as type B ARRS, additional classes of transcription factors take role in the control of cytokinin-regulated gene expression in shoot development (e.g., STM, WUS, GL1) and root development (e.g., SHY2, SCR, PLT1) (Argueso et al. 2010). Thus, it was suggested that crosstalk between cytokinin and other plant hormones at the transcriptional level is widespread.

1.5 Abscisic Acid

Abscisic acid (ABA) is a major phytohormone that regulates a broad range of events during development and adaptive stress responses in plants. It plays crucial roles in responses of vegetative tissues to abiotic stresses such as drought and high salinity (Zhu 2002). It accumulates in cells under osmotic stress, promotes stomatal closure, and regulates the expression of various protective or adaptive genes. ABA and coordinated action of different hormonal signaling pathways control maintenance of root growth, regulation of stress-responsive gene expression, accumulation of osmocompatible solutes, and synthesis of dehydrins and late embryogenesis abundant (LEA) proteins under environmental stress (Zhu 2002; Sharp et al. 2004; Verslues et al. 2006). Recently, role of ABA in response to biotic stress has been reviewed as well (Ton et al. 2009). ABA might be providing resistance to pathogens and disease via inhibition of pathogen entry through stomata or via increasing susceptibility by crosstalk with other signaling pathways.

Mutants altered in phytohormone sensitivity have led to identification of physiological receptors for auxin (Dharmasiri et al. 2005; Kepinski and Leyser 2005), gibberellins (Ueguchi-Tanaka et al. 2005), and other phytohormones. However, similar genetic screens for mutants have not directly yielded ABA receptors. On the other hand, ABA perception and signal transduction have been studied extensively. Microinjection into cytosol or treatment with ABA or its analogs has suggested multiple ABA receptors at various locations including cytosol and plasma membrane. Though controversy exists, flowering time control protein FCA (Razem et al. 2006), G-protein-coupled receptor 2 (GCR2) (Liu et al. 2007), GCR-type G-protein 1 (GTG1) and GTG2 (Pandey et al. 2009), and Mg-chelatase H subunit (ChlH) (Shen et al. 2006) were identified as ABA receptors. Among these putative receptors, FCA was later shown to be not binding ABA (Risk et al. 2008). It was indicated that the filter-based ligand-binding assays employed in receptor studies are prone to artifacts because of incomplete removal of nonprotein-bound ABA. Similar concerns were raised for ABA-binding ability of ChlH and GCR2 (Risk et al. 2009; Guo et al. 2008). Alternative techniques like affinity chromatography were employed to reinforce the hypothesis that ChlH can bind to ABA in *Arabidopsis thaliana* (Wu et al. 2009). Although GCRs and ChlH were proposed to play important roles in ABA responses, their physiological and molecular connections to well-known signaling factors such as type 2C protein phosphatases (PP2C) and sucrose nonfermenting (SNF) 1-related protein kinase 2 (SnRK2) remained unclear.

Negative regulatory system employed in ABA signaling cascade is composed of PP2C phosphatases and SnRK2 kinases which act as negative and positive regulators, respectively (Fig. 1.2). Mutants of *Arabidopsis*, insensitive to ABA, were used for identification of two genes, ABA-insensitive1 (ABI1) and ABI2, encoding group A PP2Cs (Leung et al. 1994, 1997; Meyer et al. 1994). Discovery of these phosphatases has led to isolation or characterization of various other regulators of ABA signaling including protein kinases. Members of SnRK2 family such as ABA-activated protein kinase (AAPK) from *Vicia faba* (Li et al. 2000) and *Arabidopsis* SRK2E/Open stomata 1 (OST1)/SnRK2.6 (Mustilli et al. 2002; Yoshida et al. 2002) were determined as positive regulators in ABA signaling. Gene encoding ABA-induced protein kinase 1 (PKABA1), which is a serine–threonine type protein kinase, was isolated from wheat (Anderberg and Walker-Simmons 1992). In the absence of ABA, PP2C inactivates SnRK2 by direct dephosphorylation. On the other hand, in response to environmental or developmental cues, ABA promotes inhibition of PP2C and accumulation of phosphorylated SnRK2. Active SnRK2 subsequently phosphorylates ABA-responsive element (ABRE)-binding factors (AREBs/ABFs) and initiates ABA-regulated gene expression.

ABA signaling model was updated with the discovery of pyrabactin resistance 1/pyrabactin resistance 1-like/regulatory component of ABA Receptor (PYR/PYL/RCAR) proteins as a new type of soluble ABA receptor (Ma et al. 2009; Park et al. 2009). Furthermore, protein phosphatase–kinase complexes (PP2C–SnRK2) were identified as downstream components of PYR/PYL/RCARs (Umezawa et al. 2009; Vlad et al. 2009). After these major findings, several studies offered a double-negative regulatory system for ABA signaling which consists of four components: ABA receptors (PYR/PYL/RCAR), protein phosphatases (PP2C), protein kinases (SnRK2), and their downstream targets (Fujii et al. 2009; Umezawa et al. 2009) (Fig. 1.2). In the presence of ABA, interaction of PYR/PYL/RCAR and PP2C is promoted, resulting in PP2C inhibition and SnRK2 activation. Besides direct interactions between PYR/PYL/RCARs, PP2Cs, and SnRK2s, the interaction between other ABA-binding receptors (e.g., GCRs, GTGs, and ChlH) and any component of signaling (e.g., PP2Cs, SnRK2s, and AREBs/ABFs) is unknown.

The double-negative regulatory system provided by signaling complex of PYR/PYL/RCARs, group A PP2Cs, and subclass III SnRK2s is very simple yet sophisticated. The system probably varies widely in plant cells, tissues, and organs at various developmental stages. There are 14 PYR/PYL/RCARs, 9 PP2Cs, 3 SnRK2s, and 9 AREB/ABFs in *A. thaliana* alone to regulate transcription (Ma et al. 2009; Park et al. 2009; Umezawa et al. 2009; Uno et al. 2000), increasing number of possible combinations of the signaling complex to more than 3,000 (Umezawa et al. 2010). Fine tuning of ABA responses in plant cells is probably provided by multiple determinants, like spatial or temporal limitations, stress-responsive gene expression patterns, subcellular localization, and preferences in protein–protein interactions (Umezawa et al. 2010).

Downstream targets of the PYR/PYL/RCAR–PP2C–SnRK2 complex should be determined to clarify the details of ABA signaling. These include proteins that interact with PP2C and SnRK2. Several bZIP transcription factors (AREBs/ABFs) and some membrane proteins have been identified as substrates for SnRK2 phosphorylation. In guard cells SRK2E/OST1/SnRK2.6, homologue of SRK2D/SnRK2.2 and SRK2I/SnRK2.3 acts as positive regulator of stomatal closure (Mustilli et al. 2002). It activates anion channel SLAC1 and inhibits cation channel KAT1 which is essential for K^+ uptake during stomatal opening (Geiger et al. 2009; Raghavendra et al. 2010). ABA- and PYR/PYL/RCAR-mediated inactivation of PP2C allows activation of SLAC1 which has a central role in guard cells (Fig. 1.2).

It is well known that abiotic stress conditions like drought and salinity activate ABA-dependent gene expression systems involving various transcription factors like AREBs/ABFs, MYC/MYB, C-repeat binding factors (CBFs)/drought-responsive element (DRE)-binding proteins (DREBs), and NAC family proteins. On the other hand, cold stress regulates gene expression in an ABA-independent manner through some CBFs/DREBs (Agarwal and Jha 2010). Large-scale transcriptome analyses, which provided valuable information on ABA-mediated regulation of transcription, have shown that ABA dramatically alters genomic expression (Hoth et al. 2002; Seki et al. 2002). These genome-wide expression studies not only revealed key components of ABA signaling but also contributed in identification of novel downstream target genes. Key regulators of ABA-mediated gene expression are AREBs/ABFs with ABI5 as a typical representative. Several SnRK2s regulate AREB/ABFs in ABA signaling in response to water stress (Fujii and Zhu 2009). Wheat SnRK2 ortholog, PKABA1, phosphorylates the wheat AREB1 ortholog, TaABF, and the rice SnRK2 orthologs, SAPK8, SAPK9, and SAPK10, phosphorylate the AREB1 ortholog TRAB1, in vitro (Johnson et al. 2002; Kagaya et al. 2002; Kobayashi et al. 2005). OsABI5 from rice showed transcript upregulation by ABA and high salinity and downregulation by drought and cold. Its overexpression enhanced salinity tolerance (Zou et al. 2008).

The AREBs/ABFs encode bZIP transcription factors and belong to the group A subfamily, which is composed of nine homologs in the *Arabidopsis* genome

(Jakoby et al. 2002). The AREBs/ABFs were isolated by using ABRE sequences as bait in yeast one-hybrid screening method (Choi et al. 2000). The bZIP transcription factors interact as dimers with ABREs (PyACGTGGC), which are ACGT containing G-box-like *cis*-elements in promoter regions. ABA response usually requires a combination of an ABRE with a coupling element (CE), which is similar to an ABRE or a DRE (Himmelbach et al. 2003). ABRE-binding AREBs/ABFs, DRE-binding AP2-type transcription factors, and other transcriptional regulators such as viviparous1 (VP1)/ABI3 also contribute to ABA-mediated gene expression. ABI3 binds to ABI5 and enhances its action. ABI4, an AP2-type transcription factor, and a number of additional *trans*-acting factors including MYC/MYB family proteins act as positive ABA response regulators (Yamaguchi-Shinozaki and Shinozaki 2006). ZmABI4 interacts specifically with CE and functions in ABA signaling during germination and in sugar sensing in maize (Niu et al. 2002).

Among the group A bZIP subfamily, AREB1/ABF2, AREB2/ABF4, and ABF3 are induced by dehydration, high salinity, and ABA treatment in vegetative tissues (Uno et al. 2000; Kim et al. 2004; Fujita et al. 2005). In *Arabidopsis*, four cDNA sequences of ABFs (ABF1, ABF2, ABF3, and ABF4) similar to AREB1 and AREB2 were identified. ABF1 expression was induced by cold, ABF2 and ABF3 by high salt and ABF4 by cold, drought, and high salt (Choi et al. 2000). Recently, an *areb1/areb2/abf3* triple mutant was generated (Yoshida et al. 2010). Transcriptome analysis of triple mutant revealed novel AREB/ABF downstream genes in response to water stress, including many LEA class and group A PP2C genes and transcription factors. These results indicate that AREB1, AREB2, and ABF3 are master transcription factors that cooperatively regulate ABRE-dependent gene expression in ABA signaling under stress conditions (Yoshida et al. 2010). Various bZIP transcription factor genes of different groups were identified from soybean (*Glycine max*). It was found that GmbZIP44, GmbZIP62, and GmbZIP78 belonging to subgroup S, C, and G, respectively, were also involved in salt and freezing stresses. These proteins bind to ABRE and couple of other *cis*-acting elements with differential affinity and improve stress tolerance in transgenic *Arabidopsis* by upregulating ERF5, KIN1, COR15A, COR78A, and P5CS1 and downregulating DREB2A and COR47 (Liao et al. 2008).

Orthologs of AREBs/ABFs have also been reported in barley (Casaretto and Ho 2003) and rice (Lu et al. 2009; Amir Hossain et al. 2010). OsbZIP72 was shown to be an ABRE-binding factor in rice using the yeast hybrid systems. Transgenic rice overexpressing OsbZIP72 was hypersensitive to ABA and showed elevated levels of expression of ABA response genes such as LEAs. Transgenic rice plants displayed an enhanced ability of drought tolerance (Lu et al. 2009). Expression of OsABF1 was found to be induced by various abiotic stress treatments such as anoxia, salinity, drought, oxidative stress, and cold (Amir Hossain et al. 2010). In cultivated tomato (*Solanum lycopersicum*), two members of AREBs/ABFs, namely, SlAREB1 and SlAREB2, were identified. Expression of SlAREB1 and SlAREB2 was induced by drought and salinity in both leaves and root tissues. Microarray and cDNA-amplified fragment length polymorphism (AFLP) analyses were employed in order to identify SlAREB1 target genes responsible for the enhanced tolerance in SlAREB1-overexpressing lines. Genes encoding oxidative stress-related proteins, lipid-transfer proteins (LTPs), transcription regulators, and LEA proteins were found among the upregulated genes in transgenic lines (Orellana et al. 2010). ABA regulation of gene expression in *Arabidopsis* guard cells was investigated using microarrays. Global transcriptomes of guard cells were compared to gene expression in leaves and other tissues, and approximately 300 genes showing ABA regulation unique to guard cells were determined (Wang et al. 2011).

1.6 Jasmonic Acid

Lipid-derived jasmonic acid (JA) and its metabolites, collectively known as jasmonates, are important plant signaling molecules that mediate plant responses to environmental stress and function in various aspects of growth and development (Wasternack 2007; Balbi and Devoto 2008). Phytohormones regulate development not via linear pathways, but through complex interconnections between different signaling pathways. Extensive crosstalk occurs between JAs and salicylic acid (SA), another signaling molecule with an important function in plant defense responses (Beckers and Spoel 2006). In higher plants, after synthesis via the octadecanoid pathway, JA can be conjugated to amino acids, preferentially to isoleucine (Ile) to form Ile–JA or converted to methyl jasmonate (Me–JA) or other metabolites (Wasternack 2007). Ile–JA has been proposed to be the active form of the hormone. Pathogens, mechanical wounding, water deficit, and some other abiotic stresses trigger a rapid increase in JA levels. In general, JAs help to modulate the competitive allocation of energy to defense or growth.

Mutants of *Arabidopsis* were utilized for determination of key components of JA-signaling pathway. Central roles of an F-box protein, coronatine-insensitive 1 (COI1) (Xie et al. 1998), and negative transcriptional regulators, Jasmonate ZIM-domain (JAZ) (Thines et al. 2007; Chini et al. 2007) proteins, were defined in *Arabidopsis*. Taken together, these results suggested SCF^{COI1}-dependent degradation of JAZ repressors via 26 S proteasome following perception of Ile–JA. As in the case of GA and auxin signaling pathways, Ile–JA, active hormone, relieves the repression by JAZ, transcriptional regulator (Fig. 1.1). Moreover, coronatine, which is a phytotoxin that is structurally related to JA, binds to COI1–JAZ complexes with high affinity, which strongly suggests that COI1 functions as a receptor (Melotto et al. 2008; Katsir et al. 2008; Gfeller et al. 2010). However, direct binding of Ile–JA to COI1 has not been shown yet. Thus, crystal structure analyses of COI–JAZ complexes, identification of new JAZ targets, and determination of JA-responsive genes will help to clarify the JA-signaling pathway.

JAZ proteins directly interact with MYC2, repressing its activity in the absence of Ile–JA (Fig. 1.1). MYC2 encodes a bHLH transcription factor and induces JA-mediated responses such as wounding, inhibition of root growth, JA biosynthesis, oxidative stress adaptation, and anthocyanin biosynthesis (Boter et al. 2004).

MYC2 binds to the G-box (CACGTG) or T/G-box (AACGTG) in the promoters of JA-regulated genes (Chini et al. 2007). Ethylene response factor 1 (ERF1) and other ERFs integrate JA and ethylene signals and regulate some of the MYC2-modulated responses in an opposite fashion (Lorenzo et al. 2003). Recently, involvement of additional transcriptional factors belonging to different families such as NAC (e.g., ANAC019, ANAC055) and WRKY (e.g., WRKY70, WRKY18) have been reported (Bu et al. 2008; Xu et al. 2006; Fonseca et al. 2009).

Research has been concentrated on role of JA and its metabolites in defense response against biotic stresses. Response often implies changes in the content of several phytohormones, which correlate with changes in the expression of genes involved in their biosynthesis and the responses they regulate. Local or systemic responses at the level of gene expression have been investigated using high-density microarrays (López-Ráez et al. 2010; Schlink 2010; Lewsey et al. 2010). On the other hand, JA was reported to take part in responses to some abiotic stresses such as salinity, drought, and boron toxicity. Desiccation response was shown to involve the regulation of JA-responsive genes in barley leaf segments (Lehmann et al. 1995). Exogenous application of JA to salt-stressed rice seedlings improved recovery, suggesting a role for JA during response to salinity stress (Kang et al. 2005). In barley, induction of genes involved in JA biosynthesis or known as JA-responsive genes was reported as a key feature of response to salinity (Walia et al. 2006). JA was hypothesized to be involved in the adaptation of barley to salt stress. Treatment with JA before salt application partially alleviated photosynthetic inhibition caused by salinity stress. Expression profiling after a short-term exposure to salinity stress indicated a considerable overlap between genes regulated by salinity stress and JA application. It was suggested that three JA-regulated genes, arginine decarboxylase, ribulose 1,5-bisphosphate carboxylase/oxygenase (Rubisco) activase, and apoplastic invertase, were possibly involved in salinity tolerance mediated by JA (Walia et al. 2007). In a global transcriptome analysis of response to boron toxicity using microarrays, it was shown that high concentrations of boric acid treatment resulted in upregulation of JA-biosynthetic and JA-induced genes in barley leaves. Induction of JA-related genes was found to be an important late response to boron toxicity (Öz et al. 2009). In maize developing kernels, expression patterns of some genes in several stress response-associated pathways, including ABA and JA, were examined, and these specific genes were responsive to drought stress positively (Luo et al. 2010).

1.7 Ethylene

When key components of ethylene signaling from membrane receptors to nuclear activators were investigated in *Arabidopsis*, five membrane receptors, ethylene response 1 (ETR1), ETR2, ethylene response sensor 1 (ERS1), ERS2, and ethylene insensitive 4 (EIN4), were determined. These receptors act as negative regulators through genetically identified negative regulator, constitutive triple response

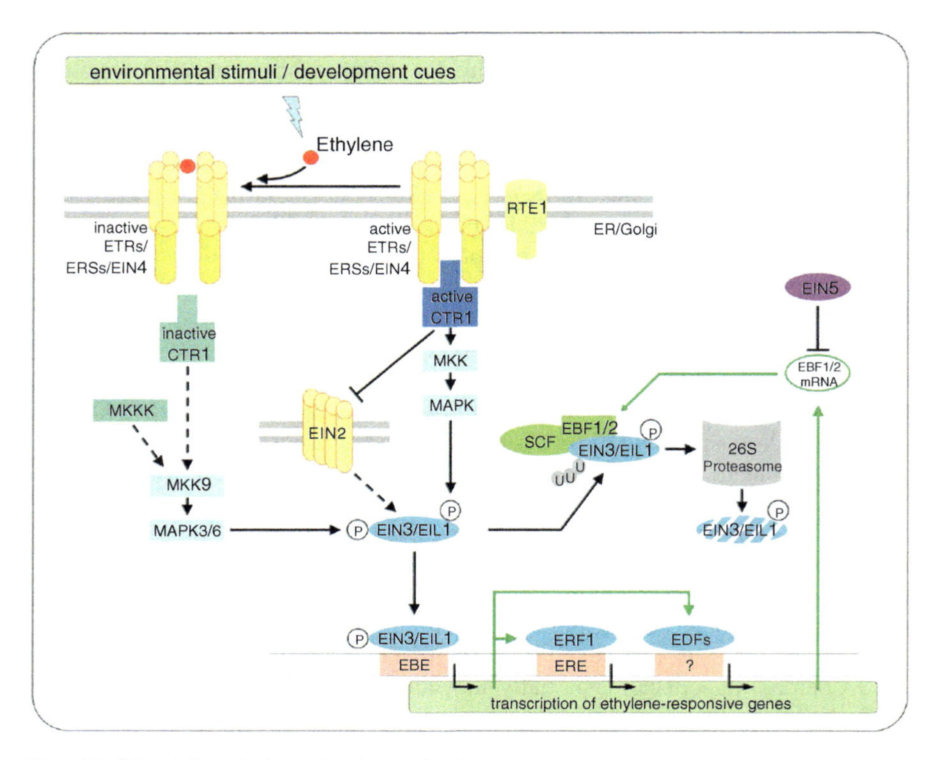

Fig. 1.3 Model for ethylene signal transduction pathway. Accumulation of ethylene triggers cellular events involving kinases or phosphatases to induce transcription of ethylene-responsive genes. *Arrows* and *T-bars* indicate activation and inhibition, respectively. *Dashed arrows* or *T-bars* indicate possible interactions

(CTR1), encoding a putative Raf-like MAPK kinase kinase (MKKK) (Kieber et al. 1993). Another membrane protein EIN2 has a pivotal role by regulating the availability of key transcription factor, EIN3, in ethylene signaling downstream of CTR1 (Fig. 1.3). The mechanism how EIN2 regulated the EIN3 is still unknown.

EIN3 is a plant-specific transcription factor mediating ethylene-regulated gene expression (Chao et al. 1997). It belongs to a multigene family in *Arabidopsis*, including EIN3, EIN3-Like 1 (EIL1), EIL2, EIL3, EIL4, and EIL5, in which EIN3 and EIL1 are the most closely related homologs. It is supposed that EIN3 and EIL1 are the major transcription factors in mediating ethylene responses.

Biochemical studies showed that EIN3 and EIL1 can directly bind to the promoter of ERF1 (ethylene response factor 1), which belongs to the EREBP (ethylene-responsive element-binding protein) family of transcription factors (Solano et al. 1998).

Ethylene response factors (ERFs), the first member of which was identified in tobacco, act at the last step of ethylene signaling pathways (Ohme-Takagi and Shinshi 1995). To date, in different plant species, ERFs have been found to be involved not only in growth, development, and regulation of metabolism (van der Fits and Memelink 2000; van der Graaff et al. 2000; Banno et al. 2001) but also in the response to biotic and abiotic stresses (Stockinger et al. 1997; Liu et al. 1998; Yamamoto et al. 1999; Fujimoto et al. 2000; Gu et al. 2000; Berrocal-Lobo et al. 2002; Gu et al. 2002; Dubouzet et al. 2003; Aharoni et al. 2004; Broun 2004; Zhang et al. 2005).

ERF4, *Arabidopsis* ERF1, ERF5, CBF1, DREB1, and DREB2, periwinkle ORCA2 and ORCA3, and tomato Pti4, Pti5, Tsi1, and JERF3 act as transcriptional activators that, when overexpressed, lead to the activation of downstream genes (Stockinger et al. 1997; Zhou et al. 1997; Liu et al. 1998; Solano et al. 1998; Menke et al. 1999; Fujimoto et al. 2000; Ohta et al. 2000; van der Fits and Memelink 2000; Park et al. 2001; Wang et al. 2004). Ethylene affects the expression of a group of genes related to pathogen attack, wounding, extreme temperatures, and drought stress.

It was indicated that overexpression of ERF1 rescued only a subset of *ein3* phenotypes. This result suggested that EIN3 regulates additional target genes in mediating distinct ethylene responses (Solano et al. 1998). Since mRNA levels were rapidly accumulated upon ethylene treatment, and knockout mutants resulted in partial ethylene insensitivity, four novel transcription factors EDF1–4 (ethylene-responsive DNA-binding factors) were suggested as potential target genes of EIN3 (Alonso et al. 2003). Collectively, a transcriptional cascade from EIN3/EIL1 to ERF1 and EDF is involved in the ethylene response pathway (Fig. 1.3).

It was demonstrated that EBF1 and EBF2 play a negative role in ethylene signaling by targeting EIN3 for degradation (Fig. 1.3). Interestingly, ethylene treatment results in an increase in the transcription level of EBF2, suggesting that there exists a negative feedback mechanism in ethylene signaling (Guo and Ecker 2003; Potuschak et al. 2003). When the ethylene signal is enhanced, the EIN3 protein becomes stabilized, which, in turn, induces the expression of EBF2. The accumulation of EBF2 is likely to suppress the high level of EIN3 protein to its basal level, thus restoring plant responsiveness to ethylene again (Guo and Ecker 2003; Cho and Yoo 2009).

It was shown that EIN2, EIN5, and EIN6 are positive regulators of EIN3 action (Li and Guo 2007). It was shown that *ein5* and *ein6* mutants were weakened in ethylene-induced EIN3 accumulation, but in *ein2* mutants, EIN3 accumulation was inhibited (Guo and Ecker 2003).

AP2/EREBP (APETALA2/ethylene-responsive element-binding protein) is a large family of transcription factor genes. The AP2/EREBP gene family has been divided into four subfamilies: AP2, RAV (related to ABI3/VP1), dehydration-responsive element-binding protein (DREB), and ERF (Sakuma et al. 2002). After identification of the ERF domain as a conserved motif in four DNA-binding proteins from tobacco (Ohme-Takagi and Shinshi 1995), many ERF-like genes have been identified from various plant species, such as *Arabidopsis* and rice (Nakano et al. 2006), tomato (Gu et al. 2000), soybean (Zhang et al. 2008), sugarcane (Trujillo et al. 2008), and two fruit crops, apple (Wang et al. 2007c) and plum (El-Sharkawy et al. 2009). To date, different members of plant ERF genes have been found to be mainly involved in response to biotic and abiotic stresses (Kizis et al. 2001; Agarwal et al. 2006; Trujillo et al. 2008; Zhang et al. 2009). Transcription factors encoded by genes in the DREB subfamily play an important role in the resistance of plants to abiotic stresses by recognizing the dehydration-responsive element (DRE), which has a core motif of A/GCCGAC (Liu et al. 1998). ERF and DREB subfamily transcription factors have been identified in various plant species, including rice (Cao et al. 2006), *Arabidopsis* (Liu et al. 1998), and cotton (Jin and Liu 2008). The roles of ERF and DREB proteins in the plant response to biotic and abiotic stresses have also been extensively documented (Agarwal et al. 2006, 2010).

Both DREB1 and DREB2 factors are induced by water stress or cold. Their transcripts accumulate at high levels shortly after initiation of the stress treatment. It was shown that DREB1 genes are induced by low temperature, whereas the DREB2 homologues are induced by drought and high salt stresses (Kizis et al. 2001). The increase in ethylene production occurred after wounding of tomato leaves (O'Donnell et al. 1996). Some genes, including ACC oxidase (ACO1, formerly TOM13) and PR genes, were induced by mechanical wounding (Pastuglia et al. 1997). Although ethylene alone is not sufficient to induce wound-responsive gene expression, it is required for activation of proteinase inhibitor genes by the wound response pathway (O'Donnell et al. 1996, 1998). Environmental stresses including drought, desiccation, and low temperature increased significantly the expression level of putative repressor LeERF3b, but markedly reduced the expression level of putative activator Pti4 (Chen et al. 2008).

Tobacco plants expressing JERF3 showed enhanced adaptation to drought, freezing, and osmotic stress during germination and seedling development. JERF3 activates the expression of genes through transcription, resulting in decreased accumulation of ROS and, in turn, enhanced adaptation to drought, freezing, and salt in tobacco (Wu et al. 2008).

A global analysis of transcriptional regulation in ethylene responses was performed with DNA microarrays. RNA levels of more than 22,000 genes in response to exogenous ethylene treatment or in various ethylene response mutants in *Arabidopsis* were examined. The expression levels of 628 genes were significantly altered by ethylene treatment, among which, 244 were induced and 384 were repressed (Alonso et al. 2003). When an EST-based microarray containing about 6,000 unique *Arabidopsis* genes has been examined, nearly 7% of the genes have been identified as ethylene-regulated (Zhong and Burns 2003). A kinetic analysis of the early response to ethylene using a cDNA microarray uncovered significant differences in gene expression among wild-type, *ctr1-1*, and *ein2-1* mutants (De Paepe et al. 2004). It was also found from these studies that overlap of genes regulated by ethylene and other signals, including JA, auxin, ABA, and sugar, suggested that many hormonal and signaling interactions might lie in the coordinated regulation of gene expression and ultimately will form a complex regulatory network (Schenk et al. 2000; De Paepe et al. 2004; Li and Guo 2007).

1.8 Salicylic Acid

Salicylic acid (SA), a phenolic secondary metabolite, plays a central role in defense response. It regulates both local disease resistance mechanisms, including host cell death and defense gene expression, and systemic acquired resistance (SAR) (Vlot et al. 2009). SA or its derivates function in diverse plant processes such as seed germination, seedling establishment, respiration, stomatal responses, senescence, thermotolerance, nodulation, and abiotic stress (Rajjou et al. 2006; Alonso-Ramirez et al. 2009; Norman et al. 2004; Manthe et al. 1992; Rao et al. 2002; Clarke et al. 2004; Stacey et al. 2006; Metwally et al. 2003). Moreover, genetic mutant studies in *Arabidopsis* suggest that SA is involved in modulating cell growth and trichome development (Rate et al. 1999; Traw and Bergelson 2003). However, its effects on most of these processes are minor and may be indirect because SA is excessively involved in crosstalk with other phytohormones or alter their biosynthesis (Pieterse et al. 2009).

Infection of the plants by viral, bacterial, or fungal pathogens results in an increase in SA levels. Accumulation of SA or its derivatives, mainly methyl salicylate (Me–SA) has been observed both at the site of infection and in distant tissues. Recognition of pathogen-associated molecular patterns (PAMPs) results in PAMP-triggered immunity (PTI, formerly called basal resistance) that prevents pathogen colonization. However, during the competition between pathogen and plants, pathogens have evolved effectors to suppress PAMP-triggered signals, and host plants, in turn, have evolved resistance (R) proteins to detect the presence of pathogen effectors and induce effector-triggered immunity (ETI, formerly termed R gene-mediated resistance) (Vlot et al. 2009; An and Mou 2011). One of the most important aspects of SA signaling is its role in SAR (Durrant and Dong 2004). SAR is a defense pathway that provides systemic protection to a broad range of pathogens. Pathogen attack results in an increase in SA levels both at the site of infection and at distant tissues. The response appears to require the synthesis of the volatile compound Me–SA at the infection site. Me–SA moves to other parts of the plant, where it is converted to SA by the protein SA-binding protein 2 (SABP2) (Durrant and Dong 2004; Santner et al. 2009).

Many components of SA signaling, including signal perception, have not been revealed yet (Santner et al. 2009). However, it is known that SA or its signaling is associated with the accumulation of reactive oxygen species (ROS) and the activation of diverse groups of defense-related genes, including those encoding pathogenesis-related (PR) proteins (Vlot et al. 2009). Moreover, nonexpresser of PR genes 1 (NPR1) protein and transcription factors such as TGACG-motif-binding factors (TGAs) and WRKYs have been identified as key components of SA response (Dong 2004; Boyle et al. 2009). NPR1 contains an ankyrin-repeat motif and a BTB/POZ domain. In the absence of SA or pathogen challenge, NPR1 is retained in the cytoplasm as an oligomer which is held together by intermolecular disulphide bridges. Increase in SA levels shifts the cellular redox state, and as a result, two cysteine residues (Cys82 and Cys216) are reduced by thioredoxin-H5 (TRX-H5) or TRX-H3 (Tada et al. 2008). NPR1 monomers are subsequently translocated into the nucleus where they promote the transcription of a large family of PR genes. Some PR proteins have antimicrobial activity, but in general, the function of these proteins has not been clearly defined. Besides redox state-controlled regulation of NPR1, regulation by protein degradation was also proposed for SA signaling. Since NPR1 is a member of the BTB domain family of proteins, it was suggested to be a subunit in an E3 ligase, which implies that SA action also involves regulated protein degradation (Gingerich et al. 2005).

NPR1 itself does not have DNA-binding capability (An and Mou 2011). However, it regulates transcription through interaction with TGA transcription factors. The TGA family of bZIP transcription factors can directly interact with the SAR marker gene PR-1 through binding to the activation sequence-1 (as-1) in its promoter region (Lebel et al. 1998). TGA factors that interact with NPR1 differentially regulate PR-1 expression in *Arabidopsis* (Kesarwani et al. 2007). TGA2 and NPR1 are activators of SAR and PR-1 in *A. thaliana*. TGA2 is a transcriptional repressor required for basal repression of PR-1, but during SAR, TGA2 recruits NPR1 as part of an enhanceosome (Boyle et al. 2009). Interaction between NPR1 and TGA1 or TGA4 was detected only upon SA treatment of leaves. The interaction depends on SA-induced changes to the redox environment that results in the reduction of two cysteine residues (Cys260 and Cys266) that are conserved in TGA1 and TGA4 (Despres et al. 2003). NPR1 and TGA1 are key redox-controlled regulators where NPR1 monomers interact with the reduced form of TGA1. Nitric oxide, another important messenger in plant defense signaling, was suggested to be a redox regulator of the NPR1/TGA1 system (Lindermayr et al. 2010). Besides TGAs, WRKY transcription factors have been suggested to play negative or positive regulatory roles in controlling PR gene expression. WRKY is a large family of proteins with up to 100 members in *Arabidopsis*. Overexpression of WRKY70 leads to constitutive PR gene expression, indicating that this transcription factor might be a positive regulator of PR genes. Expression of WRKY70 was shown to be activated by SA and repressed by JA (Li et al. 2004).

Microarray analysis was used to examine the role of NPR1 in the overall defense network. Hierarchical clustering of microarray data revealed that the expression of SA-mediated genes and of a much larger group of genes, whose expression requires JA and ethylene signaling, was affected in the *npr1-1* mutant (Glazebrook et al. 2003). NPR1 is a key regulatory component that is positioned at the crossroads of multiple defense pathways. Vlot et al. (2009) emphasized that induction of cell death by SA is in close cooperation with ROS and NO. Furthermore, the SA defense signal is potentiated by positive feedback loops of SA with NO, ROS, and couple of related gene products.

1.9 Brassinosteroids

The presence of brassinosteroids (BRs) in almost all tissues of a plant and in almost every species of plant kingdom has been demonstrated. BRs occur in free form and conjugated to sugars and fatty acids. They play critical roles in a range of developmental processes and in responses to environmental stress including abiotic constraints (Krishna 2003; Bajguz and Hayat 2009). Coordinated regulation of development in response to the stress requires an extensive crosstalk between phytohormones. Different components in the signaling network involving transcription, protein–protein interactions, and targeted protein destruction are nodes in crosstalk. One of the key players in a complex network of crosstalk is BRs. Crosstalk includes alternation in the expression of hormone biosynthetic genes and various signaling components (Bajguz and Hayat 2009).

It was shown that BRs increased ethylene production in mung bean epicotyl segments and increased effects of GA in azuki bean epicotyls (Arteca et al. 1983; Mayumi and Shibaoka 1995). Synergistic interaction of BRs with GA and auxin has been shown in *A. thaliana* seedlings during hypocotyl elongation (Tanaka et al. 2003). Furthermore, ABA is a known antagonist of BR signaling. Expression of proteins named BR enhanced expression (BEE1, BEE2, and BEE3) was repressed by ABA treatment. BEEs are members of bHLH transcription factors required for BR response in *Arabidopsis* (Friedrichsen et al. 2002). Stimulation of proline synthesis by ABA and salt stress was correlated with increase in expression of P5CS1, rate-limiting enzyme in proline biosynthesis. Both ABA and salt induction of P5CS1 transcription were inhibited by BRs in light-grown *Arabidopsis* plants. Thus, it was suggested that BRs might be negatively regulating proline accumulation which is a common salt and ABA response pathway (Abrahám et al. 2003). Expression of 12-oxo-phytodienoic acid reductase 3 (OPR3) gene, encoding an enzyme functioning in JA biosynthesis, was induced by BR treatment. This indicates a potential link between BR action and JA biosynthesis (Müssig et al. 2000). It was shown that exogenous application of BRs modified activities of antioxidant enzymes and cellular levels of nonenzymatic antioxidants in plants under different stress conditions (Nunez et al. 2003; Özdemir et al. 2004).

BRs had a stimulatory effect on the growth of drought-tolerant and drought-susceptible wheat varieties under stress conditions. Application of BR resulted in increased relative water content, nitrate reductase activity, chlorophyll content, and photosynthesis under both conditions (Sairam 1994). BR application relieved the salinity-induced inhibition of seed germination and seedling growth in rice. Moreover, BRs restored the level of chlorophylls and increased nitrate reductase activity under salt stress (Anuradha and Rao 2003).

Genetic screening for BR-signaling mutants in *Arabidopsis* resulted in the identification of BR-insensitive 1 (BRI1), encoding a leucine-rich repeat (LRR) receptor-like kinase (RLKs) (Li and Chory 1997). BRI1, localized to the plasma membrane, was demonstrated to be a critical component of a receptor complex for BRs. Direct binding of active BRs to BRI1 was shown using a biotin-tagged photoaffinity castasterone, a biosynthetic precursor of BR. Furthermore, minimal binding domain of BRI1 was determined using binding assays and recombinant BRI1 fragments (Kinoshita et al. 2005). In other plant species, it was also shown that mutations in BRI1 homologs were responsible for the BR-insensitive dwarf phenotype (Yamamuro et al. 2000; Bishop and Koncz 2002). The *Arabidopsis* BRI1-associated receptor kinase 1 (BAK1) was identified by a yeast two-hybrid assay. It was hypothesized that BRI1 and BAK1 function together, most likely through heterodimerization, to initiate BR signaling (Nam and Li 2002). Several other LRR-RLKs have been identified in *Arabidopsis* (Zhou et al. 2004; Cano-Delgado et al. 2004).

Other BR-signaling components including the cytoplasm-localized BR-insensitive 2 (BIN2) and three nuclear proteins, brassinazole-resistant 1 (BZR1), *bri1*-EMS-suppressor 1 (BES1), and *bri1* suppressor 1 (BSU1), are all members of different gene families. Binding of BRs to the extracellular domain of receptor kinase, BRI1 activates the receptor. Kinase activity of BRI1 releases the inhibitory BRI1 kinase inhibitor 1 (BKI1) protein from the plasma membrane and increases the affinity of BRI1 to BAK1 (Kinoshita et al. 2005; Wang and Chory 2006). Then BRI1 phosphorylates the BR-signaling kinases (BSKs), which, in turn, activate the downstream signaling cascades (Nam and Li 2002; Wang et al. 2008; Tang et al. 2008). The interaction between BRI1 and BAK1 leads to the dephosphorylation, dimerization, and consequent DNA binding of nuclear-localized BES1/BZR1 transcriptional factors, which, in turn, control the genomic response of BRs (Yin et al. 2005). The dephosphorylation of BES1/BZR1 is brought out by combination of inactivation of the BIN2 and activation of the phosphatase BSU1. In addition, 14-3-3 protein is required for the regulation of nucleocytoplasmic shuttling of phosphorylated BR transcription factors. It was also indicated that endocytosis of BRI1 is essential for the BR signaling (Russinova et al. 2004).

To identify genes that are subject to direct BR regulation, expression profiles of genes have been investigated with microarray analysis in either BR-deficient or BR-treated plants. Oxidative stress-related genes encoding monodehydroascorbate reductase and thioredoxin, the cold and drought stress response genes COR47 and COR78, and the heat stress-related genes HSP83, HSP70, HSF3, Hsc70-3, and Hsc70-G7 have been identified in transcriptome analysis (Müssig et al. 2002). Enhanced oxidative stress resistance was demonstrated in *det2* mutant, which is blocked in the biosynthetic pathway of BRs and has a loss-of-function mutation in DET2 gene (Cao et al. 2005).

1.10 Nitric Oxide

Nitric oxide (NO) has been shown to be involved in several plant functions, including defense response (Delledonne et al. 2001), growth and development (Beligni and Lamattina 2000), iron homeostasis (Murgia et al. 2002), and response to stresses such as drought (Garcia-Mata and Lamattina 2001), salt

(Zhao et al. 2004, 2007), and heat (Uchida et al. 2002). Both biotic and abiotic stresses alter NO production, additionally external application of NO donors enhances plant tolerance to specific stresses (Delledonne et al. 1998; Zhao et al. 2009). Another important role of NO in abiotic stress responses relies on its properties as a signaling molecule. NO is involved in the signaling pathway downstream of JA synthesis and upstream of H_2O_2 synthesis and regulates the expression of some genes involved in abiotic stress tolerance (Wendehenne et al. 2004).

Considerable efforts have been made to understand the response to NO at the molecular level. The physiological effects of NO signaling are tightly correlated to the modification of gene expression through NO-dependent processes. Several medium- and large-scale transcriptomic analyses have provided the identity of hundreds of putative NO-regulated genes (Huang et al. 2002; Polverari et al. 2003; Palmieri et al. 2008; Badri et al. 2008; Besson-Bard et al. 2009). Inhibitor of NO synthesis was helpful to understand the processes underlying NO signaling in plant cells and gene transcript accumulation. It was shown in the report of Palmieri et al. (2008) that NO in *Arabidopsis* induces several transcripts involved in signal transduction and basic metabolism. They have performed an *in silico* search for common transcription factor binding sites (TFBS) in the promoter regions of the selected genes. Eight families of TFBS occurred at least 15% more often in the promoter region of the candidate genes. Most of the TFBS correspond to the binding elements of stress-related transcriptional activators such as bZIP, WRKY transcription factors, strengthening a role of NO as a component of biotic or abiotic stress-related signaling pathways.

Parani et al. (2004) showed that EREBPs were induced by NO up to 13-fold over control expression. NO has also been reported to affect the DNA-binding property of transcription factors with zinc finger motifs (Kroncke et al. 2001). Most NO-modulated genes were also shown to be affected in abiotic stress-related conditions (Polverari et al. 2003). In the study of Parani et al. (2004), upregulation of transcripts of zinc finger proteins after treatment with 1.0 mM SNP, a donor of NO, was observed. NO treatment also induced transcripts coding for DREB1, DREB2, and LEA. These proteins are related to cold and drought tolerance in plants. It was also reported that MYB-related transcription factor, NAC domain protein, and WRYK-type transcription factor WRYK46 were upregulated. Other interesting transcripts induced were coding for oxidative stress-related proteins such as GSTs, ABC transporters, iron homeostasis proteins, and signal transduction factors (Parani et al. 2004). The activities of a variety of nuclear regulatory proteins are affected dramatically by NO. The formation of S-nitrosylated proteins seems to be an important mechanism in the regulation of the function/activity of transcription factors. In the case of the transcription factor AtMYB2, nitrosylation of Cys53 inhibits DNA binding, providing a functional link between S-nitrosylation and NO-dependent gene expression (Serpa et al. 2007).

1.11 Polyamines

Polyamines, including putrescine, spermidine, and spermine, are group of phytohormone-like aliphatic amine natural compounds with aliphatic nitrogen structure. Polyamines have been involved in many physiological processes, such as organogenesis, embryogenesis, and abiotic and biotic plant stress responses (Kumar et al. 1997; Walden et al. 1997; Bouchereau et al. 1999; Alcázar et al. 2006b; Kusano et al. 2008). It was shown that in response to a variety of abiotic stresses, changes in polyamine metabolism occur in plants (Bouchereau et al. 1999; Alcázar et al. 2006b; Groppa and Benavides 2008). Moreover, it has been noted that genetic transformation with polyamine biosynthetic genes encoding arginine decarboxylase (ADC), ornithine decarboxylase (ODC), S-adenosylmethionine decarboxylase (SAMDC), spermine synthase (SPMS), or spermidine synthase (SPDS) improved environmental stress tolerance in various plant species (Gill and Tuteja 2010). Polyamines could also inhibit DNA methylation, which permits expression of specific genes responsible for the synthesis of stress proteins (Kuznetsov and Shevyakova 2007).

Although it was indicated that, the levels of putrescine may account for 1.2% of the dry matter, representing at least 20% of the nitrogen in stressed plants (Galston 1991), the physiological significance of increased polyamine levels in abiotic stress responses is still unclear (Alcázar et al. 2006b; Kusano et al. 2008; Gill and Tuteja 2010).

Studies in the literature indicated that polyamines may act as cellular signals in complex crosstalk with hormonal pathways, including ABA regulation of abiotic stress responses. Transcript profiling by using quantitative real-time RT-PCR has revealed that the expression of ADC2, SPDS1, and SPMS genes was induced under water stress (Alcázar et al. 2006a). ABA treatment induced the expression of some of these genes (Perez-Amador et al. 2002; Urano et al. 2003). When the expression of ADC2, SPDS1, and SPMS was analyzed in the ABA-deficient (*aba2-3*) and ABA-insensitive (*abi1-1*) mutants subjected to water stress (Alcázar et al. 2006a), these three genes displayed reduced transcriptional induction compared to the wild type, indicating that ABA modulates polyamine metabolism at the transcription level by upregulating the expression of ADC2, SPDS1, and SPMS genes under water stress conditions (Alcázar et al. 2006a).

In addition, putrescine accumulation in the *aba2-3* and *abi1-1* mutants occurred under drought conditions when compared to wild-type plants. Metabolomic studies supported this result by showing polyamine responses to dehydration are also impaired in *nced3* mutants (Urano et al. 2009). These results brought the conclusion that upregulation of polyamine biosynthetic genes and accumulation of putrescine under water stress are mainly ABA-dependent responses.

Under salt stress conditions, it was indicated that there was a rapid increase in the expression of ADC2 and SPMS, resulting an increase in putrescine and spermine levels (Urano et al. 2003). The induction of both ADC2 and SPMS brings the idea that polyamine responses to salt stress are also ABA dependent.

It was also shown that stress-responsive, low temperature-responsive (LTR), drought-responsive (DRE), and ABA-responsive elements (ABRE- and/or ABRE-related motifs) are present in the promoters of the polyamine biosynthetic genes (Alcázar et al. 2006b). These results indicate that the expressions of some of the genes involved in polyamine biosynthesis are regulated by ABA under drought and salt treatments.

Transcript profiling has also revealed that cold enhances the expression of ADC1, ADC2, and SAMDC2 genes (Urano et al. 2003; Cuevas et al. 2008, 2009). While the levels of free spermidine and spermine remain constant or even decrease in response to cold treatment, free putrescine levels are increased on cold treatment with the induction of ADC genes (Cuevas et al. 2008). The *adc1* and *adc2* mutations caused higher sensitivity to freezing conditions, in both acclimated and nonacclimated plants, while addition of putrescine complemented this stress sensitivity (Cuevas et al. 2008, 2009). It was suggested that putrescine and ABA are integrated in a positive feedback loop, in which ABA and putrescine reciprocally promote biosynthesis in response to abiotic stress. When transcriptomic analysis of an ADC2 overexpressor line was performed, downregulation of several genes encoding transcription factors belonging to the AP2/ERF domain family, which are involved in salt, cold, and dehydration responses, (e.g., DREB1C, DREB2A) was observed (Alcázar et al. 2005).

The effect of polyamines on gene expression at transcriptional level was demonstrated, and it was proposed that this effect is determined by the direct interaction of polyamines with DNA and/or *trans*-acting protein factors (Lindemose et al. 2005). It was shown that polyamines activate protein phosphorylation and increase activities of certain protein kinases in plants (Tassoni et al. 1998). It was indicated in the report of Takahashi et al. (2004) that when spermine was exogenously applied, the expression of five hypersensitive response marker genes (e.g., SR203J, HMGR, HSR201, HSR515, and harpin-induced 1) was increased in tobacco leaves. On the other hand, spermine induces mitochondrial dysfunction and activation of SA-induced protein kinase (SIPK) and wound-induced protein kinase (WIPK). These two kinases are involved in the regulation of both defense gene expression and hypersensitive response-like cell death. Taken together, it might be suggested that all these components might be part of the same signaling pathway, or they might be key components of crosstalk between various pathways.

1.12 Crosstalk

Plants are frequently exposed to diverse biotic and abiotic stresses throughout their life cycle. In stress-induced growth processes, the initial stimulus seems to be translated into a hormone response that changes the hormonal regime in specific tissues or even in the whole plant. Because of intensive crosstalk, the increase in one hormone level can decrease the response to another; for example, stress hormones such as JA, ABA, and ethylene seem to negatively affect the levels of growth-promoting hormones, such as auxin and GA. Conversely, increased auxin levels during favorable conditions can reduce stress responses (Wolters and Jürgens 2009). One phytohormone regulates various developmental processes or stress responses, whereas a specific process may be coordinated by multiple phytohormones. In general, phytohormones modulate the competitive allocation of energy to stress response or growth in a complex network with a high degree of crosstalk.

Common themes in phytohormone signaling have emerged with the characterization of critical components involved in the phytohormone signaling. These themes are ubiquitin-dependent protein degradation by the 26 S proteasome, feedback regulatory loops for precise control of phytohormone response, and complex network of crosstalk between signaling pathways (Xiong et al. 2009). One of the most important routes, phytohormones exert their effects, is through the regulation of gene expression. A set of responsive downstream genes are induced or repressed upon perception of a specific hormone. However, these groups of genes are not unique. An important feature of hormone-responsive data sets is the frequent occurrence of genes associated with other hormone signaling pathways. A given physiological process can be regulated by different phytohormones through controlling the expression of a common set of downstream genes. Transcriptional repression on some of JA-responsive genes is relieved by DELLA proteins, the key negative regulator of GA signaling (Hou et al. 2010). Binding of DELLA to JAZ removes the repression on MYC2, and subsequently, the downstream JA-responsive genes are expressed.

Another node of crosstalk is at the level of biosynthesis, metabolism, or transport. Response of one phytohormone might affect the metabolism of another. Cytokinin-treated *Arabidopsis* seedlings displayed decreased expressions of GA biosynthetic genes (Brenner et al. 2005). It was also shown that GA metabolism genes were regulated by auxin signaling in *Arabidopsis* (Frigerio et al. 2006). Putative transcription factor BREVIS RADIX (BRX) provides molecular details for a negative feedback loop to maintain homeostasis between the BR and auxin pathways (Mouchel et al. 2006). Moreover, different signaling pathways could share common signaling components, leading to a more complicated phytohormone signaling than expected. It has been shown that the LRR kinase from tomato not only perceives BRs and initiates BR signaling but also binds to systemin and triggers the systemic wound responses (Montoya et al. 2002). Recent studies showed that BAK1 is able to interact with the bacterial flagellin receptor, flagellin-sensitive 2 (FLS2) and induce the burst of ROS and plant defense response to bacterial pathogens (Chinchilla et al. 2007). Molecular studies revealed that the crosstalk between different phytohormones represents a precisely coordinated web of nodes and lines. Considering the crosstalk among different hormone signaling pathways, the roles of hormone signaling in regulating expression of the genome seem very complex.

1.13 Conclusion

Recently, scientific understanding of the molecular mechanisms of phytohormone biosynthesis, perception, and signaling has been improved dramatically. Receptors, regulators, transcription factors, as well as downstream responsive genes and proteins have been identified. However, there are still major challenges and questions concerning interactions or crosstalk between different phytohormones and fine tuning of gene expression, especially in crops, under various environmental conditions. It is also crucial to address how phytohormone signaling and changes in gene expression are integrated into phenotype and specific agronomic traits. Further integration of molecular, biochemical, and physiological studies will help us answer these questions. Understanding phytohormone signaling in molecular and cellular details in crop plants will provide innovative tools for improving agricultural practices.

References

Abel S, Theologis A (1996) Early genes and auxin action. Plant Physiol 111:9–17

Abel S, Oeller PW, Theologis A (1994) Early auxin-induced genes encode short-lived nuclear proteins. Proc Natl Acad Sci USA 91:326–330

Abrahám E, Rigó G, Székely G, Nagy R, Koncz C, Szabados L (2003) Light-dependent induction of proline biosynthesis by abscisic acid and salt stress is inhibited by brassinosteroid in Arabidopsis. Plant Mol Biol 51:363–372

Achard P, Vriezen WH, Van Der Straeten D, Harberd NP (2003) Ethylene regulates Arabidopsis development via the modulation of DELLA protein growth repressor function. Plant Cell 15:2816–2825

Achard P, Cheng H, De Grauwe L, Decat J, Schoutteten H, Moritz T, Van Der Straeten D, Peng J, Harberd NP (2006) Integration of plant responses to environmentally activated phytohormonal signals. Science 311:91–94

Agarwal PK, Jha J (2010) Transcription factors in plants and ABA dependent and independent abiotic stress signaling. Biologia Plant 54:201–212

Agarwal PK, Agarwal P, Reddy MK, Sopory SK (2006) Roles of DREB transcription factors in biotic and abiotic stress tolerance in plants. Plant Cell Rep 25:1263–1274

Agarwal P, Agarwal PK, Joshi AJ, Sopory SK, Reddy MK (2010) Overexpression of PgDREB2A transcription factor enhances abiotic stress tolerance and activates downstream stress-responsive genes. Mol Biol Rep 37:1125–1135

Aharoni A, Dixit S, Jetter R, Thoenes E, van Arkel G, Pereira A (2004) The SHINE clade of AP2 domain transcription factors activates wax biosynthesis, alters cuticle properties, and confers drought tolerance when overexpressed in Arabidopsis. Plant Cell 16:2463–2480

Alcázar R, García-Martínez JL, Cuevas JC, Tiburcio AF, Altabella T (2005) Overexpression of ADC2 in Arabidopsis induces dwarfism and late-flowering through GA deficiency. Plant J 43:425–436

Alcázar R, Cuevas JC, Patrón M, Altabella T, Tiburcio AF (2006a) Abscisic acid modulates polyamine metabolism under water stress in Arabidopsis thaliana. Physiol Plant 128:448–455

Alcázar R, Marco F, Cuevas JC, Patrón M, Ferrando A, Carrasco P, Tiburcio AF, Altabella T (2006b) Involvement of polyamines in plant response to abiotic stress. Biotechnol Lett 28:1867–1876

Aleman L, Kitamura J, Abdel-mageed H, Lee J, Sun Y, Nakajima M, Ueguchi-Tanaka M, Matsuoka M, Allen RD (2008) Functional analysis of cotton orthologs of GA signal transduction factors GID1 and SLR1. Plant Mol Biol 68:1–16

Alonso JM, Stepanova AN, Solano R, Wisman E, Ferrari S, Ausubel FM, Ecker JR (2003) Five components of the ethylene response pathway identified in a screen for weak ethylene insensitive mutants in Arabidopsis. Proc Natl Acad Sci USA 100:2992–2997

Alonso-Ramirez A, Rodriguez D, Reyes D, Jimenez JA, Nicolas G, Lopez-Climent M, Gomez-Cadenas A, Nicolas C (2009) Crosstalk between gibberellins and salicylic acid in early stress responses in Arabidopsis thaliana seeds. Plant Signal Behav 4:750–751

Alvarez S, Marsh EL, Schroeder SG, Schachtman DP (2008) Metabolomic and proteomic changes in the xylem sap of maize under drought. Plant Cell Environ 31:325–340

Amir Hossain M, Lee Y, Cho JI, Ahn CH, Lee SK, Jeon JS, Kang H, Lee CH, An G, Park PB (2010) The bZIP transcription factor OsABF1 is an ABA responsive element binding factor that enhances abiotic stress signaling in rice. Plant Mol Biol 72:557–566

An C, Mou Z (2011) Salicylic acid and its function in plant immunity. J Integr Plant Biol 53:412–428

Anderberg RJ, Walker-Simmons MK (1992) Isolation of a wheat cDNA clone for an abscisic acid-inducible transcript with homology to protein kinases. Proc Natl Acad Sci USA 89:10183–10187

Anuradha S, Rao SSR (2003) Application of brassinosteroids to rice seeds (Oryza sativa L.) reduced the impact of salt stress on growth, prevented photosynthetic pigments loss and increased nitrate reductase activity. Plant Growth Regul 40:29–32

Argueso CT, Ferreira FJ, Kieber J (2009) Environmental perception avenues: the interaction of cytokinin and environmental response pathways. Plant Cell Environ 32:1147–1160

Argueso CT, Raines T, Kieber JJ (2010) Cytokinin signaling and transcriptional networks. Curr Opin Plant Biol 13:533–539

Argyros RD, Mathews DE, Chiang YH, Palmer CM, Thibault DM, Etheridge N, Argyros DA, Mason MG, Kieber JJ, Schaller GE (2008) Type B response regulators of Arabidopsis play key roles in cytokinin signaling and plant development. Plant Cell 20:2102–2116

Arnaud N, Grain T, Sorefan K, Fuentes S, Wood TA, Lawrenson T, Sablowski R, Ostergaard L (2010) Gibberellins control fruit patterning in Arabidopsis thaliana. Gene Dev 24:2127–2132

Arteca RN, Tsai DS, Schlagnhaufer C, Mandava NB (1983) The effect of brassinosteroids on auxin-induced ethylene production by etiolated mung bean segments. Physiol Plant 59:539–544

Asakura Y, Hagino T, Ohta Y, Aoki K, Yonekura-Sakakibara K, Deji A, Yamaya T, Sugiyama T, Sakakibara H (2003) Molecular characterization of His-Asp phosphorelay signaling factors in maize leaves: implications of the signal divergence by cytokinin-inducible response regulators in the cytosol and the nuclei. Plant Mol Biol 52:331–341

Badri DV, Loyola-Vargas VM, Du J, Stermitz FR, Broeckling CD, Iglesias-Andreu L, Vivanco JM (2008) Transcriptome analysis of Arabidopsis roots treated with signaling compounds: a focus on signal transduction, metabolic regulation and secretion. New Phytol 179:209–223

Bajguz A, Hayat S (2009) Effects of brassinosteroids on the plant responses to environmental stresses. Plant Physiol Biochem 47:1–8

Balbi V, Devoto A (2008) Jasmonate signalling network in Arabidopsis thaliana: crucial regulatory nodes and new physiological scenarios. New Phytol 177:301–318

Ballas N, Wong LM, Theologis A (1993) Identification of the auxin-responsive element, AuxRE, in the primary indole acetic acid-inducible gene, PS-IAA4/5, of pea (Pisum sativum). J Mol Biol 233:580–596

Banno H, Ikeda Y, Niu QW, Chua NH (2001) Overexpression of Arabidopsis ESR1 induces initiation of shoot regeneration. Plant Cell 13:2609–2618

Bano A, Hansen H, Dörffling K, Hahn H (1994) Changes in the contents of free and conjugated abscisic acid, phaseic acid and cytokinins in xylem sap of drought stressed sunflower plants. Phytochemistry 37:345–347

Bari R, Jones JD (2009) Role of plant hormones in plant defence responses. Plant Mol Biol 69:473–488

Beckers GJ, Spoel SH (2006) Fine-tuning plant defence signalling: salicylate versus jasmonate. Plant Biol (Stuttg) 8:1–10

Beligni MV, Lamattina L (2000) Nitric oxide stimulates seed germination and de-etiolation, and inhibits hypocotyl elongation, three light-inducible responses in plants. Planta 210:215–221

Berleth T, Krogan NT, Scarpella E (2004) Auxin signals-turning genes on and turning cells around. Curr Opin Plant Biol 7:553–563

Berrocal-Lobo M, Molina A, Solano R (2002) Constitutive expression of ETHYLENE-RESPONSE-FACTOR1 in Arabidopsis confers resistance to several necrotrophic fungi. Plant J 29:23–32

Besson-Bard A, Gravot A, Richaud P, Auroy P, Duc C, Gaymard F, Taconnat L, Renou JP, Pugin A, Wendehenne D (2009) Nitric oxide contributes to cadmium toxicity in Arabidopsis by promoting cadmium accumulation in roots and by up-regulating genes related to iron uptake. Plant Physiol 149:1302–1315

Bishop GJ, Koncz C (2002) Brassinosteroid insensitive 1 and plant steroid signaling. Plant Cell 14: S97–S110

Bolle C (2004) The role of GRAS proteins in plant signal transduction and development. Planta 218:683–692

Boter M, Ruiz-Rivero O, Abdeen A, Prat S (2004) Conserved MYC transcription factors play a key role in jasmonate signaling both in tomato and Arabidopsis. Genes Dev 18:1577–1591

Bouchereau A, Aziz A, Larher F, Martin-Tanguy J (1999) Polyamines and environmental challenges: recent development. Plant Sci 140:103–125

Boyle P, Le Su E, Rochon A, Shearer HL, Murmu J, Chu JY, Fobert PR, Despres C (2009) The BTB/POZ domain of the Arabidopsis disease resistance protein NPR1 interacts with the repression domain of TGA2 to negate its function. Plant Cell 21:3700–3713

Brady SM, Sarkar SF, Bonetta D, McCourt P (2003) The ABSCISIC ACID INSENSITIVE 3 (ABI3) gene is modulated by farnesylation and is involved in auxin signaling and lateral root development in Arabidopsis. Plant J 34:67–75

Brandstatter I, Kieber JJ (1998) Two genes with similarity to bacterial response regulators are rapidly and specifically induced by cytokinin in Arabidopsis. Plant Cell 10:1009–1020

Braun N, Wyrzykowska J, Muller P, David K, Couch D, Perrot-Rechenmann C, Fleming AJ (2008) Conditional repression of AUXIN BINDING PROTEIN1 reveals that it coordinates cell division and cell expansion during postembryonic shoot development in Arabidopsis and tobacco. Plant Cell 20:2746–2762

Brenner WG, Romanov GA, Kollmer I, Burkle L, Schmulling T (2005) Immediate-early and delayed cytokinin response genes of Arabidopsis thaliana identified by genome-wide expression profiling reveal novel cytokinin-sensitive processes and suggest cytokinin action through transcriptional cascades. Plant J 44:314–333

Broun P (2004) Transcription factors as tools for metabolic engineering in plants. Curr Opin Plant Biol 7:202–209

Brugiere N, Jiao S, Hantke S, Zinselmeier C, Roessler JA, Niu X, Jones RJ, Habben JE (2003) Cytokinin oxidase gene expression in maize is localized to the vasculature, and is induced by cytokinins, abscisic acid, and abiotic stress. Plant Physiol 132:1228–1240

Bu Q, Jiang H, Li CB, Zhai Q, Zhang J, Wu X, Sun J, Xie Q, Li C (2008) Role of the Arabidopsis thaliana NAC transcription factors ANAC019 and ANAC055 in regulating jasmonic acid-signaled defense responses. Cell Res 18:756–767

Cano-Delgado A, Yin Y, Yu C, Vafeados D, Mora-García S, Cheng JC, Nam KH, Li J, Chory J (2004) BRL1 and BRL3 are novel brassinosteroid receptors that function in vascular differentiation in Arabidopsis. Development 131:5341–5351

Cao S, Xu Q, Cao Y, Qian K, An K, Zhu Y, Binzeng H, Zhao H, Kua B (2005) Loss-of function mutations in DET2 gene lead to an enhanced resistance to oxidative stress in Arabidopsis. Physiol Plant 123:57–66

Cao Y, Song F, Goodman RM, Zhong Z (2006) Molecular characterization of four rice genes encoding ethylene-responsive transcriptional factors and their expressions in response to biotic and abiotic stress. J Plant Physiol 163:1167–1178
Casaretto J, Ho TH (2003) The transcription factors HvABI5 and HvVP1 are required for the ABA induction of gene expression in barley aleurone cells. Plant Cell 15:271–284
Chandler PM, Marion-Poll A, Ellis M, Gubler F (2002) Mutants at the Slender1 locus of Barley cv. Himalaya. Molecular and physiological characterization. Plant Phys 129:181–190
Chandler PM, Harding CA, Ashton AR, Mulcair MD, Dixon NE, Mander LN (2008) Characterization of gibberellin receptor mutants of barley (Hordeum vulgare L.). Mol Plant 1:285–294
Chao Q, Rothenberg M, Solano R, Roman G, Terzaghi W, Ecker JR (1997) Activation of the ethylene gas response pathway in Arabidopsis by the nuclear protein ETHYLENE-INSENSITIVE3 and related proteins. Cell 89:1133–1144
Chapman EJ, Estelle M (2009) Mechanism of auxin-regulated gene expression in plants. Annu Rev Genet 43:265–285
Chen G, Hu Z, Grierson D (2008) Differential regulation of tomato ethylene responsive factor LeERF3b, a putative repressor, and the activator Pti4 in ripening mutants and in response to environmental stresses. J Plant Physiol 165:662–670
Chinchilla D, Zipfel C, Robatzek S et al (2007) A flagellin-induced complex of the receptor FLS2 and BAK1 initiates plant defence. Nature 448:497–500
Chini A, Fonseca S, Fernández G, Adie B, Chico JM, Lorenzo O, García-Casado G, López-Vidriero I, Lozano FM, Ponce MR, Micol JL, Solano R (2007) The JAZ family of repressors is the missing link in jasmonate signalling. Nature 448:666–671
Cho YH, Yoo SD (2009) Emerging complexity of ethylene signal transduction. J Plant Biol 52:283–288
Choi H, Hong J, Ha J, Kang J, Kim SY (2000) ABFs, a family of ABA responsive element binding factors. J Biol Chem 275:1723–1730
Clarke SM, Mur LAJ, Wood JE, Scott IM (2004) Salicylic acid dependent signaling promotes basal thermotolerance but is not essential for acquired thermotolerance in *Arabidopsis thaliana*. Plant J 38:432–437
Cuevas JC, Lopez-Cobollo R, Alcázar R, Zarza X, Koncz C, Altabella T, Salinas J, Tiburcio AF, Ferrando A (2008) Putrescine is involved in Arabidopsis freezing tolerance and cold acclimation by regulating abscisic acid levels in response to low temperature. Plant Physiol 148:1094–1105
Cuevas JC, Lopez-Cobollo R, Alcazar R, Zarza X, Koncz C, Altabella T, Salinas J, Tiburcio AF, Ferrando A (2009) Putrescine as a signal to modulate the indispensable ABA increase under cold stress. Plant Signal Behav 4:219–220
Dai C, Xue HW (2010) Rice early flowering1, a CKI, phosphorylates DELLA protein SLR1 to negatively regulate gibberellin signalling. EMBO J 29:1916–1927
De Paepe A, Vuylsteke M, Van Hummelen P, Zabeau M, Van Der Straeten D (2004) Transcriptional profiling by cDNA-AFLP and microarray analysis reveals novel insights into the early response to ethylene in Arabidopsis. Plant J 39:537–559
Delledonne M, Xia Y, Dixon RA, Lamb C (1998) Nitric oxide functions as a signal in plant disease resistance. Nature 394:585–588
Delledonne M, Zeier J, Marocco A, Lamb C (2001) Signal interactions between nitric oxide and reactive oxygen intermediates in the plant hypersensitive disease resistance response. Proc Natl Acad Sci USA 98:13454–13459
Despres C, Chubak C, Rochon A, Clark R, Bethune T, Desveaux D, Fobert PR (2003) The Arabidopsis NPR1 disease resistance protein is a novel cofactor that confers redox regulation of DNA binding activity to the basic domain/leucine zipper transcription factor TGA1. Plant Cell 15:2181–2191
Dharmasiri N, Dharmasiri S, Estelle M (2005) The F-box protein TIR1 is an auxin receptor. Nature 435:441–445

Dill A, Jung HS, Sun TP (2001) The DELLA motif is essential for gibberellin-induced degradation of RGA. Proc Natl Acad Sci USA 98:14162–14167
Dong X (2004) NPR1, all things considered. Curr Opin Plant Biol 7:547–552
Dubouzet JG, Sakuma Y, Ito Y, Kasuga M, Dubouzet EG, Miura S, Seki M, Shinozaki K, Yamaguchi-Shinozaki K (2003) OsDREB genes in rice, Oryza sativa L., encode transcription activators that function in drought-, high-salt- and cold-responsive gene expression. Plant J 33:751–763
Durrant WE, Dong X (2004) Systemic acquired resistance. Annu Rev Phytopathol 42:185–209
Ellis CM, Nagpal P, Young JC, Hagen G, Gulifoyle TJ, Reed JW (2005) AUXIN RESPONSE FACTOR1 and AUXIN RESPONSE FACTOR2 regulate senescence and floral organ abscission in Arabidopsis thaliana. Development 132:4563–4574
El-Sharkawy I, Sherif S, Mila I, Bouzayen M, Jayasankar S (2009) Molecular characterization of seven genes encoding ethylene-responsive transcriptional factors during plum fruit development and ripening. J Exp Bot 60:907–922
Feng S, Martinez C, Gusmaroli G, Wang Y, Zhou J, Wang F, Chen L, Yu L, Iglesias-Pedraz JM, Kircher S, Schäfer E, Fu X, Fan LM, Deng XW (2008) Coordinated regulation of Arabidopsis thaliana development by light and gibberellins. Nature 451:475–479
Fonseca S, Chico JM, Solano R (2009) The jasmonate pathway: the ligand, the receptor and the core signalling module. Curr Opin Plant Biol 12:539–547
Friedrichsen DM, Nemhauser J, Muramitsu T, Maloof JN, Alonso J, Ecker JR, Furuya M, Chory J (2002) Three redundant brassinosteroid early response genes encode putative bHLH transcription factors required for normal growth. Genetics 162:1445–1456
Frigerio M, Alabadi D, Perez-Gomez J et al (2006) Transcriptional regulation of gibberellin metabolism genes by auxin signaling in Arabidopsis. Plant Physiol 142:553–563
Fu X, Harberd NP (2003) Auxin promotes Arabidopsis root growth by modulating gibberellin response. Nature 421:740–743
Fu X, Richards DE, Ait-Ali T, Hynes LW, Ougham H, Peng J, Harberd NP (2002) Gibberellin-mediated proteasome-dependent degradation of the barley DELLA protein SLN1 repressor. Plant Cell 14:3191–3200
Fujii H, Zhu J (2009) Arabidopsis mutant deficient in 3 abscisic acid-activated protein kinases reveals critical roles in growth, reproduction, and stress. Proc Natl Acad Sci USA 106:8380–8385
Fujii H, Chinnusamy V, Rodrigues A, Rubio S, Antoni R, Park SY, Cutler SR, Sheen J, Rodriguez PL, Zhu JK (2009) In vitro reconstitution of an abscisic acid signalling pathway. Nature 462:660–664
Fujimoto SY, Ohta M, Usui A, Shinshi H, Ohme-Takagi M (2000) Arabidopsis ethylene-responsive element binding factors act as transcriptional activators or repressors of GCC box-mediated gene expression. Plant Cell 12:393–405
Fujita Y, Fujita M, Satoh R, Maruyama K, Parvez MM, Seki M, Hiratsu K, Ohme-Takagi M, Shinozaki K, Yamaguchi-Shinozaki K (2005) AREB1 is a transcription activator of novel ABRE dependent ABA signaling that enhances drought stress tolerance in Arabidopsis. Plant Cell 17:3470–3488
Fukaki H, Nakao Y, Okushima Y, Theologis A, Tasaka M (2005) Tissue specific expression of stabilized SOLITARY-ROOT/IAA14 alters lateral root development in Arabidopsis. Plant J 44:382–395
Galston AW (1991) On the trail of a new regulatory system in plants. New Biol 3:450–453
Garcia-Mata C, Lamattina L (2001) Nitric oxide induces stomatal closure and enhances the adaptive plant responses against drought stress. Plant Physiol 126:1196–1204
Gazarian IG, Lagrimini LM, Mellon FA, Naldrett MJ, Ashby GA, Thorneley RN (1998) Identification of skatolyl hydroperoxide and its role in the peroxidase-catalysed oxidation of indol-3-yl acetic acid. Biochem J 333:223–232
Geiger D, Scherzer S, Mumm P, Stange A, Marten I, Bauer H, Ache P, Matschi S, Liese A, Al-Rasheid KA, Romeis T, Hedrich R (2009) Activity of guard cell anion channel SLAC1 is controlled by drought-stress signaling kinase–phosphatase pair. Proc Natl Acad Sci USA 106:21425–21430
Gfeller A, Liechti R, Farmer EE (2010) Arabidopsis jasmonate signaling pathway. Sci Signal 3:cm4
Gill SS, Tuteja N (2010) Polyamines and abiotic stress tolerance in plants. Plant Signal Behav 51:26–33

Gingerich DJ, Gagne JM, Salter DW, Hellmann H, Estelle M, Ma L, Vierstra RD (2005) Cullins 3a and 3b assemble with members of the broad complex/tramtrack/bric-a-brac (BTB) protein family to form essential ubiquitin-protein ligases (E3s) in Arabidopsis. J Biol Chem 280:18810–18821
Glazebrook J, Chen W, Estes B, Chang HS, Nawrath C, Metraux JP, Zhu T, Katagiri F (2003) Topology of the network integrating salicylate and jasmonate signal transduction derived from global expression phenotyping. Plant J 34:217–228
Goda H, Sasaki E, Akiyama K, Maruyama-Nakashita A, Nakabayashi K et al (2008) The AtGenExpress hormone and chemical treatment data set: experimental design, data evaluation, model data analysis and data access. Plant J 55:526–542
Gomi K, Sasaki A, Itoh H, Ueguchi-Tanaka M, Ashikari M, Kitano H, Matsuoka M (2004) GID2, an F-box subunit of the SCF E3 complex, specifically interacts with phosphorylated SLR1 protein and regulates the gibberellin-dependent degradation of SLR1 in rice. Plant J 37:626–634
Gray WM (2004) Hormonal regulation of plant growth and development. PLoS Biol 2:E311
Gray WM, Kepinski S, Rouse D, Leyser O, Estelle M (2001) Auxin regulates SCFTIR1-dependent degradation of AUX/IAA proteins. Nature 414:271–276
Griffiths J, Murase K, Rieu I, Zentella R, Zhang ZL, Powers SJ, Gong F, Phillips AL, Hedden P, Sun TP, Thomas SG (2006) Genetic characterization and functional analysis of the GID1 gibberellin receptors in Arabidopsis. Plant Cell 18:3399–3414
Groppa MD, Benavides MP (2008) Polyamines and abiotic stress: recent advances. Amino Acids 34:35–45
Gu YQ, Yang C, Thara VK, Zhou J, Martin GB (2000) Pti4 is induced by ethylene and salicylic acid, and its product is phosphorylated by the Pto kinase. Plant Cell 12:771–786
Gu YQ, Wildermuth MC, Chakravarthy S, Loh YT, Yang CM, He XH, Han Y, Martin GB (2002) Tomato transcription factors Pti4, Pti5, and Pti6 activate defense responses when expressed in Arabidopsis. Plant Cell 14:817–831
Gubler F, Raventos N, Keys M, Watts R, Mundy J, Jacobsen JV (1999) Target genes and regulatory domains of the GAMYB transcriptional activator in cereal aleurone. Plant J 17:1–9
Gubler F, Chandler PM, White RG, Llewellyn DJ, Jacobsen JV (2002) Gibberellin signaling in barley aleurone cells. Control of SLN1 and GAMYB expression. Plant Physiol 129:191–200
Guilfoyle TJ, Hagen G (2007) Auxin response factors. Curr Opin Plant Biol 10:453–460
Guo H, Ecker JR (2003) Plant responses to ethylene gas are mediated by SCFEBF1/EBF2-dependent proteolysis of EIN3 transcription factor. Cell 115:667–677
Guo J, Zeng Q, Emami M, Ellis BE, Chen JG (2008) The GCR2 gene family is not required for ABA control of seed germination and early seedling development in Arabidopsis. PLoS One 3: e2982
Hagen G, Guilfoyle TJ (2002) Auxin-responsive gene expression: genes, promoters and regulatory factors. Plant Mol Biol 49:373–385
Hamann T, Benkova E, Baurle I, Kientz M, Jürgens G (2004) The Arabidopsis BODENLOS gene encodes an auxin response protein inhibiting MONOPTEROS-mediated embryo patterning. Gene Dev 16:1610–1615
Hannah MA, Heyer AG, Hincha DK (2005) A global survey of gene regulation during cold acclimation in *Arabidopsis thaliana*. PLoS Genet 1:179–196
Harberd NP, Belfield E, Yasumura Y (2009) The angiosperm gibberellin-GID1-DELLA growth regulatory mechanism: how an "inhibitor of an inhibitor" enables flexible response to fluctuating environments. Plant Cell 21:1328–1339
Hare PD, Cress WA, van Staden J (1997) The involvement of cytokinins in plant responses to environmental stress. Plant Growth Regul 23:79–103
Hartweck LM (2008) Gibberellin signaling. Planta 229:1–13
Heinekamp T, Strathmann A, Kuhlmann M, Froissard M, Müller A, Perrot-Rechenmann C, Dröge-Laser W (2004) The tobacco bZIP transcription factor BZI-1 binds the GH3 promoter in vivo and modulates auxin-induced transcription. Plant J 38:298–309
Himmelbach A, Yang Y, Grill E (2003) Relay and control of abscisic acid signaling. Curr Opin Plant Biol 6:470–479

Hoth S, Morgante M, Sanchez JP, Hanafey MK, Tingey SV, Chua NH (2002) Genome-wide gene expression profiling in Arabidopsis thaliana reveals new targets of abscisic acid and largely impaired gene regulation in the abi1-1 mutant. J Cell Sci 115:4891–4900
Hou X, Hu WW, Shen L, Lee LY, Tao Z, Han JH, Yu H (2008) Global identification of DELLA target genes during Arabidopsis flower development. Plant Physiol 147:1126–1142
Hou X, Lee LY, Xia K, Yan Y, Yu H (2010) DELLAs modulate jasmonate signaling via competitive binding to JAZs. Dev Cell 19:884–894
Huang X, von Rad U, Durner J (2002) Nitric oxide induces the nitric oxide-tolerant alternative oxidase in Arabidopsis suspension cells. Planta 215:914–923
Huerta L, Forment J, Gadea J, Fagoaga C, Peña L, Pérez-Amador MA, García-Martínez JL (2008) Gene expression analysis in citrus reveals the role of gibberellins on photosynthesis and stress. Plant Cell Environ 31:1620–1633
Husbands A, Bell EM, Shuai B, Smith HM, Springer PS (2007) LATERAL ORGAN BOUNDARIES defines a new family of DNA-binding transcription factors and can interact with specific bHLH proteins. Nucleic Acids Res 35:6663–6671
Hussain A, Peng J (2003) DELLA Proteins and GA Signalling in Arabidopsis. J Plant Growth Regul 22:134–140
Hussain A, Cao D, Cheng H, Wen Z, Peng J (2005) Identification of the conserved serine/threonine residues important for gibberellin-sensitivity of Arabidopsis RGL2 protein. Plant J 44:88–99
Hwang I, Sheen J (2001) Two-component circuitry in Arabidopsis signal transduction. Nature 413:383–389
Ikeda A, Ueguchi-Tanaka M, Sonoda Y, Kitano H, Koshioka M, Futsuhara Y, Matsuoka M, Yamaguchi J (2001) Slender rice, a constitutive gibberellin response mutant, is caused by a null mutation of the SLR1 gene, an ortholog of the height-regulating gene GAI/RGA/RHT/D8. Plant Cell 13:999–1010
Ishida K, Yamashino T, Yokoyama A, Mizuno T (2008) Three type-B response regulators, ARR1, ARR10 and ARR12, play essential but redundant roles in cytokinin signal transduction throughout the life cycle of *Arabidopsis thaliana*. Plant Cell Physiol 49:47–57
Itoh H, Sasaki A, Ueguchi-Tanaka M, Ishiyama K, Kobayashi M, Hasegawa Y, Minami E, Ashikari M, Matsuoka M (2005) Dissection of the phosphorylation of rice DELLA protein, SLENDER RICE1. Plant Cell Physiol 46:1392–1399
Iwakawa H, Ueno Y, Semiarti E, Onouchi H, Kojima S, Tsukaya H, Hasebe M, Soma T, Ikezaki M, Machida C, Machida Y (2002) The ASYMMETRIC LEAVES2 gene of Arabidopsis thaliana, required for formation of a symmetric flat leaf lamina, encodes a member of a novel family of proteins characterized by cysteine repeats and a leucine zipper. Plant Cell Physiol 43:467–478
Jain M, Khurana JP (2009) Transcript profiling reveals diverse roles of auxin-responsive genes during reproductive development and abiotic stress in rice. FEBS J 276:3148–3162
Jain M, Tyagi AK, Khurana JP (2006) Molecular characterization and differential expression of cytokinin-responsive type-A response regulators in rice (Oryza sativa). BMC Plant Biol 6:1
Jain M, Ghanashyam C, Bhattacharjee A (2010) Comprehensive expression analysis suggests overlapping and specific roles of rice glutathione S-transferase genes during development and stress responses. BMC Genomics 11:73
Jakoby M, Weisshaar B, Dröge-Laser W, Vicente-Carbajosa J, Tiedemann J, Kroj T, Parcy F (2002) bZIP transcription factors in Arabidopsis. Trends Plant Sci 7:106–111
Jin LG, Liu JY (2008) Molecular cloning, expression profile and promoter analysis of a novel ethylene responsive transcription factor gene GhERF4 from cotton (Gossypium hirstum) Plant Physiol Bioch 46:46–53
Johnson MA, Perez-Amador MA, Lidder P, Green PJ (2000) Mutants of Arabidopsis defective in a sequence-specific mRNA degradation pathway. Proc Natl Acad Sci USA 97:13991–13996
Johnson RR, Wagner RL, Verhey SD, Walker-Simmons MK (2002) The abscisic acid-responsive kinase PKABA1 interacts with a seed-specific abscisic acid response element-binding factor, TaABF, and phosphorylates TaABF peptide sequences. Plant Physiol 130:837–846

Kagaya Y, Hobo T, Murata M, Ban A, Hattori T (2002) Abscisic acid-induced transcription is mediated by phosphorylation of an abscisic acid response element binding factor, TRAB1. Plant Cell 14:3177–3189

Kang DJ, Seo YJ, Lee JD, Ishii R, Kim KU, Shin DH, Park SK, Jang SW, Lee IJ (2005) Jasmonic acid differentially affects growth, ion uptake and abscisic acid concentration in salt-tolerant and salt-sensitive rice cultivars. J Agro Crop Sci 191:273–282

Katsir L, Schilmiller AL, Staswick PE, He SY, Howe GA (2008) COI1 is a critical component of a receptor for jasmonate and the bacterial virulence factor coronatine. Proc Natl Acad Sci USA 105:7100–7105

Kepinski S, Leyser O (2005) The Arabidopsis F-box protein TIR1 is an auxin receptor. Nature 435:446–451

Kesarwani M, Yoo J, Dong X (2007) Genetic interactions of TGA transcription factors in the regulation of pathogenesis-related genes and disease resistance in Arabidopsis. Plant Physiol 144:336–346

Kieber JJ, Rothenberg M, Roman G, Feldmann KA, Ecker JR (1993) CTR1, a negative regulator of the ethylene response pathway in Arabidopsis, encodes a member of the raf family of protein kinases. Cell 72:427–441

Kim J, Harter K, Theologis A (1997) Protein–protein interactions among the Aux/IAA proteins. Proc Natl Acad Sci USA 94:11786–11791

Kim S, Kang JY, Cho DI, Park JH, Kim SY (2004) ABF2, an ABRE-binding bZIP factor, is an essential component of glucose signaling and its overexpression affects multiple stress tolerance. Plant J 40:75–87

Kinoshita T, Cano-Delgado A, Seto H, Hiranuma S, Fujioka S (2005) Binding of brassinosteroids to the extracellular domain of plant receptor kinase BRI1. Nature 433:167–171

Kizis D, Lumbreras V, Pagès M (2001) Role of AP2/EREBP transcription factors in gene regulation during abiotic stress. FEBS Lett 498:187–189

Kobayashi Y, Murata M, Minami H, Yamamoto S, Kagaya Y, Hobo T, Yamamoto A, Hattori T (2005) Abscisic acid-activated SnRK2 protein kinases function in the gene-regulation pathway of ABA signal transduction by phosphorylating ABA response element-binding factors. Plant J 44:939–949

Kovtun Y, Chiu WL, Zeng W, Sheen J (1998) Suppression of auxin signal transduction by a MAPK cascade in higher plants. Nature 395:716–720

Kovtun Y, Chiu WL, Tena G, Sheen J (2000) Functional analysis of oxidative stress-activated mitogen-activated protein kinase cascade in plants. Proc Natl Acad Sci USA 97:2940–2945

Krishna P (2003) Brassinosteroid-mediated stress responses. J Plant Growth Regul 22:289–297

Kroncke KD, Fehsel K, Suschek C, Kolb-Bachofen V (2001) Inducible nitric oxide synthase-derived nitric oxide in gene regulation, cell death and cell survival. Int Immunopharmacol 1:1407–1420

Kumar A, Altabella T, Taylor M, Tiburcio AF (1997) Recent advances in polyamine research. Trends Plant Sci 2:124–130

Kusano T, Berberich T, Tateda C, Takahashi Y (2008) Polyamines: essential factors for growth and survival. Planta 228:367–381

Kuznetsov Vl, Shevyakova NI (2007) Polyamines and stress tolerance of plants. Plant Stress 1:50–71

Lebel E, Heifetz P, Thorne L, Uknes S, Ryals J, Ward E (1998) Functional analysis of regulatory sequences controlling PR-1 gene expression in Arabidopsis. Plant J 16:223–233

Lee HW, Kim NY, Lee DJ, Kim J (2009) LBD18/ASL20 regulates lateral root formation in combination with LBD16/ASL18 downstream of ARF7 and ARF19 in Arabidopsis. Plant Physiol 151:1377–1389

Lehmann J, Atzorn R, Brückner C, Reinbothe S, Leopold J, Wasternack C, Parthier B (1995) Accumulation of jasmonate, abscisic acid, specific transcripts and proteins in osmotically stressed barley leaf segments. Planta 197:156–162

Leung J, Bouvier-Durand M, Morris PC, Guerrier D, Chefdor F, Giraudat J (1994) Arabidopsis ABA response gene ABI1: features of a calcium-modulated protein phosphatase. Science 264:1448–1452
Leung J, Merlot S, Giraudat J (1997) The Arabidopsis ABSCISIC ACID-INSENSITIVE2 (ABI2) and ABI1 genes encode homologous protein phosphatases 2 C involved in abscisic acid signal transduction. Plant Cell 9:759–771
Lewsey MG, Murphy AM, Maclean D, Dalchau N, Westwood JH, Macaulay K, Bennett MH, Moulin M, Hanke DE, Powell G, Smith AG, Carr JP (2010) Disruption of two defensive signaling pathways by a viral RNA silencing suppressor. Mol Plant Microbe Interact 23:835–845
Li J, Chory J (1997) A putative leucine-rich repeat receptor kinase involved in brassinosteroid signal transduction. Cell 90:929–938
Li H, Guo H (2007) Molecular basis of the ethylene signaling and response pathway in Arabidopsis. J Plant Growth Regul 26:106–117
Li J, Wang X, Watson MB, Assmann SM (2000) Regulation of abscisic acid-induced stomatal closure and anion channels by guard cell AAPK kinase. Science 287:300–303
Li J, Brader G, Palva ET (2004) The WRKY70 transcription factor: a node of convergence for jasmonate-mediated and salicylate mediated signals in plant defense. Plant Cell 16:319–331
Liao Y, Zou HF, Wei W, Hao YJ, Tian AG, Huang J, Liu YF, Zhang JS, Chen SY (2008) Soybean GmbZIP44, GmbZIP62 and GmbZIP78 genes function as negative regulator of ABA signaling and confer salt and freezing tolerance in transgenic Arabidopsis. Planta 228:225–240
Lindermayr C, Sell S, Müller B, Leister D, Durner J (2010) Redox regulation of the NPR1-TGA1 system of Arabidopsis thaliana by nitric oxide. Plant Cell 22:2894–2907
Lindemose S, Nielson PE, Mallegaard NE (2005) Polyamines preferentially interact with bent adenine trancts in double stranded DNA. Nucleic Acids Res 33:1790–1803
Liu ZB, Ulmasov T, Shi X, Hagen G, Guilfoyle TJ (1994) Soybean GH3 promoter contains multiple auxin-inducible elements. Plant Cell 6:645–657
Liu Q, Kasuga M, Sakuma Y, Abe H, Miura S, Yamaguchi-Shinozaki K, Shinozaki K (1998) Two transcription factors, DREB1 and DREB2, with an EREBP/AP2 DNA binding domain separate two cellular signal transduction pathways in drought- and low-temperature-responsive gene expression, respectively, in Arabidopsis. Plant Cell 10:1391–1406
Liu KD, Kang BC, Jiang H (2005) A GH3-like gene, CcGH3, isolated from Capsicum chinense L. fruit is regulated by auxin and ethylene. Plant Mol Biol 58:447–464
Liu X, Yue Y, Li B, Nie Y, Li W, Wu WH, Ma L (2007) A G protein-coupled receptor is a plasma membrane receptor for the plant hormone abscisic acid. Science 315:1712–1716
López-Ráez JA, Verhage A, Fernández I, García JM, Azcón-Aguilar C, Flors V, Pozo MJ (2010) Hormonal and transcriptional profiles highlight common and differential host responses to arbuscular mycorrhizal fungi and the regulation of the oxylipin pathway. J Exp Bot 61:2589–2601
Lorenzo O, Piqueras R, Sanchez-Serrano JJ, Solano R (2003) ETHYLENE RESPONSE FACTOR1 integrates signals from ethylene and jasmonate pathways in plant defense. Plant Cell 15:165–178
Lu G, Gao C, Zheng X, Han B (2009) Identification of OsbZIP72 as a positive regulator of ABA response and drought tolerance in rice. Planta 229:605–615
Luo M, Liu J, Lee RD, Scully BT, Guo B (2010) Monitoring the expression of maize genes in developing kernels under drought stress using oligo-microarray. J Integr Plant Biol 52: 1059–1074
Ma Y, Szostkiewicz I, Korte A, Moes D, Yang Y, Christmann A, Grill E (2009) Regulators of PP2C phosphatase activity function as abscisic acid sensors. Science 324:1064–1068
Magome H, Yamaguchi S, Hanada A, Kamiya Y, Oda K (2004) dwarf and delayed-flowering 1, a novel Arabidopsis mutant deficient in gibberellin biosynthesis because of overexpression of a putative AP2 transcription factor. Plant J 37:720–729
Manthe B, Schulz M, Schnabl H (1992) Effects of salicylic acid on growth and stomatal movement of Vicia faba L.: Evidence for salicylic acid metabolization. J Chem Ecol 18:1525–1539

Mason MG, Mathews DE, Argyros DA, Maxwell BB, Kieber JJ, Alonso JM, Ecker JR, Schaller GE (2005) Multiple type-B response regulators mediate cytokinin signal transduction in Arabidopsis. Plant Cell 17:3007–3018

Mayumi K, Shibaoka H (1995) A possible double role for brassinolide in the reorientation of cortical microtubules in the epidermal cells of azuki bean epicotyls. Plant Cell Physiol 36:173–181

McGinnis KM, Thomas SG, Soule FD, Strader LC, Zale JM, Sun TP, Steber CM (2003) The Arabidopsis SLEEPY1 gene encodes a putative F-box subunit of an SCF E3 ubiquitin ligase. Plant Cell 15:1120–1130

Melotto M, Mecey C, Niu Y, Chung HS, Katsir L, Yao J, Zeng W, Thines B, Staswick P, Browse J, Howe GA, He SY (2008) A critical role of two positively charged amino acids in the Jas motif of Arabidopsis JAZ proteins in mediating coronatine- and jasmonoyl isoleucine-dependent interactions with the COI1 F-box protein. Plant J 55:979–988

Menke FL, Champion A, Kijne JW, Memelink J (1999) A novel jasmonate- and elicitor-responsive element in the periwinkle secondary metabolite biosynthetic gene Str interacts with a jasmonate- and elicitor-inducible AP2-domain transcription factor, ORCA2. EMBO J 16:4455–4463

Metwally A, Finkemeier I, Georgi M, Dietz KJ (2003) Salicylic acid alleviates the cadmium toxicity in barley seedlings. Plant Physiol 132:272–281

Meyer K, Leube MP, Grill E (1994) A protein phosphatase 2 C involved in ABA signal transduction in Arabidopsis thaliana. Science 264:1452–1455

Minglin L, Yuxiu Z, Tuanyao C (2005) Identification of genes up-regulated in response to Cd exposure in Brassica juncea L. Gene 363:151–158

Monroe-Augustus M, Zolman BK, Bartel B (2003) IBR5, a dual-specificity phosphatase-like protein modulating auxin and abscisic acid responsiveness in Arabidopsis. Plant Cell 15:2979–2991

Montoya T, Nomura T, Farrar K et al (2002) Cloning the tomato curl3 gene highlights the putative dual role of the leucine-rich repeat receptor kinase tBRI1/SR160 in plant steroid hormone and peptide hormone signaling. Plant Cell 14:3163–3176

Mouchel CF, Osmont KS, Hardtke CS (2006) BRX mediates feedback between brassinosteroid levels and auxin signalling in root growth. Nature 443:458–461

Murase K, Hirano Y, Sun TP, Hakoshima T (2008) Gibberellin-induced DELLA recognition by the gibberellin receptor GID1. Nature 456:459–463

Murgia I, Delledonne M, Soave C (2002) Nitric oxide mediates iron-induced ferritin accumulation in Arabidopsis. Plant J 30:521–528

Müssig C, Biesgen C, Lisso J, Uwer U, Weiler EW, Altmann T (2000) A novel stress inducible 12-oxophytodienoate reductase from Arabidopsis thaliana provides a potential link between brassinosteroid-action and jasmonic-acid synthesis. J Plant Physiol 157:143–152

Müssig C, Fischer S, Altmann T (2002) Brassinosteroid-regulated gene expression. Plant Physiol 129:1241–1251

Mustilli A, Merlot S, Vavasseur A, Fenzi F, Giraudat J (2002) Arabidopsis OST1 protein kinase mediates the regulation of stomatal aperture by abscisic acid and acts upstream of reactive oxygen species production. Plant Cell 14:3089–3099

Nakajima M, Shimada A, Takashi Y, Kim YC, Park SH, Ueguchi-Tanaka M, Suzuki H, Katoh E, Iuchi S, Kobayashi M, Maeda T, Matsuoka M, Yamaguchi I (2006) Identification and characterization of Arabidopsis gibberellin receptors. Plant J 46:880–889

Nakano O, Suzuki K, Fujimura T, Shinshi H (2006) Genome-wide analysis of the ERF gene family in Arabidopsis and rice. Plant Physiol 140:411–432

Nam KH, Li J (2002) BRI1/BAK1, a receptor kinase pair mediating brassinosteroid signaling. Cell 110:203–212

Navarro L, Bari R, Achard P, Lisón P, Nemri A, Harberd NP, Jones JD (2008) DELLAs control plant immune responses by modulating the balance of jasmonic acid and salicylic acid signaling. Curr Biol 18:650–655

Nemhauser JL, Hong F, Chory J (2006) Different plant hormones regulate similar processes through largely nonoverlapping transcriptional responses. Cell 126:467–475

Niu X, Helentjaris T, Bate NJ (2002) Maize ABI4 binds coupling element1 in abscisic acid and sugar response genes. Plant Cell 14:2565–2575

Norman C, Howell KA, Millar AH, Whelan JM, Day DA (2004) Salicylic acid is an uncoupler and inhibitor of mitochondrial electron transport. Plant Physiol 134:492–501

Nunez M, Mazzafera P, Mazorra LM, Siqueira WJ, Zullo MAT (2003) Influence of a brassinsteroid analogue on antioxidant enzymes in rice grown in culture medium with NaCl. Biol Plant 47:67–70

O'Donnell PJ, Calvert C, Atzorn R, Wasternack C, Leyser HMO, Bowles DJ (1996) Ethylene as a signal mediating the wound response of tomato plants. Science 274:1914–1917

O'Donnell PJ, Truesdale MR, Calvert CM, Dorans A, Roberts MR, Bowles DJ (1998) A novel tomato gene that rapidly responds to wound-and pathogen-related signals. Plant J 14:137–142

Ogawa M, Kusano T, Katsumi M, Sano H (2000) Rice gibberellin-insensitive gene homolog, OsGAI, encodes a nuclear-localized protein capable of gene activation at transcriptional level. Gene 245:21–29

Ogawa M, Hanada A, Yamauchi Y, Kuwalhara A, Kamiya Y, Yamaguchi S (2003) Gibberellin biosynthesis and response during Arabidopsis seed germination. Plant Cell 15:1591–1604

Oh E, Yamaguchi S, Hu J, Yusuke J, Jung B, Paik I, Lee HS, Sun TP, Kamiya Y, Choi G (2007) PIL5, a phytochrome-interacting bHLH protein, regulates gibberellin responsiveness by binding directly to the GAI and RGA promoters in Arabidopsis seeds. Plant Cell 19:1192–1208

Ohme-Takagi M, Shinshi H (1995) Ethylene-inducible DNA binding proteins that interact with an ethylene-responsive element. Plant Cell 7:173–182

Ohta M, Ohme-Takagi M, Shinshi H (2000) Three ethylene responsive transcription factors in tobacco with distinct transactivation functions. Plant J 22:29–38

Okushima Y, Overvoorde PJ, Arima K, Alonso JM, Chan A, Chang C, Ecker JR, Hughes B, Lui A, Nguyen D, Onodera C, Quach H, Smith A, Yu G, Theologis A (2005) Functional genomic analysis of the AUXIN RESPONSE FACTOR gene family members in Arabidopsis thaliana: unique and overlapping functions of ARF7and ARF19. Plant Cell 17:444–463

Olszewski N, Sun T, Gubler F (2002) Gibberellin signaling: Biosynthesis, catabolism, and response pathways. Plant Cell 14(Suppl):S61–S80

Orellana S, Yanez M, Espinoza A, Verdugo I, Gonzalez E, Ruiz-Lara S, Casaretto JA (2010) The transcription factor SlAREB1 confers drought, salt stress tolerance and regulates biotic and abiotic stress-related genes in tomato. Plant Cell Environ 33:2191–2208

Öz MT, Yılmaz R, Eyidoğan F, de Graaff L, Yücel M, Öktem HA (2009) Microarray analysis of late response to boron toxicity in barley (Hordeum vulgare L.) leaves. Turk J Agric Forest 33:191–202

Özdemir F, Bor M, Demiral T, Türkan I (2004) Effects of 24-epibrassinolide on seed germination, seedling growth, lipid peroxidation, proline content and antioxidative system of rice (Oryza sativa L.) under salinity stress. Plant Growth Regul 42:203–211

Palmieri MC, Sell S, Huang X, Scherf M, Werner T, Durner J, Lindermayr C (2008) Nitric oxide-responsive genes and promoters in Arabidopsis thaliana: a bioinformatics approach. J Exp Bot 59:177–186

Pandey S, Nelson DC, Assmann SM (2009) Two novel GPCR-type G proteins are abscisic acid receptors in Arabidopsis. Cell 136:136–148

Paponov I, Paponov M, Teale W, Menges M, Chakrabortee S et al (2008) Comprehensive transcriptome analysis of auxin responses in Arabidopsis. Mol Plant 1:321–337

Parani M, Rudrabhatla S, Myers R, Weirich H, Smith B, Leaman DW, Goldman SL (2004) Microarray analysis of nitric oxide responsive transcripts in Arabidopsis. Plant Biotechnol J 2:359–366

Park JM, Park CJ, Lee SB, Ham BK, Shin R, Paek KH (2001) Overexpression of the tobacco Tsi1 gene encoding an EREBP/AP2-type transcription factor enhances resistance against pathogen attack and osmotic stress in tobacco. Plant Cell 13:1035–1046

Park JE, Park JY, Kim YS, Staswick PE, Jeon J, Yun J, Kim SY, Kim J, Lee YH, Park CM (2007) GH3-mediated auxin homeostasis links growth regulation with stress adaptation response in Arabidopsis. J Biol Chem 282:10036–10046

Park SY, Fung P, Nishimura N, Jensen DR, Fujii H, Zhao Y, Lumba S, Santiago J, Rodrigues A, Chow TF, Alfred SE, Bonetta D, Finkelstein R, Provart NJ, Desveaux D, Rodriguez PL, McCourt P, Zhu JK, Schroeder JI, Volkman BF, Cutler SR (2009) Abscisic acid inhibits type 2 C protein phosphatases via the PYR/PYL family of START proteins. Science 324:1068–1071

Pastuglia M, Roby D, Dumas C, Cock JM (1997) Rapid induction by wounding and bacterial infection of an S gene family receptor-like kinase gene in Brassica oleracea. Plant Cell 9:49–60

Peng J, Carol P, Richards DE, King KE, Cowling RJ, Murphy GP, Harberd NP (1997) The Arabidopsis GAI gene defines a signaling pathway that negatively regulates gibberellin responses. Genes Dev 11:3194–3205

Peng J, Richards DE, Hartley NM, Murphy GP, Devos KM, Flintham JE, Beales J, Fish LJ, Worland AJ, Pelica F, Sudhakar D, Christou P, Snape JW, Gale MD, Harberd NP (1999) 'Green revolution' genes encode mutant gibberellin response modulators. Nature 400:256–261

Perez-Amador MA, Leon J, Green PJ, Carbonell J (2002) Induction of the arginine decarboxylase ADC2 gene provides evidence for the involvement of polyamines in the wound response in Arabidopsis. Plant Physiol 130:1454–1463

Pieterse CM, Leon-Reyes A, Van Der Ent S, Van Wees SCM (2009) Networking by small-molecule hormones in plant immunity. Nat Chem Biol 5:308–316

Polverari A, Molesini B, Pezzotti M, Buonaurio R, Marte M, Delledonne M (2003) Nitric oxide-mediated transcriptional changes in Arabidopsis thaliana. Mol Plant Microbe Interact 16:1094–1105

Pontin MA, Piccoli PN, Francisco R, Bottini R, Martinez-Zapater JM, Lijavetzky D (2010) Transcriptome changes in grapevine (Vitis vinifera L.) cv. Malbec leaves induced by ultraviolet-B radiation. BMC Plant Biol 10:224

Potters G, Pasternak TP, Guisez Y, Jansen MAK (2009) Different stresses, similar morphogenic responses: integrating a plethora of pathways. Plant Cell Environ 32:158–169

Potuschak T, Lechner E, Parmentier Y, Yanagisawa S, Grava S, Koncz C, Genschik P (2003) EIN3-dependent regulation of plant ethylene hormone signaling by two Arabidopsis F box proteins: EBF1 and EBF2. Cell 115:679–689

Raghavendra AS, Gonugunta VK, Christmann A, Grill E (2010) ABA perception and signaling. Trends Plant Sci 15:395–401

Rajjou L, Belghazi M, Huguet R, Robin C, Moreau A, Job C, Job D (2006) Proteomic investigation of the effect of salicylic acid on Arabidopsis seed germination and establishment of early defense mechanisms. Plant Physiol 141:910–923

Rao MV, Lee HI, Davis KR (2002) Ozone-induced ethylene production is dependent on salicylic acid, and both salicylic acid and ethylene act in concert to regulate ozone-induced cell death. Plant J 32:447–456

Rashotte AM, Mason MG, Hutchison CE, Ferreira FJ, Schaller GE, Kieber JJ (2006) A subset of Arabidopsis AP2 transcription factors mediates cytokinin responses in concert with a two-component pathway. Proc Natl Acad Sci USA 103:11081–11085

Rashotte AM, Carson SDB, To JPC, Kieber JJ (2003) Expression profiling of cytokinin action in Arabidopsis. Plant Physiol 132:1998–2011

Rate DN, Cuenca JV, Bowman GR, Guttman DS, Greenberg JT (1999) The gain-of-function Arabidopsis acd6 mutant reveals novel regulation and function of the salicylic acid signaling pathway in controlling cell death, defense, and cell growth. Plant Cell 11:1695–1708

Razem FA, El-Kereamy A, Abrams SR, Hill RD (2006) The RNA-binding protein FCA is an abscisic acid receptor. Nature 439:290–294

Richter R, Behringer C, Muller IK, Schwechheimer C (2010) The GATA-type transcription factors GNC and GNL/CGA1 repress gibberellin signaling downstream from DELLA proteins and PHYTOCHROME-INTERACTING FACTORS. Genes Dev 24:2093–2104

Riechmann JL, Heard J, Martin G, Reuber L, Jiang C-Z, Keddie J, Adam L, Pineda O, Ratcliffe OJ, Samaha RR, Creelman R, PilgrimM BP, Zhang JZ, Ghandehari D, Sherman BK, Yu G-L (2000) Arabidopsis transcription factors: genome-wide comparative analysis among eukaryotes. Science 290:2105–2110

Riou-Khamlichi C, Huntley R, Jacqmard A, Murray JA (1999) Cytokinin activation of Arabidopsis cell division through a D-type cyclin. Science 283:1541–1544
Risk JM, Macknight RC, Day CL (2008) FCA does not bind abscisic acid. Nature 456:E5–E6
Risk JM, Day CL, Macknight RC (2009) Reevaluation of abscisic acid-binding assays shows that G-Protein-Coupled Receptor 2 does not bind abscisic acid. Plant Physiol 150:6–11
Rivero RM, Kojima M, Gepstein A, Sakakibara H, Mittler R, Gepstein S, Blumwald E (2007) Delayed leaf senescence induces extreme drought tolerance in a flowering plant. Proc Natl Acad Sci USA 104:19631–19636
Russinova E, Borst JW, Kwaaitaal M, Caño-Delgado A, Yin Y, Chory J, de Vries SC (2004) Heterodimerization and endocytosis of Arabidopsis brassinosteroid receptors BRI1 and AtSERK3 (BAK1). Plant Cell 16:3216–3229
Sairam RK (1994) Effects of homobrassinolide application on plant metabolism and grain yield under irrigated and moisture stress conditions of two wheat varieties. Plant Growth Regul 14:173–181
Sakai H, Aoyama T, Bono H, Oka A (1998) Two-component response regulators from Arabidopsis thaliana contain a DNA-binding motif. Plant Cell Physiol 39:1232–1239
Sakuma Y, Liu Q, Dubouzet JG, Abe H, Shinozaki K, Yamaguchi-Shinozaki K (2002) DNA-binding specificity of the ERF/AP2 domain of Arabidopsis DREBs, transcription factors involved in dehydration- and cold-inducible gene expression. Biochem Biophys Res Commun 290:998–1009
Santner A, Calderon-Villalobos LIA, Estelle M (2009) Plant hormones are versatile chemical regulators of plant growth. Nature Chem Biol 5:301–307
Sasaki A, Itoh H, Gomi K, Ueguchi-Tanaka M, Ishiyama K, Kobayashi M, Jeong DH, An G, Kitano H, Ashikari M, Matsuoka M (2003) Accumulation of phosphorylated repressor for gibberellin signaling in an F-box mutant. Science 299:1896–1898
Schaller GE, Mathews DE, Gribskov M, Walker JC (2002) Two component signaling elements and histidyl-aspartyl phosphorelays. In: Somerville CR, Meyerowitz EM (eds) The Arabidopsis book. American Society of Plant Biologists, Rockville, MD
Schenk PM, Kazan K, Wilson I, Anderson JP, Richmond T, Somerville SC, Manners JM (2000) Coordinated plant defense responses in Arabidopsis revealed by microarray analysis. Proc Natl Acad Sci USA 97:11655–11660
Schlink K (2010) Down-regulation of defense genes and resource allocation into infected roots as factors for compatibility between Fagus sylvatica and Phytophthora citricola. Funct Integr Genomics 10:253–264
Schopfer P, Liszkay A, Bechtold M, Frahry G, Wagner A (2002) Evidence that hydroxyl radicals mediate auxin-induced extension growth. Planta 214:821–828
Schwechheimer C (2008) Understanding gibberellic acid signaling-are we there yet? Curr Opin Plant Biol 11:9–15
Seki M, Ishida J, Narusaka M, Fujita M, Nanjo T, Umezawa T, Kamiya A, Nakajima M, Enju A, Sakurai T, Satou M, Akiyama K, Yamaguchi-Shinozaki K, Carninci P, Kawai J, Hayashizaki Y, Shinozaki K (2002) Monitoring the expression pattern of around 7,000 Arabidopsis genes under ABA treatments using a full-length cDNA microarray. Funct Integr Genomics 2:282–291
Seo PJ, Park CM (2009) Auxin homeostasis during lateral root development under drought condition. Plant Signal Behav 4:1002–1004
Serpa V, Vernal J, Lamattina L, Grotewold E, Cassia R, Terenzi H (2007) Inhibition of AtMYB2 DNA-binding by nitric oxide involves cysteine S-nitrosylation. Biochem Biophys Res Commun 361:1048–1053
Sharp RE, Poroyko V, Hejlek LG, Spollen WG, Springer GK, Bohnert HJ, Nguyen HT (2004) Root growth maintenance during water deficits: physiology to functional genomics. J Exp Bot 55:2343–2351
Shashidhar VR, Prasad TG, Sudharshan L (1996) Hormone signals from roots to shoots of sunflower (Helianthus annuus L.) moderate soil drying increases delivery of abscisic acid and depresses delivery of cytokinins in xylem sap. Ann Bot 78:151–155

Shen YY, Wang XF, Wu FQ, Du SY, Cao Z, Shang Y, Wang XL, Peng CC, Yu XC, Zhu SY, Fan RC, Xu YH, Zhang DP (2006) The Mg-chelatase H subunit is an abscisic acid receptor. Nature 443:823–826

Shimada A, Ueguchi-Tanaka M, Nakatsu T, Nakajima M, Naoe Y, Ohmiya H, Kato H, Matsuoka M (2008) Structural basis for gibberellin recognition by its receptor GID1. Nature 456: 520–523

Silverstone AL, Tseng TS, Swain SM, Dill A, Jeong SY, Olszewski NE, Sun TP (2007) Functional analysis of SPINDLY in gibberellin signaling in Arabidopsis. Plant Physiol 143:987–1000

Solano R, Stepanova A, Chao Q, Ecker JR (1998) Nuclear events in ethylene signaling: a transcriptional cascade mediated by ETHYLENE-INSENSITIVE3 and ETHYLENE-RESPONSE-FACTOR1. Genes Dev 12:3703–3714

Stacey G, McAlvin CB, Kim SY, Olivares J, Soto MJ (2006) Effects of endogenous salicylic acid on nodulation in the model legumes Lotus japonicas and Medicago truncatula. Plant Physiol 141:1473–1481

Stockinger EJ, Gilmour SJ, Thomashow MF (1997) Arabidopsis thaliana CBF1 encodes an AP2 domain-containing transcriptional activator that binds to the C-repeat/DRE, a cis-acting DNA regulatory element that stimulates transcription in response to low temperature and water deficit. Proc Natl Acad Sci USA 94:1035–1040

Strader LC, Ritchie S, Soule JD, McGinnis KM, Steber CM (2004) Recessive-interfering mutations in the gibberellin signaling gene SLEEPY1 are rescued by overexpression of its homologue, SNEEZY. Proc Natl Acad Sci USA 101:12771–12776

Strader LC, Monroe-Augustus M, Bartel B (2008) The IBR5 phosphatase promotes Arabidopsis auxin responses through a novel mechanism distinct from TIR1-mediated repressor degradation. BMC Plant Biol 8:41

Sullivan ML, Green PJ (1996) Mutational analysis of the DST element in tobacco cells and transgenic plants: identification of residues critical for mRNA instability. RNA 2:308–315

Sun TP (2010) Gibberellin-GID1-DELLA: a pivotal regulatory module for plant growth and development. Plant Physiol 154:567–570

Sun TP (2011) The molecular mechanism and evolution of the GA-GID1-DELLA signaling module in plants. Curr Biol 21:R338–R345

Sun TP, Gubler F (2004) Molecular mechanism of gibberellin signaling in plants. Annu Rev Plant Biol 55:197–223

Suzuki M, Kao CY, Cocciolone S, McCarty DR (2001) Maize VP1 complements Arabidopsis abi3 and confers a novel ABA/auxin interaction in roots. Plant J 28:409–418

Szemenyei H, Hannon M, Long JA (2008) TOPLESS mediates auxin-dependent transcriptional repression during Arabidopsis embryogenesis. Science 319:1384–1386

Tada Y, Spoel SH, Pajerowska-Mukhtar K, Mou Z, Song J, Dong X (2008) Plant immunity requires conformational changes of NPR1 via S-nitrosylation and thioredoxins. Science 321:952–956

Takahashi Y, Uehara Y, Berberich T, Ito A, Saitoh H, Miyazaki A, Terauchi R, Kusano T (2004) A subset of hypersensitive response marker genes, including HSR203J, is the downstream target of a spermine signal transduction pathway in tobacco. Plant J 40:586–595

Tanaka K, Nakamura Y, Asami T, Yoshida S, Matsuo T, Okamoto S (2003) Physiological roles of brassinosteroids in early growth of Arabidopsis: brassinosteroids have a synergistic relationship with gibberellin as well as auxin in light-grown hypocotyl elongation. J Plant Growth Regul 22:259–271

Tang W, Kim TW, Oses-Prieto JA, Sun Y, Deng Z, Zhu S, Wang R, Burlingame AL, Wang ZY (2008) BSKs mediate signal transduction from the receptor kinase BRI1 in Arabidopsis. Science 321:557–560

Tao LZ, Cheung AY, Wu HM (2002) Plant Rac-like GTPases are activated by auxin and mediate auxin-responsive gene expression. Plant Cell 14:2745–2760

Tassoni A, Antognoni F, Battistini ML, Sanvido OA, Bani N (1998) Characterization of spremidine binding to solubilized plasma membrane proteins from Zucchini hypocotyls. Plant Physiol 117:971–977

Terol J, Domingo C, Talon M (2006) The GH3 family in plants: genome wide analysis in rice and evolutionary history based on EST analysis. Gene 371:279–290

Thines B, Katsir L, Melotto M, Niu Y, Mandaokar A, Liu G, Nomura K, He SY, Howe GA, Browse J (2007) JAZ repressor proteins are targets of the SCF(COI1) complex during jasmonate signalling. Nature 448:661–665

Tian CE, Muto H, Higuchi K, Matamura T, Tatematsu K, Koshiba T, Yamamoto KT (2004) Disruption and overexpression of auxin response factor 8 gene of Arabidopsis affect hypocotyls elongation and root growth habit, indicating its possible involvement in auxin homeostasis in light condition. Plant J 40:333–343

Tiwari SB, Hagen G, Guilfoyle T (2003) The roles of auxin response factor domains in auxin-responsive transcription. Plant Cell 15:533–543

Tjian R, Maniatis T (1994) Transcriptional activation: a complex puzzle with few easy pieces. Cell 77:5–8

To JP, Deruère J, Maxwell BB, Morris VF, Hutchison CE, Ferreira FJ, Schaller GE, Kieber JJ (2007) Cytokinin regulates type-A Arabidopsis response regulator activity and protein stability via two-component phosphorelay. Plant Cell 19:3901–3914

Ton J, Flors V, Mauch-Mani B (2009) The multifaceted role of ABA in disease resistance. Trends Plant Sci 14:310–317

Tran LS, Urao T, Qin F, Maruyama K, Kakimoto T, Shinozaki K, Yamaguchi-Shinozaki K (2007) Functional analysis of AHK1/ATHK1 and cytokinin receptor histidine kinases in response to abscisic acid, drought, and salt stress in Arabidopsis. Proc Natl Acad Sci USA 104:20623–20628

Traw MB, Bergelson J (2003) Interactive effects of jasmonic acid, salicylic acid, and gibberellin on induction of trichomes in Arabidopsis. Plant Physiol 133:1367–1375

Tromas A, Perrot-Rechenmann C (2010) Recent progress in auxin biology. C R Biol 333:297–306

Trujillo LE, Sotolongo M, Menéndez C, Ochogavía ME, Coll Y, Hernández I, Borrás-Hidalgo O, Thomma BP, Vera P, Hernández L (2008) SodERF3, a novel sugarcane ethylene responsive factor (ERF), enhances salt and drought tolerance when overexpressed in tobacco plants. Plant Cell Physiol 49:512–525

Tuteja N, Sopory SK (2008) Chemical signaling under abiotic stress environment in plants. Plant Signal Beh 3:525–536

Uchida A, Jagendorf AT, Hibino T, Takabe T, Takabe T (2002) Effects of hydrogen peroxide and nitric oxide on both salt and heat stress tolerance in rice. Plant Sci 163:515–523

Ueguchi-Tanaka M, Matsuoka M (2010) The perception of gibberellins: clues from receptor structure. Curr Opin Plant Biol 13:503–508

Ueguchi-Tanaka M, Ashikari M, Nakajima M, Itoh H, Katoh E, Kobayashi M, Chow TY, Hsing YI, Kitano H, Yamaguchi I, Matsuoka M (2005) GIBBERELLIN INSENSITIVE DWARF1 encodes a soluble receptor for gibberellin. Nature 437:693–698

Ueguchi-Tanaka M, Nakajima M, Motoyuki A, Matsuoka M (2007) Gibberellin receptor and its role in gibberellin signaling in plants. Annu Rev Plant Biol 58:183–198

Ulmasov T, Hagen G, Guilfoyle TJ (1997) ARF1, a transcription factor that binds to auxin response elements. Science 276:1865–1868

Umezawa T, Sugiyama N, Mizoguchi M, Hayashi S, Myouga F, Yamaguchi-Shinozaki K, Ishihama Y, Hirayama T, Shinozaki K (2009) Type 2 C protein phosphatases directly regulate abscisic acid-activated protein kinases in Arabidopsis. Proc Natl Acad Sci USA 106:17588–17593

Umezawa T, Nakashima K, Miyakawa T, Kuromori T, Tanokura M, Shinozaki K, Yamaguchi-Shinozaki K (2010) Molecular basis of the core regulatory network in ABA responses: sensing, signaling and transport. Plant Cell Physiol 51:1821–1839

Uno Y, Furihata T, Abe H, Yoshida R, Shinozaki K, Yamaguchi-Shinozaki K (2000) Arabidopsis basic leucine zipper transcription factors involved in an abscisic acid-dependent signal transduction pathway under drought and high-salinity conditions. Proc Natl Acad Sci USA 97:11632–11637

Urano K, Yoshiba Y, Nanjo T, Igarashi Y, Seki M, Sekiguchi F, Yamaguchi-Shinozaki K, Shinozaki K (2003) Characterization of Arabidopsis genes involved in biosynthesis of polyamines in abiotic stress responses and developmental stages. Plant Cell Environ 26:1917–1926

Urano K, Maruyama K, Ogata Y, Morishita Y, Takeda M, Sakurai N, Suzuki H, Saito K, Shibata D, Kobayashi M, Yamaguchi-Shinozaki K, Shinozaki K (2009) Characterization of the ABA-regulated global responses to dehydration in Arabidopsis by metabolomics. Plant J 57:1065–1078

van der Fits L, Memelink J (2000) ORCA3, a jasmonate-responsive transcriptional regulator of plant primary and secondary metabolism. Science 289:295–297

van der Graaff E, Dulk-Ras AD, Hooykaas PJ, Keller B (2000) Activation tagging of the LEAFY PETIOLE gene affects leaf petiole development in Arabidopsis thaliana. Development 127:4971–4980

Verslues PE, Agarwal M, Katiyar-Agarwal S, Zhu J, Zhu JK (2006) Methods and concepts in quantifying resistance to drought, salt and freezing, abiotic stresses that affect plant water status. Plant J 45:523–539

Vlad F, Rubio S, Rodrigues A, Sirichandra C, Belin C, Robert N, Leung J, Rodriguez PL, Laurière C, Merlot S (2009) Protein phosphatases 2 C regulate the activation of the Snf1-related kinase OST1 by abscisic acid in Arabidopsis. Plant Cell 21:3170–3184

Vlot AC, Dempsey DA, Klessig DF (2009) Salicylic Acid, a multifaceted hormone to combat disease. Annu Rev Phytopathol 47:177–206

Walden R, Cordeiro A, Tiburcio AF (1997) Polyamines: small molecules triggering pathways in plant growth and development. Plant Physiol 113:1009–1013

Walia H, Wilson C, Wahid A, Condamine P, Cui X, Close TJ (2006) Expression analysis of barley (Hordeum vulgare L.) during salinity stress. Funct Integr Genomics 6:143–156

Walia H, Wilson C, Condamine P, Liu X, Ismail AM (2007) Close TJ (2007) Large-scale expression profiling and physiological characterization of jasmonic acid-mediated adaptation of barley to salinity stress. Plant Cell Environ 30:410–421

Wang X, Chory J (2006) Brassinosteroids regulate dissociation of BKI1, a negative regulator of BRI1 signaling, from the plasma membrane. Science 313:1118–1122

Wang H, Huang Z, Chen Q, Zhang Z, Zhang H, Wu Y, Huang D, Huang R (2004) Ectopic overexpression of tomato JERF3 in tobacco activates downstream gene expression and enhances salt tolerance. Plant Mol Biol 55:183–192

Wang D, Pei K, Fu Y, Sun Z, Li S, Liu H, Tang K, Han B, Tao Y (2007a) Genome wide analysis of the auxin response factors (ARF) gene family in rice (Oryza sativa). Gene 394:13–24

Wang D, Pajerowska-Mukhtar K, Culler AH, Dong X (2007b) Salicylic acid inhibits pathogen growth in plants through repression of the auxin signaling pathway. Curr Biol 17:1784–1790

Wang A, Tan D, Takahashi A, Zhong LT, Harada T (2007c) MdERFs, two ethylene response factors involved in apple fruit ripening. J Exp Bot 58:3743–3748

Wang X, Kota U, He K, Blackburn K, Li J, Goshe MB, Huber SC, Clouse SD (2008) Sequential transphosphorylation of the BRI1/BAK1 receptor kinase complex impacts early events in brassinosteroid signaling. Dev Cell 15:220–235

Wang F, Zhu D, Huang X, Li S, Gong Y, Yao Q, Fu X, Fan LM, Deng XW (2009) Biochemical insights on degradation of Arabidopsis DELLA proteins gained from a cell-free assay system. Plant Cell 21:2378–2390

Wang S, Bai Y, Shen C, Wu Y, Zhang S, Jiang D, Guilfoyle TJ, Chen M, Qi Y (2010) Auxin-related gene families in abiotic stress response in Sorghum bicolor. Funct Integr Genomics 10:533–546

Wang RS, Pandey S, Li S, Gookin TE, Zhao Z, Albert R, Assmann SM (2011) Common and unique elements of the ABA-regulated transcriptome of Arabidopsis guard cells. BMC Genomics 12:216

Wasternack C (2007) Jasmonates: an update on biosynthesis, signal transduction and action in plant stress response, growth and development. Ann Bot (Lond) 100:681–697

Weijers D, Benkova E, Jäger KE, Schlereth A, Hamann T, Kientz M, Wilmoth JC, Reed JW, Jürgens G (2005) Developmental specificity of auxin response by pairs of ARF and Aux/IAA transcriptional regulators. EMBO J 24:1874–1885

Wendehenne D, Durner J, Klessig DF (2004) Nitric oxide: a new player in plant signalling and defence responses. Curr Opin Plant Biol 7:449–455

Willige BC, Ghosh S, Nill C, Zourelidou M, Dohmann EM, Maier A, Schwechheimer C (2007) The DELLA domain of GA INSENSITIVE mediates the interaction with the GA INSENSITIVE DWARF1A gibberellin receptor of Arabidopsis. Plant Cell 19:1209–1220

Wolters H, Jürgens G (2009) Survival of the flexible: hormonal growth control and adaptation in plant development. Nat Rev Genet 10:305–317

Woodward AW, Bartel B (2005) Auxin: Regulation, action, and interaction. Ann Bot 95: 707–735

Wu L, Zhang Z, Zhang H, Wang XC, Huang R (2008) Transcriptional modulation of ERF protein JERF3 in oxidative stress response enhances tolerance of tobacco seedlings to salt, drought and freezing. Plant Physiol 148:1953–1963

Wu FQ, Xin Q, Cao Z, Liu ZQ, Du SY, Mei C, Zhao CX, Wang XF, Shang Y, Jiang T, Zhang XF, Yan L, Zhao R, Cui ZN, Liu R, Sun HL, Yang XL, Su Z, Zhang DP (2009) The magnesium-chelatase H subunit binds abscisic acid and functions in abscisic acid signaling: New Evidence in Arabidopsis. Plant Physiol 150:1940–1954

Xie DX, Feys BF, James S, Nieto-Rostro M, Turner JG (1998) COI1: an Arabidopsis gene required for jasmonate-regulated defense and fertility. Science 280:1091–1094

Xiong GS, Li JY, Wang YH (2009) Advances in the regulation and crosstalks of phytohormones. Chin Sci Bull 54:4069–4082

Xu X, Chen C, Fan B, Chen Z (2006) Physical and functional interactions between pathogen-induced Arabidopsis WRKY18, WRKY40, and WRKY60 transcription factors. Plant Cell 18:1310–1326

Yamaguchi-Shinozaki K, Shinozaki K (2006) Transcriptional regulatory networks in cellular responses and tolerance to dehydration and cold stresses. Annu Rev Plant Biol 57:781–803

Yamamoto S, Suzuki K, Shinshi H (1999) Elicitor-responsive, ethylene-independent activation of GCC box-mediated transcription that is regulated by both protein phosphorylation and dephosphorylation in cultured tobacco cells. Plant J 20:571–579

Yamamuro C, Ihara Y, Wu X, Noguchi T, Fujioka S, Takatsuto S, Ashikari M, Kitano H, Matsuoka M (2000) Loss of function of a rice brassinosteroid insensitive1 homolog prevents internode elongation and bending of the lamina joint. Plant Cell 12:1591–1606

Yang GX, Jan A, Shen SH, Yazaki J, Ishikawa M, Shimatani Z, Kishimoto N, Kikuchi S, Matsumoto H, Komatsu S (2004) Microarray analysis of brassinosteroids- and gibberellin-regulated gene expression in rice seedlings. Mol Genet Genomics 271:468–478

Yang Y, Yu X, Wu P (2006) Comparison and evolution analysis of two rice subspecies LATERAL ORGAN BOUNDARIES domain gene family and their evolutionary characterization from Arabidopsis. Mol Phylogenet Evol 39:248–262

Yasumura Y, Crumpton-Taylor M, Fuentes S, Harberd NP (2007) Step by-step acquisition of the gibberellin-DELLA growth-regulatory mechanism during land-plant evolution. Curr Biol 17:1225–1230

Yin Y, Vafeados D, Tao Y, Yoshida S, Asami T, Chory J (2005) A new class of transcription factors mediates brassinosteroid-regulated gene expression in Arabidopsis. Cell 120:249–59

Yoshida R, Hobo T, Ichimura K, Mizoguchi T, Takahashi F, Aronso J, Ecker JR, Shinozaki K (2002) ABA-activated SnRK2 protein kinase is required for dehydration stress signaling in Arabidopsis. Plant Cell Physiol 43:1473–1483

Yoshida T, Fujita Y, Sayama H, Kidokoro S, Maruyama K, Mizoi J, Shinozaki K, Yamaguchi-Shinozaki K (2010) AREB1, AREB2, and ABF3 are master transcription factors that cooperatively regulate ABRE-dependent ABA signaling involved in drought stress tolerance and require ABA for full activation. Plant J 61:672–685

Zentella R, Zhang ZL, Park M, Thomas SG, Endo A, Murase K, Fleet CM, Jikumaru Y, Nambara E, Kamiya Y, Sun TP (2007) Global analysis of DELLA direct targets in early gibberellin signaling in Arabidopsis. Plant Cell 19:3037–3057

Zhang X, Zhang Z, Chen J, Chen Q, Wang XC, Huang R (2005) Expressing TERF1 in tobacco enhances drought tolerance and abscisic acid sensitivity during seedling development. Planta 222:494–501

Zhang G, Chen M, Chen X, Xu Z, Guan S, Li LC, Li A, Guo J, Mao L, Ma Y (2008) Phylogeny, gene structures, and expression patterns of the ERF gene family in soybean (Glycine max L.). J Exp Bot 59:4095–4107

Zhang G, Chen M, Li L, Xu Z, Chen X, Guo J, Ma Y (2009) Overexpression of the soybean GmERF3 gene, an AP2/ERF type transcription factor for increased tolerances to salt, drought, and diseases in transgenic tobacco. J Exp Bot 60:3781–3796

Zhang ZL, Ogawa M, Fleet CM, Zentella R, Hu J, Heo JO, Lim J, Kamiya Y, Yamaguchi S, Sun TP (2011) Scarecrow-like 3 promotes gibberellin signaling by antagonizing master growth repressor DELLA in Arabidopsis. Proc Natl Acad Sci USA 108:2160–2165

Zhao L, Zhang F, Guo J, Yang Y, Li B, Zhang L (2004) Nitric oxide functions as a signal in salt resistance in the calluses from two ecotypes of reed. Plant Physiol 134:849–857

Zhao MG, Tian QY, Zhang WH (2007) Nitric oxide synthase dependent nitric oxide production is associated with salt tolerance in Arabidopsis. Plant Physiol 144:206–217

Zhao MG, Chen L, Zhang LL, Zhang WH (2009) Nitric reductase dependent nitric oxide production is involved in cold acclimation and freezing tolerance in Arabidopsis. Plant Physiol 151:755–767

Zhong GV, Burns JK (2003) Profiling ethylene-regulated gene expression in Arabidopsis thaliana by microarray analysis. Plant Mol Biol 53:117–131

Zhou J, Tang X, Martin GB (1997) The Pto kinase conferring resistance to tomato bacterial speck disease interacts with proteins that bind a cis-element of pathogenesis related genes. EMBO J 16:3207–3218

Zhou A, Wang H, Walker JC, Li J (2004) BRL1, a leucine-rich repeat receptor-like protein kinase, is functionally redundant with BRI1 in regulating Arabidopsis brassinosteroid signaling. Plant J 40:399–409

Zhu JK (2002) Salt and drought stress signal transduction in plants. Annu Rev Plant Biol 53:247–273

Zhuang YL, Ren GJ, Zhu Y, Hou GH, Qu X, Li ZX, Yue GD, Zhang JR (2008) Transcriptional profiles of immature ears and tassels in maize at early stage of water stress. Biologia Plant 52:754–758

Zou M, Guan Y, Ren H, Zhang F, Chen F (2008) A bZIP transcription factor, OsABI5, is involved in rice fertility and stress tolerance. Plant Mol Biol 66:675–683

Chapter 2
Cross-Talk Between Phytohormone Signaling Pathways Under Both Optimal and Stressful Environmental Conditions

Marcia A. Harrison

Abstract The perception of abiotic stress triggers the activation of signal transduction cascades that interact with the baseline pathways transduced by phytohormones. The convergence points among hormone signal transduction cascades are considered cross-talk, and together they form a signaling network. Through this mechanism, hormones interact by activating either a common second messenger or a phosphorylation cascade. This chapter reviews kinase cascades as cross-talk points in hormonal networks during abiotic stress conditions. These transduction cascades lead to the regulation of gene expression that directly affects the biosynthesis or action of other hormones. Examples of stress-related hormone transduction networks are provided for drought and wounding conditions. The expression of specific genes associated with drought and wounding stress will be compared with expression changes that occur during other abiotic stress conditions. This evaluation will be used to construct a model of abiotic stress signaling that incorporates the signaling components that are most common across all abiotic stress conditions and are, therefore, relevant to developing stress tolerance in crop plants.

Abbreviations

ABA	Abscisic acid
ABF	ABA-responsive element-binding factor
ABI	ABA insensitive
ABRE	ABA-responsive element
ACO	1-Aminocyclopropane-1-carboxylate oxidase

M.A. Harrison (✉)
Department of Biological Sciences, Marshall University, One John Marshall Drive, Huntington, WV 25755, USA
e-mail: harrison@marshall.edu

N.A. Khan et al. (eds.), *Phytohormones and Abiotic Stress Tolerance in Plants*,
DOI 10.1007/978-3-642-25829-9_2, © Springer-Verlag Berlin Heidelberg 2012

ACS	1-Aminocyclopropane-1-carboxylic acid synthase
AGO1	Argonaute1
AP2C1	*Arabidopsis* Ser/Thr phosphatase of type 2C
AtHK1	*Arabidopsis* histidine kinase 1
CDPK	Calcium-dependent protein kinase
CKX	Cytokinin oxidase/dehydrogenase
CPK	Calcium-dependent protein kinase gene/protein abbreviation
EIN3	Ethylene insensitive
EREBP	Ethylene-responsive element-binding protein
ERF	Ethylene-response factor
GORK	Guard cell outward-rectifying K^+
GST	Glutathione-*S*-transferase 1
GUS	β-glucuronidase
HK	Histidine kinase
IP_3	Inositol trisphosphate
JA	Jasmonic acid
KAT	K+ channel in *Arabidopsis thaliana*
MAPK	Mitogen-activated protein kinase
MeJA	Methyl JA
miRNA	MicroRNA
MKK	MAPK kinase
MPK	MAPK gene/protein abbreviation
NO	Nitric oxide
OPR3	12-oxophytodienoat-10,11-reductase
PP2C	Protein phosphatase 2C
PYR	Pyrabactin (4-bromo-N-[pyridin-2-yl methyl] naphthalene-1-sulfonamide) resistance
RCAR	Regulatory component of ABA receptor
ROS	Reactive oxygen species
RSRE	Rapid stress response element
RWR	Rapid wound response
SEN1	Senescence-associated protein 1
SLAC1	Slow anion channel-associated 1
SnRK2	Sucrose nonfermenting 1-related protein kinase
TCH3	Touch-induced 3

2.1 Introduction

In plants, the perception of abiotic stress triggers the activation of signal transduction cascades that interact with the baseline pathways transduced by phytohormones. The convergence points among hormone signal transduction cascades are considered cross-talk, and together they form a signaling network. Through this mechanism, hormones interact by activating either a common second messenger or a phosphorylation cascade. This chapter reviews kinase cascades as cross-talk points in hormonal networks during abiotic stress conditions. These transduction cascades lead to the regulation of gene expression that directly affects the biosynthesis or action of other hormones, a process that represents an additional layer of hormonal cross-talk also addressed here.

Examples of stress-related hormone transduction networks are provided for drought and wounding conditions. The expression of specific genes associated with drought and wounding stress will be reviewed and compared with expression changes that occur during other abiotic stress conditions. This evaluation will be used to construct a basic model of abiotic stress signaling that incorporates the signaling components that are most common across all abiotic stress conditions and are, therefore, relevant to developing stress tolerance in crop plants.

2.2 Drought Stress

Information regarding the regulation of drought stress comes from experiments that directly examine either drought (induced by dehydration causing loss of turgor) or osmotic stress (induced by increasing extracellular solute concentration, which also results in turgor loss), and from inferences based on experiments involving the functions of hormones and other signals. In a meta-analysis of more than 450 papers concerning the role of drought in photosynthesis, the authors conclude that experimental data are often disjointed or not comparable, making it difficult to discern general trends (Pinheiro and Chaves 2011). However, changes in gene expression induced by drought stress certainly overlap with hormone-regulated gene expression. Therefore, this discussion will present data from studies of hormone action as well as from those that focus on events that occur during drought stress.

Stomatal closure is a rapid response to drought stress and is regulated by a complex network of signaling pathways. During drought stress, abscisic acid (ABA) is considered to be the primary phytohormone that triggers short-term responses such as stomatal closure (Zhang et al. 2006). ABA controls longer-term growth responses through the regulation of gene expression that favors maintenance of root growth, which optimizes water uptake (Zhang et al. 2006). As part of the regulation of drought stress responses, ABA may interact with jasmonic acid (JA) and nitric oxide (NO) to stimulate stomatal closure, while its regulation of gene expression includes the induction of genes associated with response to ethylene, cytokinin, or auxin. ABA-mediated regulation of signal transduction pathways during drought stress is presented here as an example to illustrate the signal transduction elements that may play roles in cross-talk associated with the response to drought stress.

2.2.1 Stomatal Closure in Response to Drought Stress

ABA undergoes a dramatic increase in concentration after drought stress (Hoad 1975), and studies indicate that ABA is the major regulatory hormone that controls drought stress-induced stomatal closure (Finkelstein and Rock 2002).

The increase in ABA concentration in guard cells is trigged by the reduction in the amount of water around the roots. During drought conditions, dehydration of root tissue is sensed by a drought-specific histidine kinase (HK) osmoreceptor. In *Arabidopsis* plants grown under normal conditions, ATHK1, a transmembrane, two-component histidine kinase, is expressed at a higher level in the roots than in the stems and leaves. Loss-of-function *athk1* mutants produced lower levels of endogenous ABA, indicating that the increase in ABA synthesis during osmotic stress is dependent upon sensing of drought conditions by ATHK1 (Wohlbach et al. 2008). *ATHK1* expression greatly increases, and its protein accumulates to a high level, in root tissue experiencing drought stress conditions (Urao et al. 1999). ATHK1 is a positive regulator of the drought stress response and activates many downstream targets, including ABA biosynthetic enzymes (Wohlbach et al. 2008) and ABA-responsive transcription factors (Tran et al. 2007).

2.2.1.1 ABA Accumulation in Guard Cells Regulates Stomatal Closure

Changes in ABA concentration in response to drought stress vary dramatically from tissue to tissue. Using an ABA-dependent reporter construct, Christmann et al. (2005) demonstrated a basic time course for drought-induced ABA accumulation in tissues of *Arabidopsis* seedlings whose roots were subjected to a −1.0 MPa water stress treatment. An increase in ABA concentration in the vascular tissue of the cotyledons was observed by 4 h after treatment (Christmann et al. 2005). After 4 h, ABA was relatively uniformly distributed in the leaf tissue, but by 8 h posttreatment, a higher concentration of ABA was present in guard cells than in other leaf tissue. However, within this 8-h time frame, virtually no change in ABA concentration was detected in the roots. Therefore, the activation of ATHK1 in root tissue may trigger a hydraulic signaling system, proposed by Christmann et al. (2007), that acts as a long-distance signal to stimulate ABA production in the vascular parenchyma and guard cells. Other studies confirm that vascular parenchyma cells contain the enzymes associated with ABA synthesis, which thus leads to the high concentration of ABA in the vascular tissue after drought stress (Koiwai et al. 2004; Endo et al. 2008).

From the vascular tissue, a specific type of ATP-binding cassette transporter exports ABA (Kuromori et al. 2010), while another transporter imports ABA into leaf tissue, including guard cells (Umezawa et al. 2010). In addition, the increase in ABA concentration may in part originate from the release of active ABA from its ABA–glucose ester conjugate, which is stored in the vacuoles of leaf cells and can also circulate in the plant (Wasilewska et al. 2008; Seiler et al. 2011).

2.2.1.2 ABA Regulates the Activity of Ion Channels

ABA regulates several types of ion channels, causing anion efflux, K^+ efflux, and the inhibition of K^+ import (Fig. 2.1). One mechanism of ABA action begins when ABA binds to receptors belonging to a protein family known as PYR/RCARs, the pyrabactin (4-bromo-N-[pyridin-2-yl methyl]naphthalene-1-sulfonamide) resistance (PYR)/regulatory component of ABA receptor (RCAR) (Hubbard et al. 2010). In *Arabidopsis*, 14 highly conserved PYR/RCARs have been identified. Upon binding to ABA, PYR/RCARs inhibit the activity of specific protein phosphatase 2Cs (PP2Cs), which are negative regulators of ABA signaling (Hubbard et al. 2010). The inactivation of PP2C allows for the phosphorylation and activation of three sucrose nonfermenting 1-related protein kinase 2s (SnRK2s), which belong

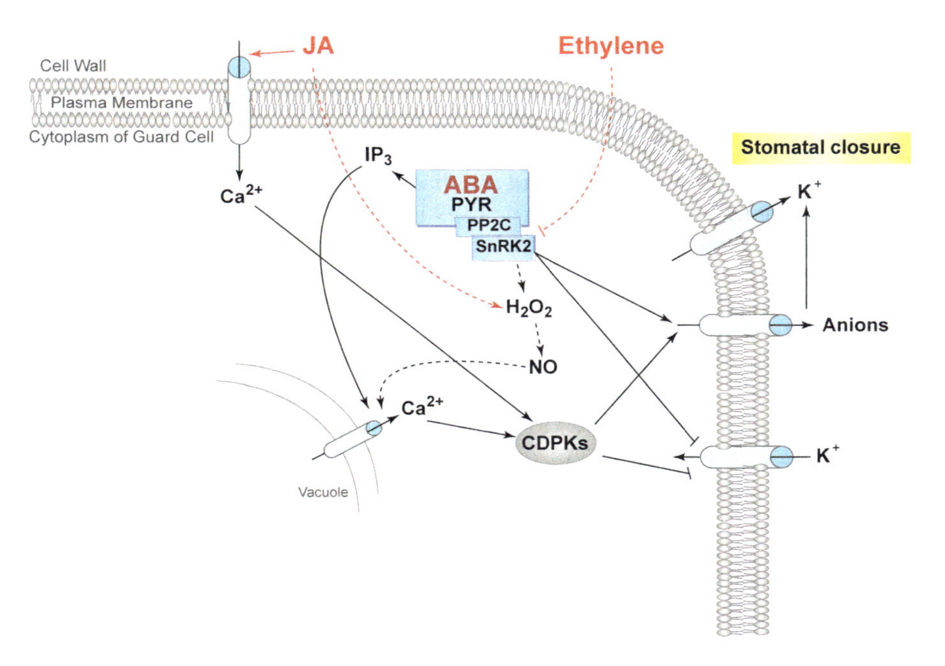

Fig. 2.1 This model illustrates the interactions between phytohormone signal pathways during ABA-induced stomatal closure as a result of drought stress. Drought triggers an increase in ABA concentration in guard cells, where it becomes bound to its receptor complex, in which SnRK2 is activated. The SnRK2 phosphorylates and inactivates a potassium import channel and activates an anion efflux channel, which in turn simulates K^+ efflux. SnRK2 also activates enzymes involved in H_2O_2 production; the resultant H_2O_2 activates NO that then triggers the influx of Ca^{2+} from the vacuole into the cytoplasm. The Ca^{2+} activates CDPKs, which then stimulate anion efflux and inhibit K^+ influx. The resulting ion loss causes water efflux, loss of turgor, and stomatal closure. In addition, Ca^{2+} influx from the vacuole into the cytoplasm is induced by ABA-mediated production of IP_3. Stress-induced JA production interacts with ABA-mediated stomatal closure by stimulating the influx of extracellular Ca^{2+} and/or by activating H_2O_2/NO signaling. Ethylene acts as a negative regulator of ABA, and thus, of this pathway. Steps inferred from experiments exploring ABA function using turgid cells are indicted with *dashed lines*

to the SnRK2 subfamily, whose members are associated with the regulation of abiotic stress (Fujita et al. 2009). While the precise mechanism of SnRK2 activation is unknown, it has been shown to autophosphorylate. SnRK2 acts as a positive regulator of several targets on the plasma membrane, including the anion exporter slow anion channel-associated 1 (SLAC1) and the K^+ channel in *Arabidopsis thaliana* (KAT1) in guard cells (Hubbard et al. 2010). SnRK2 activates SLAC1 through phosphorylation, while phosphorylation of KAT1 by SnRK2 inhibits its function, thus decreasing the influx of K^+ into the cell (Fig. 2.1). Increased SLAC1 activity causes an efflux of anions, which depolarizes the membrane and results in the loss of K^+ through the depolarization-activated K^+ efflux channel called guard cell outward-rectifying K^+ (GORK) (Jeanguenin et al. 2008) (Fig. 2.1). The collective loss of anions and K^+ ions from the guard cells causes water to move out of these cells, which results in the reduction in turgor that triggers stomatal closure in response to ABA.

2.2.1.3 ABA Activates Ca^{2+} Signaling Pathways

Ca^{2+} influx into the cytoplasm of guard cells has been observed within minutes after ABA treatment (Schroeder and Hagiwara 1990). This influx may occur through the release of the second messenger inositol 1, 4, 5-trisphosphate (IP_3), which activates Ca^{2+} channels located in the vacuole and endoplasmic reticulum (Krinke et al. 2007; Kwak et al. 2008) (Fig. 2.1). The influx of Ca^{2+} into the cytoplasm initiates a number of Ca^{2+}-dependent events associated with ABA signal transduction.

Numerous Ca^{2+}-dependent protein kinases (CDPKs) are activated during drought stress conditions and control stomatal closure through the regulation of ion channels (Fig. 2.1). In ABA-associated regulation of SLAC1, SnRK2 inhibits the phosphatase ABA insensitive 1 (ABI1), a negative regulator of CPK21. Therefore, ABI1 inactivation allows the activation of CPK21, which phosphorylates SLAC1, and thus activates anion efflux (Geiger et al. 2010). In *Arabidopsis* leaves, the concentration of another CDPK, CPK10, increases within 30 min after drought stress begins and causes the inhibition of inward K^+ currents (Zou et al. 2010). These results indicate that an increase in the cytoplasmic concentration of Ca^{2+} stimulates Ca^{2+}-dependent pathways that inhibit K^+ import while activating SLAC1, triggering the membrane depolarization that activates K^+ efflux (Fig. 2.1). CDPK pathways thus contribute to the loss of ions from guard cells, which in turn results in the loss of turgor, and ultimately to stomatal closure.

2.2.1.4 H_2O_2 and NO Are Associated with ABA in the Regulation of Stomatal Closure

An increase in oxidative stress is a common result of most abiotic stress conditions, including drought (Jaspers and Kangasjärvi 2010), and is often associated with an increase in NO production (Neill et al. 2008). While there is considerable evidence for the roles of H_2O_2 and NO in ABA-mediated stimulation of stomatal closure, their roles in signaling associated with drought stress remains unclear (Neill et al. 2008; Sirichandra et al. 2009). Studies of ABA action demonstrate that ABA-mediated regulation of stomatal closure requires NO and H_2O_2 (Bright et al. 2006). In these studies, both H_2O_2 and ABA treatments stimulated NO synthesis within 25 min after drought induction and resulted in stomatal closure in leaf tissue within 2.5 h (Bright et al. 2006). ABA-stimulated stomatal closure was reduced when NO was removed, and in mutants with impaired NO biosynthesis. Likewise, mutants that lack a key enzyme for H_2O_2 production demonstrated reduced NO production and reduced stomatal closure. In addition, an ABA-induced SnRK2 activates a guard cell NADPH oxidase that releases H_2O_2 (Sirichandra et al. 2009; Hubbard et al. 2010). These results demonstrate both the complex interactions between these molecules and that ABA regulation requires the production of H_2O_2 to stimulate NO production (Fig. 2.1).

2.2.1.5 Cross-Talk Between ABA and Ethylene Involves H_2O_2 and NO

An active interaction between ethylene and ABA has been shown to control the regulation of stomatal closure. For instance, an elevated ethylene concentration in leaves inhibits ABA-induced stomatal closure (Tanaka et al. 2005). In drought-stressed *Arabidopsis ethylene overproducer 1* mutants, stomata closed more slowly and were less sensitive to ABA; in addition, ethylene applied to ABA-treated epidermal peels from wild-type *Arabidopsis* leaves inhibited stomatal closure (Tanaka et al. 2005). When ethylene was applied independently of ABA, it induced H_2O_2 synthesis within 30 min after treatment (Desikan et al. 2006). Using ethylene-response mutants, Desikan et al. (2006) provided evidence that stomatal closure by ethylene is regulated through its signal transduction pathway, which both stimulates production and requires H_2O_2 synthesis.

Direct evidence that ethylene production increases as a result of drought stress has been inconsistent, with reports of both increased ethylene production and ethylene inhibition after drought or osmotic treatment (Abeles et al. 1992). However, most studies that report increased ethylene production use detached leaves, and thus may not reflect the response in intact plants under drought conditions (Morgan et al. 1990). In studies using intact cotton plants, ethylene production was consistently lower in drought-exposed plants than in plants that were watered daily (Morgan et al. 1990). In addition, ABA appears to inhibit ethylene production, and ABA-deficient maize seedlings have greatly increased ethylene production (Sharp 2002). Therefore, the dramatic increase in ABA concentration that occurs under drought stress probably causes a reduction in ethylene production. Conversely, ethylene concentration may increase in response to other stress conditions, which could interfere with ABA-regulated stomatal closure. Wilkinson and Davies (2009) propose that the impact of the interaction between ethylene and ABA is dependent upon the oxidative stress load of the plant. In their model, ethylene and ABA can act independently to induce the formation of reactive oxygen species (ROS), such as

H_2O_2, which then stimulate NO production and stomatal closure. However, when ethylene is present in combination with high ABA levels, ABA-regulated stomatal closure is inhibited (Fig. 2.1). Therefore, as part of the stress response, ethylene may act as a feedback mechanism to allow the influx of some CO_2 for photosynthesis, even during extreme stress conditions (Tanaka et al. 2005; Neill et al. 2008).

2.2.1.6 JA Cross-Talk Involves Protein Kinase Phosphorylation Cascades

JA biosynthesis is induced by stress conditions such as wounding and herbivory (Wasternack 2007), and many JA-associated signaling genes are regulated by drought stress (Huang et al. 2008). JA interacts with ABA-regulated stomatal closure by increasing Ca^{2+} influx, which stimulates CDPK production and the resultant signal cascade. Treatment of turgid, excised *Arabidopsis* leaves with either ABA or methyl JA (MeJA) results in stomatal aperture reduction within 10 min (Munemasa et al. 2007). Inhibition of ABA biosynthesis by using chemical inhibitors or ABA-deficient mutants suppresses MeJA-induced Ca^{2+} oscillations in guard cells and also impairs stomatal closure (Hossain et al. 2011). Therefore, MeJA-mediated regulation of stomatal closure interacts with ABA-mediated regulation of Ca^{2+} signal transduction pathways (Fig. 2.1).

Studies involving the interactions of ABA with MeJA in guard cells reveal that both induce the formation of ROS and NO, and that both are present at reduced concentrations in MeJA-insensitive plants (Munemasa et al. 2007). Munemasa et al. (2011) demonstrated that CPK6 acts downstream of NO and ROS signaling, and therefore may be a target of NO-stimulated Ca^{2+} influx into the cytoplasm. CPK6 is also required for ABA activation of cytoplasmic Ca^{2+} channels (Ca^{2+}-permeable cation channels) (Munemasa et al. 2011) and slow-type anion currents in guard cells (Munemasa et al. 2007). Therefore, JA-induced Ca^{2+} influx into the cytoplasm initiates CPK6 activation, which in turn activates slow-type anion channels, therefore interacting with ABA-induced stomatal closure NO, ROS, ethylene and JA (Munemasa et al. 2011) (Fig. 2.1).

2.2.1.7 A Comparison of ABA- and Drought-Mediated Stomatal Closure

Studies of ABA- and drought-associated regulation of stomatal closure demonstrate the complexity of overlapping transduction pathways. As indicated by Neill et al. (2008), signaling mechanisms may not necessarily be the same under both optimum and environmentally stressful conditions. Many of the studies that investigate the regulation of stomatal aperture rely on the responses displayed by turgid cells (Bright et al. 2006; Desikan et al. 2006; Hossain et al. 2011; Munemasa et al. 2007, 2011; Tanaka et al. 2005). These studies present strong evidence to support a role for NO in ABA-mediated regulation of stomatal closure (Bright et al. 2006). However, the use of turgid cells in studies that explore the interactions of JA and ethylene with ABA relative to the regulation of stomatal closure may mean that those results do not translate to situations where physiological stress is occurring. Interestingly, Garcia-Mata and Lamattina (2001) showed that an increase in NO concentration improves drought tolerance in wheat.

2.2.2 *ABA-Mediated Regulation of Gene Expression*

A continually elevated ABA concentration initiates changes in gene expression that inhibit shoot growth and maintain root growth, thus increasing the root-to-shoot ratio. In a whole-genome microarray study using *Arabidopsis*, more than 1,900 genes were found to be drought-responsive, 1,300 of which are also regulated by ABA (Huang et al. 2008). Expression of these drought-responsive genes overlapped with expression profiles of genes regulated by hormones such as JA, auxin, cytokinin, or ethylene (Huang et al. 2008). In addition, numerous microRNAs (miRNAs) are upregulated in response to osmotic stress (Liu et al. 2008).

2.2.2.1 ABA-Dependent Gene Expression During Drought Stress

Longer-term physiological responses to abiotic stress conditions are caused by changes in gene regulation through ABA-mediated regulation of transcription factors that bind to ABA-responsive elements (ABREs) on ABA-regulated genes (Fig. 2.2). In addition to signaling stomatal closure, phosphorylation cascades also lead to changes in ABA-regulated transcription factors. For example, the ABA-responsive transcription factors (ABFs) ABF1 and ABF4 are activated when they are phosphorylated by CPK4 or CPK11 (which are ABA-inducible kinases) (Zhu et al. 2007).

Zhu et al. (2007) also demonstrated that other kinases can phosphorylate ABF1 and ABF4. For example, ABF1 was reported to be a target of some SnRK2 members (Fujita et al. 2009), and CPK32 phosphorylates ABF4 (Choi et al. 2005). Therefore, these transcription factors are probably affected by several signals and pathways (Zhu et al. 2007) and are thus important cross-talk points.

2.2.2.2 MAPK-Directed Phosphorylation Cascades Also Act as Cross-Talk Points

Osmosensing and subsequent signal transduction require phosphorylation of key intermediates through the activity of mitogen-activated protein kinases (MAPKs) (Boudsocq and Laurière 2005). Current models indicate that NO is a central signaling molecule that triggers phosphorylation events through MAPKs as an early response to drought stress (Courtois et al. 2008; Neill et al. 2008) (Fig. 2.2). In drought-stressed wheat seedlings, elevated NO concentrations enhance the transcription of specific drought-induced genes (Garcia-Mata and Lamattina 2001).

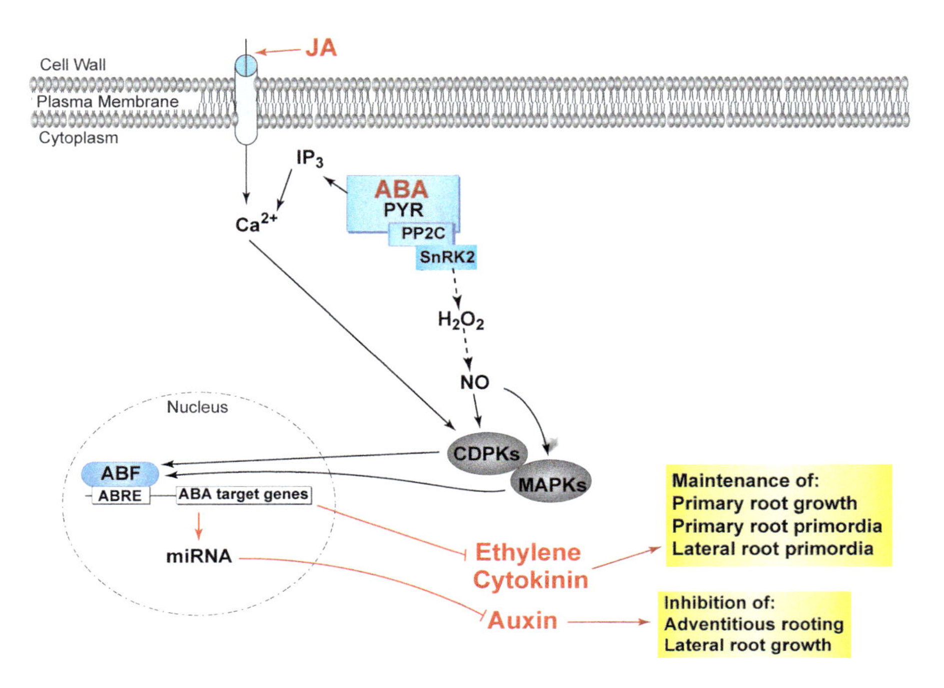

Fig. 2.2 This model illustrates drought-induced gene regulation that controls root growth and architecture. Elevated ABA concentration activates H_2O_2, NO, and Ca^{2+} signaling as discussed in Fig. 2.1. As shown, specific CDPKs and MAPKs activate ABA-regulated transcription factors, which control the expression of ABA-responsive genes. These genes inhibit the ethylene response via the downregulation of several ethylene-responsive and ethylene biosynthetic genes; they also reduce cytokinin levels by triggering an increase in the expression of enzymes that oxidize cytokinin. An increase in the expression of miRNAs associated with auxin signaling downregulates auxin responses. Collectively, these interactions favor the maintenance of the root primordia of primary roots and existing lateral roots while inhibiting lateral root elongation and adventitious root formation. Steps inferred from experiments exploring ABA function using turgid cells are indicted with *dashed lines*

In *Arabidopsis*, MPK6 and its associated MAPK cascade components are activated by drought stress (Boudsocq and Laurière 2005). MPK6 targets two ethylene biosynthetic enzymes, 1-aminocyclopropane-1-carboxylic acid [ACC] synthase 2 (ACS2) and ACS6 (Liu and Zhang 2004; Jaspers and Kangasjärvi 2010), and is also involved in ethylene signaling that leads to the regulation of ethylene-responsive genes (Hahn and Harter 2009). In addition, both *ACS2* and *ACS6* contain ABRE on their promoters (PlantCARE, Lescot et al. 2002). Expression profiles from the roots of 18-day-old wild-type *Arabidopsis* plants show that *ACS2* expression increases within 30 min of 300 mM mannitol treatment to the roots (*Arabidopsis* eFP Browser, Winter et al. 2007). *ACS2* is highly induced by osmotic stress by 3 h of treatment and remains elevated for 24 h. However, when osmotic treatment is compared to drought caused by exposing roots to a 15-min stream of air, causing the plant a water loss that totals 10% of the fresh weight, a different *ACS2* expression pattern emerges. Under drought conditions, *ACS2* expression is transiently upregulated in the roots for 6–12 h after dehydration, and shoots displayed no expression change from the control level (eFP Browser, Winter et al. 2007). In addition, evaluation of the promoter regions of other ethylene biosynthetic genes shows that in addition to ACS2 and ACS6, ACS7, ACS11, and ACC oxidase 4 (ACO4) have at least one ABRE on their promoters and therefore are candidates for ABA-mediated regulation (PlantCARE, Lescot et al. 2002). Thus, ethylene production may be differentially regulated by various stress conditions, and the responses to water loss and osmotic stress are likely to have differing regulatory components. While drought conditions may transiently upregulate specific ACS genes in root tissue, genes involved in ethylene biosynthesis and/or the response to ethylene tend to be downregulated in response to drought stress (Huang et al. 2008). For example, Huang et al. (2008) noted that ACC oxidase (ACO), which regulates the final step in ethylene biosynthesis, is downregulated by dehydration; its expression increased after rehydration. Therefore, these studies support a reduced, but perhaps transient, expression of ethylene during drought stress (Fig. 2.2).

The negative regulation of the MAPKs may serve as an essential feedback component of ABA-mediated signaling. Two forms of MAP kinase phosphatase, PP2C5 and each its related homolog, *Arabidopsis* Ser/Thr phosphatase of type 2C (AP2C1), each acts as negative regulators of MPK3 and MPK6 by dephosphorylation and each is upregulated by treatment with ABA (Schweighofer et al. 2007; Brock et al. 2010). Interestingly, Brock et al. (2010) provide evidence that these phosphatases are positive regulators of ABA-mediated signaling. Loss-of-function mutants (*ap2c1* or *pp2c5*) displayed reduced expression of selected ABA-responsive genes. Huang et al. (2008) also reported the increased expression of several PP2Cs during drought stress. Thus, increased dephosphorylation of MAPKs may provide a feedback mechanism for multiple regulatory pathways associated with drought responses.

2.2.2.3 Maintenance of Primary Root Growth During Water Stress

When *Arabidopsis* plants are exposed to osmotic (100–400 mM sorbitol) stress, mutants deficient in the osmotic sensor ATHK1 have shorter primary roots compared to wild-type plants, indicating that root growth regulation involves perception by an osmoreceptor (Wohlbach et al. 2008). This hypothesis is supported by studies in which mutants that overexpress *ATHK1* maintain root elongation under osmotic conditions (Wohlbach et al. 2008). Under drought conditions, the region near the root tip continues to grow, while regions away from the tip are inhibited or cease to grow (Sharp 2002). The tip region of the primary root is also the area where ABA and ROS accumulate and where a decrease in ethylene production is noted 3.5 h after drought stress (Yamaguchi and Sharp 2010). Therefore, during drought stress, an increase in ABA concentration inhibits ethylene production and downregulates genes that respond to the presence of ethylene, thus maintaining root growth (Sharp 2002) (Fig. 2.2).

Cytokinin is considered to be a negative regulator of root growth and branching, and root-specific degradation of cytokinin may also contribute to the primary root growth and branching induced by drought stress (Werner et al. 2010). Using *Arabidopsis* plants that display increased expression of cytokinin oxidase/dehydrogenase (CKX) genes under the control of a root-specific reporter, Werner et al. (2010) demonstrated that an increase in cytokinin degradation in the roots results in an increase in both primary root length and lateral root formation during drought conditions. An analysis of the *Arabidopsis* CKX promoter indicates that the region contains three ABREs (PlantCARE, Lescot et al. 2002), which implies potential ABA-based regulation of cytokinin degradation in the roots (Fig. 2.2). This activity would stimulate root growth and lateral root production; thus, increasing the root-to-shoot is associated with drought conditions and with increased drought tolerance (Werner et al. 2010).

2.2.2.4 Regulation of Lateral and Adventitious Root Development During Water Stress

While auxin is considered to be the primary hormone involved in the initiation and growth of lateral and adventitious roots, ABA may play a role in regulating lateral root growth under stress conditions (De Smet et al. 2006). During drought conditions, *Arabidopsis* produces specialized, short lateral roots that remain in a dormant or nongrowing condition while the plant is under stress (Wasilewska et al. 2008). These roots replace dehydrated lateral roots once drought conditions are relieved. Therefore, for *Arabidopsis*, the additional ABA produced in response to drought stress inhibits the outgrowth of lateral roots from existing meristems (De Smet et al. 2006).

Several miRNAs that are associated with auxin signaling are upregulated by drought or osmotic stress and may play important roles in the regulation of root architecture. For example, in *Arabidopsis*, both *miR167* and *miR168* are involved in the regulation of auxin response factors and both contain ABREs, indicating their own regulation by ABA signaling (Liu et al. 2008). *miR167* targets auxin response factors *ARF6* and *ARF8*, which act redundantly to regulate responses to auxin, and are positive regulators of adventitious root development from shoot tissue (Gutierrez et al. 2009). *miR168* targets *ARGONAUTE1* (*AGO1*), which encodes a RNA Slicer enzyme that downregulates the expression of *ARF17*, which is a negative regulator of root formation. *Arabidopsis* plants that overproduce ARF17 or are deficient in ARF6 or ARF8 have fewer adventitious roots than wild-type plants (Gutierrez et al. 2009). Interestingly, *miR167* and *miR168* are differentially regulated after osmotic stress treatment (Liu et al. 2008). *miR168* expression is upregulated in *Arabidopsis* seedlings 2–6 h after stress induced by 200 mM mannitol, while *miR167* expression increases 6–24 h after this treatment (Liu et al. 2008).

Data from the *Arabidopsis* eFP Browser indicate that osmotic (300 mM mannitol) treatment to roots downregulates *ARF8* at 3, 12, and 24 h after treatment (Winter et al. 2007). A microarray analysis supports the overall downregulation of genes involved in auxin-regulated responses as a consequence of root dehydration (Huang et al. 2008).

Gutierrez et al. (2009) proposed a model in which miRNAs fine-tune auxin-mediated regulation of adventitious rooting by controlling ARF transcription. In this case, during drought stress, ABA-induced production of *miR167* (which targets *ARF6* and *ARF8*) would inhibit adventitious root growth, and increased expression of *miR168* reduces *AGO1* expression, which would result in an increase in *ARF17* concentration, again leading to a decrease in adventitious root production. Considering that these miRNA genes appear to be differentially controlled, the level of auxin production could be nimbly up- or downregulated in response to changing stress conditions, particularly through the regulation of these transcription factors. Such a system would allow a dynamic response, where, for instance, under drought conditions, it would transiently favor decreased adventitious rooting while also increasing overall primary root growth.

In contrast to the adjustment in adventitious rooting, there were no changes in either the length or number of lateral roots for either loss-of-function mutants or overexpressers of *ARF6*, *ARF8*, or *ARF17* (Gutierrez et al. 2009). In a separate study, *miR167* was found to regulate the elongation, but not the initiation, of lateral roots in *Arabidopsis* (Gifford et al. 2008). Current models support the differential regulation of lateral and adventitious roots by miRNAs and their associated targets (Meng et al. 2010).

2.2.2.5 A Summary of Drought Regulation of Root Growth and Development

Drought-associated changes in root architecture are directed by cross-talk between ABA with ethylene, auxin, and cytokinin (Fig. 2.2). When *Arabidopsis* is exposed to drought stress, an increase in ABA in the roots inhibits ethylene production and stimulates cytokinin degradation, which together result in the maintenance of primary root growth. ABA-controlled regulation of genes (such as those that encode ARFs and miRNAs) associated with auxin-regulated root production inhibits the formation of adventitious roots and the outgrowth of lateral roots from existing primordia. However, research on other species demonstrates that the regulation of lateral root formation is highly adaptive, and in rice, for instance, ABA stimulates lateral root formation (Chen et al. 2006). Thus, the complex balance between stimulation and feedback inhibition produced by interacting hormone signaling pathways may result in different scenarios that are fine-tuned by the specific stress conditions and plant species.

2.3 Hormonal Cross-Talk Associated with Wounding

Mechanical wounding of leaf tissue is associated with increases in ethylene, JA, and ABA, which collectively trigger cellular senescence at the wounding site. Cross-talk points that affect gene regulation after wounding also trigger defense mechanisms associated with biotic signaling.

2.3.1 Ethylene Production Increases After Mechanical Wounding

Mechanical wounding causes a transient increase in ethylene production above the normally low, basal level in most plant tissues (Saltveit and Dilley 1978). While there is considerable variation between the levels of basal and peak wound-induced ethylene production based on species and tissue types, the timing of the induction and peak rate of wound-induced ethylene production remain consistent within tissues of a given species (Saltveit and Dilley 1978). For example, in etiolated corn seedlings, wound-induced ethylene increases after a lag of 20 min, with peak production occurring at 58 min, while light-grown plants such as *Forsythia* had a 30-min lag and peak ethylene production at 133 min (Saltveit and Dilley 1978). This transient increase in ethylene production is considered to be a short-term response to wounding that leads to subsequent long-term growth and developmental responses (León et al. 2001; Telewski 2006).

Many ethylene biosynthetic enzymes, including ACS and ACO, are wound inducible (Reymond et al. 2000; Tsuchisaka and Theologis 2004). Analysis of the *Arabidopsis* ACS isoforms composed of ACS promoters fused to a reporter gene demonstrates that *ACS2*, *ACS4*, *ACS6*, *ACS7*, and *ACS8* are each locally upregulated in tissue around the site of wounding (Tsuchisaka and Theologis 2004) (Fig. 2.3), and an increase in phosphorylation of both ACS2 and ACS6 by MPK6 has been measured within 15 min after wounding of *Arabidopsis* seedlings (Liu and Zhang 2004). This indicates that in addition to stimulating expression of *ACS* genes, wounding activates a MAPK kinase cascade, which activates and stabilizes ACS2 and ACS6, resulting in an increase in ethylene production (Liu and Zhang 2004).

2.3.2 Rapid Signaling Events Associated with Mechanical Wounding

The most rapid cellular change associated with wounding or mechanical stimulation (e.g., by touch) is the increase in concentration of cytoplasmic Ca^{2+} that occurs within seconds of stimulation (Haley et al. 1995; Monshausen et al. 2009).

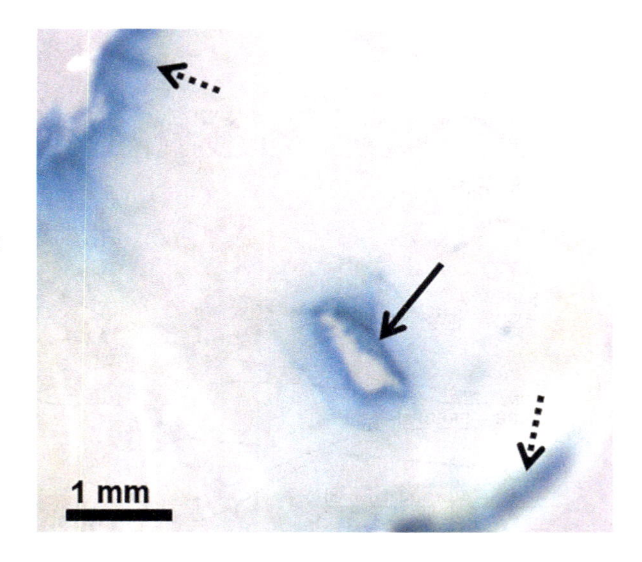

Fig. 2.3 This image illustrates an example of a localized gene expression pattern associated with wound-inducible *ACS* genes. Reporter expression in a transgenic *Arabidopsis* leaf carrying *ACS6 promoter::GUS* constructs occurs around the area punctured with a dissecting needle (*arrow*). GUS staining also occurs at the cut edge where the leaf was excised from the plant (*dashed arrows*)

This cytosolic Ca^{2+} interacts with Ca^{2+}-binding or Ca^{2+}-responsive proteins whose expression is also rapidly upregulated after wounding (Walley et al. 2007).

An increase in ROS production occurs within 2 min of wounding of *Arabidopsis* leaf tissue (Miller et al. 2009). ROS production rapidly increases after treatment with Ca^{2+} ionophores that release Ca^{2+} ions, and is inhibited by Ca^{2+}-transport inhibitors (Monshausen et al. 2009). Monshausen et al. (2009) propose that the influx of Ca^{2+} into the cytoplasm stimulates NADPH oxidation to form superoxide in the cell wall. Superoxide is rapidly converted to H_2O_2, which can easily enter the cell and trigger the regulation of gene expression (Monshausen et al. 2009) (Fig. 2.4).

Microarray studies have revealed that numerous genes are expressed within minutes of wounding; for instance, Walley et al. (2007) found that 162 genes were induced within 5 min of wounding of *Arabidopsis* rosette leaves. The promoters of 47 of these rapid wound-responsive (RWR) genes contain a rapid stress response element (RSRE) (Walley et al. 2007). In transgenic *Arabidopsis* plants containing four RSREs fused to a reporter, wound-induced reporter expression occurred within 5 min in an area local to the wound site, indicating that the response element is capable of detecting wounding stimulus in a manner similar to that observed for the RWR genes (Walley et al. 2007). Ca^{2+} and ROS are both considered candidates for the regulation of RSRE-containing genes, but currently there is no direct evidence confirming that they regulate proteins that bind to RSREs (Walley and Dehesh 2010) (Fig. 2.4). However, a number of RWRs, such as calmodulin-related protein and touch-induced 3 (TCH3), are Ca^{2+}-activated genes (Walley et al. 2007), and the RSRE sequence is similar to one which binds a calmodulin-binding transcription activator, thus supporting a role for Ca^{2+} in the regulation of RWR genes (Walley and Dehesh 2010).

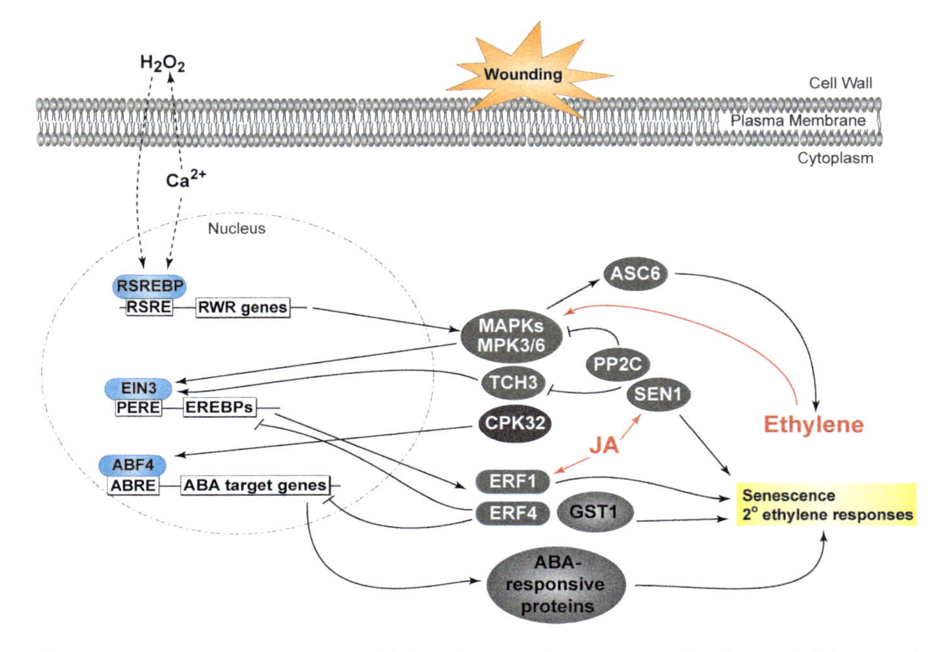

Fig. 2.4 This model illustrates wound-induced changes in gene expression that result in increased leaf senescence. Wounding causes rapid Ca^{2+} oscillations in the cell, which stimulate H_2O_2 production. Ca^{2+} and H_2O_2 regulate the expression of rapid wound-responsive (RWR) genes through the activation of transcription factors (RSRE binding proteins, RSREBPs), which include MAPKs associated with the activation of the ethylene signaling component, EIN3. EIN3 regulates the expression of ethylene-responsive element-binding proteins (EREBPs), including ERF1 and GST1, which trigger ethylene-mediated stimulation of senescence. ERF4, which is also an EREBP, is a negative regulator of both ABA- and ethylene-mediated responses. Other RWRs include the ethylene biosynthetic enzyme ACS6, senescence-associated gene SEN1, and CPK32. CPK32 activates the ABA transcription factor ABF4 to initiate ABA-mediated responses. Another RWR, the phosphatase PP2C, acts to dephosphorylate and inactivate other RWRs such as MAPKs and TCH3. Wound-induced JA synergistically interacts through its activation of ERF1 and SEN1. Rapid wound-responsive genes products are indicated by *shaded symbols with white lettering*

Early signaling events include rapid upregulation of components of the ethylene-response pathway. Transcription profiles show rapid upregulation of kinase cascade components, including the MAPK kinase, MKK9, and its target MPK3, supporting an important role for increased phosphorylation in the early responses to wounding (An et al. 2010). MPK3 phosphorylation targets transcription activator ethylene insensitive 3 (EIN3) (An et al. 2010). EIN3 binds to promoter regions of ethylene-responsive element-binding proteins (EREBPs) and regulates transcription of both the senescence-associated gene glutathione-S-transferase 1(GST1) and ethylene-response factor 1 (ERF1), a RWR that regulates the secondary gene targets of ethylene signaling such as those involved in the defense response and cellular inhibition (Schaller and Kieber 2002). Therefore, the phosphorylation and stimulation of EIN3 activity would explain the rapid increase in the expression of primary ethylene-responsive genes, such as ERF1 and ERF4 (Fig. 2.4). In addition, MPK3 also phosphorylates ACS6 (a RWR), which stimulates ethylene biosynthesis (An et al. 2010) (Fig. 2.4). Therefore, the expression of many of the RWR genes serves to initiate early tissue senescence at the wound site.

Interestingly, PP2C-type phosphatases (including AP2C1) that dephosphorylate MAPKs are also identified as RWRs (Walley et al. 2007). Plants carrying the AP2C1 promoter fused to a GUS (β-glucuronidase) reporter gene demonstrate an increase in AP2C1 expression localized around the site of wounding (Schweighofer et al. 2007). Transgenic *Arabidopsis* plants that overexpress AP2C1 display both greatly reduced MPK6 activity and reduced ethylene production after wounding (Schweighofer et al. 2007). In addition, TCH3 is also negatively regulated by dephosphorylation, and may be a target of the RWR PP2C-type phosphatases (Wright et al. 2002). Therefore, feedback regulation by dephosphorylation is also a rapid response to wounding (Fig. 2.4)

2.3.3 Gene Expression Patterns Associated with Mechanical Wounding

Expression studies of wound-induced genes in *Arabidopsis* demonstrated the upregulation of many genes associated with JA and with ethylene biosynthesis and response (Cheong et al. 2002). Microarray studies that profiled gene expression in wounded *Arabidopsis* leaves revealed that many genes have a similar expression pattern, in which induction occurs by 15 min and expression peaks at 1–2 h after wounding; this pattern is similar to the ethylene profile described above (Reymond et al. 2000). Genes that display this type of expression pattern include JA- and ethylene-associated genes as well as several RWRs, including calmodulin (*TCH1*), calmodulin-related proteins (*TCH2* and *TCH3*), *MPK3*, and *GST1* (Reymond et al. 2000).

ABA also accumulates in the tissue around a wound site, probably in response to dehydration associated with the damage caused by wounding (León et al. 2001). The expression of CPK32, which phosphorylates (and thus activates) the ABA-responsive transcription factor ABF4 (Choi et al. 2005), greatly increases from 5 to 30 min after wounding (Chotikacharoensuk et al. 2006). ABA is known to promote leaf senescence (Finkelstein and Rock 2002), and thus, the increase in ABA-responsive transcription factors suggests that ABA may contribute to the stimulation of senescence around a wound area.

JA production is also wound inducible and stimulates leaf senescence in plant tissues (Wasternack 2007). JA synthesis follows a similar pattern of transient increase as that of ethylene and wound-induced gene expression (Chotikacharoensuk et al. 2006; Schweighofer et al. 2007). JA is activated independently from ethylene, but JA treatment upregulates numerous *ERF* genes (McGrath et al. 2005). These include the transcriptional activator ERF1 and transcriptional repressor ERF4, both of which are RWR proteins (Walley et al. 2007). Interestingly, ERF4 is a negative regulator of ABA and ethylene responses (Yang et al. 2005) (Fig. 2.4). Senescence-associated protein 1 (SEN1) is also potentially regulated by JA (The Arabidopsis Information Resource, Swarbreck et al. 2008) (Fig. 2.4).

2.3.4 Hormonal Responses to Mechanical Wounding Are Rapid and May Also Be Involved in Biotic Stress Response

Mechanical wounding stimulates rapidly induced events that involve increases in both Ca^{2+} and ROS, which then activate the expression of MAPK and CDPK cascade components. These protein kinases activate transcription factors and proteins that are part of response pathways that involve the phytohormones ethylene, JA, and ABA (Fig. 2.4). Thus, the response pathways can be stimulated downstream of the point where an increase in the hormone itself would be necessary for inciting a response. Subsequent increases in hormone production may act to sustain the wound-induced response.

Many RWR genes are known to be upregulated under both abiotic and biotic stress conditions (Walley et al. 2007). Walley et al. (2007) examined both abiotically induced RWR expression as well as profiles for various biotic stresses and found considerable overlap in the expression of genes associated with early responses to pathogenic infections, indicating that RWR genes may participate in the regulation of the initial defense mechanism. JA is a key regulatory signal associated with both the local and systemic response to pathogen attack (Wasternack 2007). It is known that mechanical wounding stimulates JA biosynthesis (León et al. 2001), and it has been proposed that JA also contributes to the systemic response that stimulates immunity to subsequent herbivory (Wasternack 2007).

2.4 Signaling During Abiotic Stress Conditions

Several signaling components discussed earlier are common across many abiotic stress conditions. ABA is considered the primary hormone involved in the signaling of many abiotic stresses, including drought, salt, and cold. Therefore, the stress conditions that result in elevated ABA concentrations may be part of a general stress-regulation network that shares many of the same signaling components described for the regulation of drought stress. Likewise, mechanical stresses would stimulate the production of RWRs and follow a common mechanosensory-regulated response mechanism. RWR genes are also associated with several abiotic responses. Therefore, rapid gene expression changes may be characteristic to most abiotic stress responses. In addition, wounding damage to tissue causes dehydration around the wound, so some aspects of that pathway will converge with the drought-stress response.

While many components of the pathways act in a similar manner, and many of the same genes are induced by several abiotic stresses, it should be noted that the expression pattern of any specific gene varies in its timing and level of expression within specific tissues and under different experimental conditions. For example, MKK9, which is upregulated by osmotic stress, dehydration, salt, wounding, and cold treatments (Ma and Bohnert 2007), is highly upregulated in roots from 1 to 24 h when salt-stressed but during this same time period is expressed most strongly in the shoots during osmotic stress (eFP Browser, Winter et al. 2007). While the variation may occur because the treatments vary in their severity and duration, the hormonal status of the plants under the different stress conditions may also affect cross-talk and the complex feedback regulatory mechanisms associated with the interacting pathways.

2.4.1 Common Regulatory Elements of Abiotic Stress Responses

Many of the genes that are affected by wounding are associated with several abiotic stressors, such as cold, drought, heat, salt, and ABA treatment (Cheong et al. 2002). For example, several RWR genes (e.g., *ACS6*, *ERF1*, and several other *ERFs*, Ca^{2+}-signaling components, *TCH3*, and *MPK3*) are rapidly induced in roots not only by wounding but also by salt stress (Ma and Bohnert 2007) (Fig. 2.5). In addition, 47% of the RWR genes are also activated by cold stress, and other genes associated with several abiotic responses are RWRs, providing evidence that the rapid regulation of these genes is a common component of abiotic stresses (Walley et al. 2007) (Fig. 2.5).

Cellular dehydration is a result of many abiotic stress conditions, especially drought, salt, and cold stress, but also occurs in wounded tissue. Microarray studies have provided substantial evidence of overlapping gene expression associated with abiotic stress conditions that involve dehydration. For example, a comparison of changes in gene expression caused by drought or high salt conditions found that 154 genes were upregulated and 104 were downregulated, indicating a high correlation with gene regulation during the same two stress conditions in *Brassica rapa* (Chen et al. 2010). In another study, 197 genes were identified as being induced by several different stress conditions (wounding, cold, salt, osmotic, biotic, and chemical) (Ma and Bohnert 2007). The functions of the gene products include signaling components that are involved in ROS metabolism, Ca^{2+} signaling (such as CPK32, which regulates ABF4), ethylene signaling (including ACS6 and numerous ERFs), and MAPK cascade components (e.g., MKK9, which leads to ACS6 activation). Therefore, Ma and Bohnert (2007) concluded that both ROS and ethylene signaling pathways are central to the coordination of stress signals while ABF4 binds to promoter regions of ABA-responsive genes, thereby regulating them (Zhu et al. 2007).

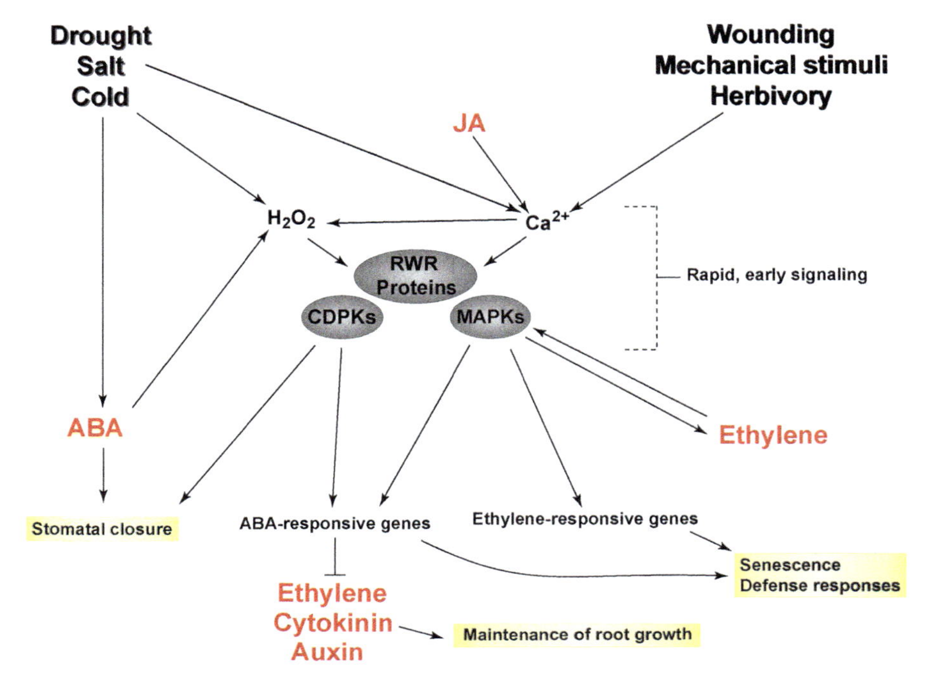

Fig. 2.5 This model of hormonal cross-talk integrates elements involved in the responses to numerous abiotic and biotic stress conditions. Negative regulators and regulatory feedback steps have been omitted for clarity

Kinase cascades are associated with several abiotic stress conditions, and certain components may serve as hubs. In *Arabidopsis*, MPK4 and MPK6 (which activate ACS2 and ACS6) are induced by cold, salt, and osmotic stress conditions, indicating that these MAPK forms are central hubs in abiotic stress conditions (Boudsocq and Laurière 2005). Jaspers and Kangasjärvi (2010) note that conditions (drought, cold, salt, temperature extremes, etc.) that lead to oxidative stress and subsequent H_2O_2 production activate a key MAP kinase cascade whose targets include transcription factors involved in ABA and ethylene signaling. Therefore, ROS such as H_2O_2 convey information from a variety of upstream signals and transduce the information to targeted genes to generate a response.

Stress conditions often increase the expression of certain miRNAs, including *miR167*, *miR168*, and *miR393*, which are involved in auxin regulation. *miR168* is upregulated during cold, osmotic, and salt stresses, and *miR167* is upregulated by cold and salt (Liu et al. 2008), while *miR393* expression increases during cold, drought, salt, and UV irradiation conditions (Martin et al. 2010). The promoters of both *miR167* and *miR168* contain response elements for both auxin and ABA (Liu et al. 2008). As discussed previously, *miR167* and *miR168* have been shown to control the regulation of auxin-mediated responses.

The *miR393* gene contains an ABRE (Liu et al. 2008) and regulates the expression of the auxin-binding protein transport inhibitor response 1 (*TIR1*), which enables the expression of auxin-regulated genes. Increased levels of TIR1 favor lateral root formation (Pérez-Torres et al. 2008); therefore, the degradation of *TIR1* transcripts would result in less lateral root formation under abiotic stress conditions (Martin et al. 2010). This is consistent with a general downregulation of auxin-responsive genes found in microarray studies of plants experiencing abiotic stress (Cheong et al. 2002). In addition to the ABRE, *miR393*'s promoter also contains response elements for ethylene, auxin, and JA, indicating potential regulation by several hormones (Liu et al. 2008).

As with the histidine kinase osmoreceptor, AHK1, the cytokinin receptors AHK2 and AHK3 were induced within 10 min of stress treatment in *Arabidopsis* plants (Tran et al. 2007). AHK2 was upregulated after dehydration, and AHK3 production was stimulated by multiple abiotic stresses (dehydration, salt, and cold, as well as by ABA treatment) (Tran et al. 2007). Loss-of-function mutants *ahk2* and *ahk3* exhibit an increased tolerance to drought and salt, which indicates that AHK2 and AHK3 act as negative regulators of osmotic stress responses (Tran et al. 2007).

As previously discussed, JA biosynthesis is induced by wounding (Figs. 2.1 and 2.2). In addition, the expression of a JA biosynthetic enzyme, 12-oxophytodienoat-10,11-reductase (OPR3), increased by 3 h after treatment with salt or Ca^{2+} (Chotikacharoensuk et al. 2006). Ludwig et al. (2005) found that abiotic stress conditions stimulated Ca^{2+} signaling and CDPK2 activity resulting in elevated levels of OPR3 and JA. In addition, the expression of 197 genes was found to be common to both drought and JA-associated signaling (Huang et al. 2008). An analysis of transcription regulation sites determined that several genes involved in the JA biosynthetic pathway are upregulated after NO treatment, indicating a potential regulation of JA signaling through the ROS/NO pathway (Palmieri et al. 2008). However, the role of NO in the signaling of abiotic stresses remains unclear. While NO may play a role in ABA regulation of stomatal closure during non-stressed conditions, and may be involved in JA signaling, abiotic stress conditions may preferentially trigger a Ca^{2+} pulse and ROS burst that both activate RWRs, inducing the rapid stress-responsive pathway (Fig. 2.5).

2.4.2 Cross-Talk Between Abiotic and Biotic Stresses

A model by Ludwig et al. (2005) indicates that both abiotic and biotic stress conditions stimulate CDPK and MAPK pathways and that ethylene acts as a critical cross-talk point between the two pathways. Many of the RWRs are also associated with biotic stress conditions and may also be considered part of the pathogen defense response, especially in association with tissue wounding and through the activation of ethylene biosynthetic and signaling components (Walley et al. 2007) (Fig. 2.5). In addition, JA synthesis results from both wounding and biotic stresses (León et al. 2001), thus indicating its involvement in ethylene regulation and the regulation of stress responses such as senescence (Fig. 2.5).

2.4.3 Stress Tolerance

The breeding of plants to increase their tolerance to abiotic stress is an ongoing and important agricultural effort. Plants tolerant to abiotic stress conditions would possess such desirable traits as an increase in sensitivity to ABA, which would result in drought-tolerant traits such as increases in stomatal closure, root growth, and lateral branching of roots. Current laboratory methods used to produce transgenic plants allow researchers to manipulate genes much more precisely than by traditional breeding methods. However, the complexity of native stress regulation, which involves numerous proteins and cross-talk with phytohormone pathways, makes precise regulation difficult. Understanding the hormonal cross-talk that occurs during the regulation of stress conditions may provide insight into useful gene targets that would allow the production of stress-tolerant plants. Unfortunately, current public and political opinion of transgenic plants blunts the effectiveness of this approach. Nonetheless, specific genes could be targeted using strategies that involve miRNAs or transcription factors, which might be more acceptable to the general public. The combination of traditional breeding with modern genetic approaches could be used to improve screening for stress-tolerant cultivars.

2.4.3.1 Gene Targets

Evaluation of loss-of-function and gain-of-function mutants has identified numerous genes that confer stress tolerance to several abiotic conditions. Loss-of-function mutants associated with components of ABA signaling generally result in plants with increased sensitivity to abiotic stress conditions. Overexpression of these genes often results in tolerance. For example, plants that overexpress the osmoreceptor ATHK1 demonstrate increased tolerance to drought (Tran et al. 2007). Specific types of CDPKs (e.g., CPK10 in *Arabidopsis* and OsCDPK7 in *Oryza*) have been found to increase stress tolerance in plants in which they are overexpressed (Saijo et al. 2001; Zou et al. 2010). Conversely, loss-of-function mutants of the negative regulators of ABA signaling tend to result in increased stress tolerance. For example, loss-of-function mutants for the cytokinin receptors AHK2 and AKH3 display increased tolerance to salt and drought (Tran et al. 2007). Mutants of miRNA genes have also been shown to alter the regulation of transcription factors as well as other signaling components, making them potential targets for engineering stress-tolerant crops.

2.4.3.2 Strategies for Developing Stress Tolerance in Plants

The major cross-talk points that involve both ABA and ethylene signaling pathway components have been considered as primary targets for manipulation to improve the response of crops to multiple stress conditions. Other important targets include the gene products that protect plants from the increased oxidative stress created by many abiotic stressors (Neill et al. 2008). However, given the complexity of interactions associated with abiotic and biotic stress responses, a targeted strategy is preferred.

One strategy that provides a more targeted approach involves manipulating the expression of transcription factors, which would only affect a subset of stress-responsive genes (Umezawa et al. 2006). For example, the overexpression of an ethylene-responsive element-binding protein (EREBP) increases ROS detoxification and ROS-induced cell death (Ogawa et al. 2005). In a shared response to various stresses, many RWR genes that are transcription factors are upregulated early and in a transient manner; this feature may be useful in identifying plants resistant to multiple stress conditions. Walley et al. (2007) suggested that using RSREs to drive expression of specific genes could be a useful tool in engineering stress tolerance in plants. Interestingly, Capiati et al. (2006) demonstrated that wounding induces salt tolerance in tomatoes, indicating that focusing on the rapid responses to wounding that show significant cross-talk with other abiotic stress conditions could be a useful approach. The regulation of specific miRNAs may also provide a method for more precisely modulating the stress response through the regulation of specific transcription factors. However, transcription factors and miRNAs have multiple targets, and altered regulation that improves certain aspects of plant growth may also result in unfavorable traits.

Another approach to improve the targeting of specific regulatory components is to direct expression through the use of tissue-specific or stress-inducible promoter constructs (Hardy 2010). For example, the overexpression of OsCDPK7 in rice is localized to the root primordial and vascular tissues and may protect those tissues during abiotic stress conditions (Saijo et al. 2001). Another target could be the root-specific reduction of cytokinin levels through overexpression of cytokinin oxidase/dehydrogenase, which increases stress tolerance while maintaining adequate shoot growth and development in both *Arabidopsis* and tobacco (Werner et al. 2010). The use of stress-inducible promoters allows for normal plant growth and development since the gene(s) will only be expressed under stress conditions (Hardy 2010).

Gene expression profiles of *Arabidopsis* transcripts have led to a wealth of knowledge concerning the genes associated with that plant's response to abiotic stress. Increased computational capability, such as the clustering methods used by Ma and Bohnert (2007) have allowed more sophisticated data analysis, including an improved ability to sort genes into functional groups. Data from *Arabidopsis* is useful for identifying key components but is not readily transferable to crops, especially the cereal species (Tester and Bacic 2005). As pointed out by Langridge et al. (2006), a functional genomic approach can help improve our knowledge of stress tolerance in cereal species, but the application of this information to field use will require additional time. As more crop genomes are sequenced, it will be determined whether the approaches used for *Arabidopsis* may be effectively applied to crop species. Identification of the key regulatory components involved in tissue specificity and in the responses to various stressors could provide a mechanism for identifying important genes and critical components of stress-related signaling pathways. Such information could be used to develop transgenic plants or to identify natural variants, perhaps by using improved molecular-based screening methods (Tester and Bacic 2005).

Acknowledgments I thank Susan Weinstein for her careful reading and helpful advice concerning this manuscript. I also thank Jamie Lau for assistance with the graphics and Richard Pitaniello for editing assistance.

References

Abeles FB, Morgan PW, Saltveit ME (1992) Ethylene in plant biology, 2nd edn. San Diego, California

An F, Zhao Q, Ji Y, Li W, Jiang Z, Yu X, Zhang C, Han Y, He W, Liu Y, Zhang S, Ecker JR, Guo H (2010) Ethylene-induced stabilization of ETHYLENE INSENSITIVE3 and EIN3-LIKE1 is mediated by proteasomal degradation of EIN3 Binding F-Box 1 and 2 that requires EIN2 in *Arabidopsis*. Plant Cell 22:2384–2401

Boudsocq M, Laurière C (2005) Osmotic signaling in plants. Multiple pathways mediated by emerging kinase families. Plant Physiol 138:1185–1194

Bright J, Desikan R, Hancock JT, Weir IS, Neill SJ (2006) ABA induced NO generation and stomatal closure in *Arabidopsis* are dependent on H_2O_2 synthesis. Plant J 45:113–122

Brock AK, Willmann R, Kolb D, Grefen L, Lajunen HM, Bethke G, Lee J, Nürnberger T, Gust AA (2010) The *Arabidopsis* mitogen-activated protein kinase phosphatase PP2C5 affects seed germination, stomatal aperture, and abscisic acid-inducible gene expression. Plant Physiol 153:1098–1111

Capiati DA, País SM, Téllez-Iñón MT (2006) Wounding increases salt tolerance in tomato plants: evidence on the participation of calmodulin-like activities in cross-tolerance signaling. J Exp Bot 57:2391–2400

Chen CW, Yang YW, Lur HS, Tsai YG, Chang MC (2006) A novel function of abscisic acid in the regulation of rice (*Oryza sativa* L.) root growth and development. Plant Cell Physiol 47:1–13

Chen L, Ren F, Zhong H, Feng Y, Jiang W, Li X (2010) Identification and expression analysis of genes in response to high-salinity and drought stresses in *Brassica napus*. Acta Biochim Biophys Sin (Shanghai) 42:154–164

Cheong YH, Chang HS, Gupta R, Wang X, Zhu T, Luan S (2002) Transcriptional profiling reveals novel interactions between wounding, pathogen, abiotic stress, and hormonal responses in *Arabidopsis*. Plant Physiol 129:661–677

Choi HI, Park HJ, Park JH, Kim S, Im MY, Seo HH, Kim YW, Hwang I, Kim SY (2005) *Arabidopsis* calcium-dependent protein kinase AtCPK32 interacts with ABF4, a transcriptional regulator of abscisic acid-responsive gene expression, and modulates its activity. Plant Physiol 139:1750–1761

Chotikacharoensuk T, Arteca RN, Arteca JM (2006) Use of differential display for the identification of touch-induced genes from an ethylene-insensitive *Arabidopsis* mutant and partial characterization of these genes. J Plant Physiol 163:1305–1320

Christmann A, Hoffmann T, Teplova I, Grill E, Műller A (2005) Generation of active pools of abscisic acid revealed by in vivo imaging of water-stressed *Arabidopsis*. Plant Physiol 137:209–219

Christmann A, Weiler EW, Steudle E, Grill E (2007) A hydraulic signal in root-to-shoot signalling of water shortage. Plant J 52:167–174

Courtois C, Besson A, Dahan J, Bourque S, Dobrowolska G, Pugin A, Wendehenne D (2008) Nitric oxide signaling in plants: interplays with Ca^{2+} and protein kinases. J Exp Bot 59:155–163

De Smet I, Zhang HM, Inzé D, Beeckman T (2006) A novel role for abscisic acid emerges from underground. Trends Plant Sci 11:434–439

Desikan R, Last K, Harrett-Williams R, Tagliavia C, Harter K, Hooley R, Hancock JT, Neill SJ (2006) Ethylene-induced stomatal closure in *Arabidopsis* occurs via AtrbohF-mediated hydrogen peroxide synthesis. Plant J 47:907–916

Endo A, Sawada Y, Takahashi H, Okamoto M, Ikegami K, Koiwai H, Seo M, Toyomasu T, Mitsuhashi W, Shinozaki K et al (2008) Drought induction of *Arabidopsis* 9-*cis*-epoxycarotenoid dioxygenase occurs in vascular parenchyma cells. Plant Physiol 147:1984–1993

Finkelstein RR, Rock CD (2002) Abscisic acid biosynthesis and response. The Arabidopsis Book 1:e0058. doi:10.1199/tab.0058

Fujita Y, Nakashima K, Yoshida T, Katagiri T, Kidokoro S, Kanamori N et al (2009) Three SnRK2 protein kinases are the main positive regulators of abscisic acid signaling in response to water stress in *Arabidopsis*. Plant Cell Physiol 50:2123–2132

Garcia-Mata C, Lamattina L (2001) Nitric oxide induces stomatal closure and enhances the adaptive plant responses against drought stress. Plant Physiol 126:1196–1204

Geiger D, Scherzer S, Mumm P, Marten I, Ache P, Matschi S et al (2010) Guard cell anion channel SLAC1 is regulated by CDPK protein kinases with distinct Ca^{2+} affinities. Proc Natl Acad Sci USA 107:8023–8028

Gifford ML, Dean A, Gutierrez RA, Coruzzi GM, Birnbaum KD (2008) Cell-specific nitrogen responses mediate developmental plasticity. Proc Natl Acad Sci USA 105:803–808

Gutierrez L, Bussell JD, Păcurar DI, Schwambach J, Păcurar M, Bellinia C (2009) Phenotypic plasticity of adventitious rooting in *Arabidopsis* is controlled by complex regulation of AUXIN RESPONSE FACTOR transcripts and microRNA abundance. Plant Cell 21:3119–3132

Hahn A, Harter K (2009) Mitogen-activated protein kinase cascades and ethylene: Signaling, biosynthesis, or both? Plant Physiol 149:1207–1210

Haley A, Russell AJ, Wood N, Allan AC, Knight M, Campbell AK, Trewavas AJ (1995) Effects of mechanical signaling on plant cell cytosolic calcium. Proc Natl Acad Sci USA 92:4124–4128

Hardy A (2010) Candidate stress response genes for developing commercial drought tolerant crops. Basic Biotechnol 6:54–58

Hoad GV (1975) Effect of osmotic stress on abscisic acid levels in xylem sap of sunflower (*Helianthus annuus* L). Planta 124:25–29

Hossain MA, Munemasa S, Uraji M, Nakamura Y, Mori IC, Murata Y (2011) Involvement of endogenous abscisic acid in methyl jasmonate-induced stomatal closure in *Arabidopsis*. Plant Physiol 156:430–438

Huang D, Wu W, Abrams SR, Adrian J, Cutler AJ (2008) The relationship of drought-related gene expression in *Arabidopsis* thaliana to hormonal and environmental factors. J Exp Bot 59:2991–3007

Hubbard KE, Nishimura N, Hitomi K, Getzoff ED, Schroeder JI (2010) Early abscisic acid signal transduction mechanisms: newly discovered components and newly emerging questions. Genes Dev 24:1695–1708

Jaspers P, Kangasjärvi J (2010) 200 Reactive oxygen species in abiotic stress signaling. Physiol Plant 138:405–413

Jeanguenin L, Lebaudy A, Xicluna J, Alcon C, Hosy E, Duby G, Michard E, Lacombe B, Dreyer I, Thibaud J-B (2008) Heteromerization of *Arabidopsis* Kv channel α-subunits. Plant Signal Behav 3:622–625

Koiwai H, Nakaminami K, Seo M, Mitsuhashi W, Toyomasu T, Koshiba T (2004) Tissue-specific localization of an abscisic acid biosynthetic enzyme, AAO3, in *Arabidopsis*. Plant Physiol 134:1697–1707

Krinke O, Novotná Z, Valentová O, Martinec J (2007) Inositol trisphosphate receptor in higher plants: is it real? J Exp Bot 58:361–376

Kuromori T, Miyaji T, Yabuuchi H, Shimizu H, Sugimoto E, Kamiya A et al (2010) ABC transporter AtABCG25 is involved in abscisic acid transport and responses. Proc Natl Acad Sci USA 107:2361–2366

Kwak JM, Mäser P, Schroeder JI (2008) The clickable guard cell, version II: Interactive model of guard cell signal transduction mechanisms and pathways. The Arabidopsis Book 6:e0114. doi:10.1199/tab.0114

Langridge P, Paltridge N, Fincher G (2006) Functional genomics of abiotic stress tolerance in cereals. Brief Funct Genom Proteom 4:343–354

León J, Rojo E, Sánchez-Serrano JJ (2001) Wound signalling in plants. J Exp Bot 52:1–9

Lescot M, Déhais P, Thijs G, Marchal K, Moreau Y, Van de Peer Y, Rouzé P, Rombauts S (2002) PlantCARE, a database of plant cis-acting regulatory elements and a portal to tools for in silico analysis of promoter sequences. Nucleic Acids Res 30:325–327

Liu Y, Zhang S (2004) Phosphorylation of 1-aminocyclopropane-1-carboxylic acid synthase by MPK6, a stress-responsive mitogen-activated protein kinase, induces ethylene biosynthesis in *Arabidopsis*. Plant Cell 16:3386–3399

Liu HH, Tian X, Li YJ, Wu CA, Zheng CC (2008) Microarray-based analysis of stress-regulated microRNAs in *Arabidopsis thaliana*. RNA 14:836–843

Ludwig AA, Saitoh H, Felix G, Freymark G, Miersch O, Wasternack C, Boller T, Jones JDG, Romeis T (2005) Ethylene-mediated cross-talk between calcium-dependent protein kinase and MAPK signaling controls stress responses in plants. Proc Natl Acad Sci USA 102:10736–10741

Ma S, Bohnert HJ (2007) Integration of *Arabidopsis thaliana* stress-related transcript profiles, promoter structures, and cell specific expression. Genome Biol 8:R49. doi:10.1186/gb-2007-8-4-r49

Martin RC, Liu P-P, Goloviznina NA, Nonogaki H (2010) microRNA, seeds, and Darwin?: diverse function of miRNA in seed biology and plant responses to stress. J Exp Bot 61:2229–2234

McGrath KC, Dombrecht B, Manners JM, Schenk PM, Edgar CI, Maclean DJ, Scheible WR, Udvardi MK, Kazan K (2005) Repressor -and activator-type ethylene response factors functioning in jasmonate signaling and disease resistance identified via a genome-wide screen of *Arabidopsis* transcription factor gene expression. Plant Physiol 139:949–959

Meng Y, Mab X, Chen D, Wub P, Chen M (2010) MicroRNA-mediated signaling involved in plant root development. Biochem Biophys Res Commun 393:345–349

Miller G, Schlauch K, Tam R, Cortes D, Torres MA, Shulaev V, Dangl JL, Mittler R (2009) The plant NADPH oxidase RBOHD mediates rapid systemic signaling in response to diverse stimuli. Sci Signal 2:ra45

Monshausen GB, Bibikova TN, Weisenseel MH, Gilroy S (2009) Ca^{2+} regulates reactive oxygen species production and pH during mechanosensing in *Arabidopsis* roots. Plant Cell 21:2341–2356

Morgan PW, He CJ, De Greef JA, De Proft MP (1990) Does water deficit stress promote ethylene synthesis by intact plants? Plant Physiol 94:1616–1624

Munemasa S, Oda K, Watanabe-Sugimoto M, Nakamura Y, Shimoishi Y, Murata Y (2007) The *coronatine-insensitive 1* mutation reveals the hormonal signaling interaction between abscisic acid and methyl jasmonate in *Arabidopsis* guard cells. Specific impairment of ion channel activation and second messenger production. Plant Physiol 143:1398–1407

Munemasa S, Hossain MA, Nakamura Y, Mori IC, Murata Y (2011) The *Arabidopsis* calcium-dependent protein kinase, CPK6, functions as a positive regulator of methyl jasmonate signaling in guard cells. Plant Physiol 155:553–561

Neill S, Barros R, Bright J, Desikan R, Hancock J, Harrison J, Morris P, Ribeiro D, Wilson I (2008) Nitric oxide, stomatal closure, and abiotic stress. J Exp Bot 59:165–176

Ogawa T, Pan L, Kawai-Yamada M, Yu LH, Yamamura S, Koyama T, Kitajima S, Ohme-Takagi M, Sato F, Uchimiya H (2005) Functional analysis of *Arabidopsis* ethylene-responsive element binding protein conferring resistance to Bax and abiotic stress-induced plant cell death. Plant Physiol 138:1436–1445

Palmieri MC, Sell S, Huang X, Scherf M, Werner T, Durner J, Lindermay C (2008) Nitric oxide-responsive genes and promoters in *Arabidopsis thaliana*: a bioinformatics approach. J Exp Bot 59:177–186

Pérez-Torres CA, López-Bucio J, Cruz-Ramírez A, Ibarra-Laclette E, Dharmasiri S, Estelle M, Herrera-Estrellab L (2008) Phosphate availability alters lateral root development in *Arabidopsis* by modulating auxin sensitivity via a mechanism involving the TIR1 auxin receptor. Plant Cell 20:3258–3272

Pinheiro C, Chaves MM (2011) Photosynthesis and drought: Can we make metabolic connections from available data? J Exp Bot 62:869–82

Reymond P, Weber H, Damond M, Farmer EE (2000) Differential gene expression in response to mechanical wounding and insect feeding in *Arabidopsis*. Plant Cell 12:707–720

Saijo Y, Kinoshita N, Ishiyama K, Hata S, Kyozuka J, Hayakawa T, Nakamura T, Shimamoto K, Yamaya T, Izui K (2001) A Ca^{2+}-dependent protein kinase that endows rice plants with cold- and salt-stress tolerance functions in vascular bundles. Plant Cell Physiol 42:1228–1233

Saltveit ME Jr, Dilley DR (1978) Rapidly induced wound ethylene from excised segments of etiolated *Pisum sativum* L., cv. Alaska: I. Characterization of the response. Plant Physiol 61:447–450

Schaller E, Kieber JJ (2002) Ethylene. The Arabidopsis Book 1:e0071. doi:10.1199/tab.0071

Schroeder JI, Hagiwara S (1990) Repetitive increases in cytosolic Ca^{2+} of guard cells by abscisic acid activation of non-selective Ca^{2+} permeable channels. Proc Natl Acad Sci USA 87:9305–9309

Schweighofer A, Kazanaviciute V, Scheikl E, Teige M, Doczi R, Hirt H, Schwanninger M, Kant M, Schuurink R, Mauch F, Buchala A, Cardinale F, Meskiene I (2007) The PP2C-type phosphatase AP2C1, which negatively regulates MPK4 and MPK6, modulates innate immunity, jasmonic acid, and ethylene levels in *Arabidopsis*. Plant Cell 19:2213–2224

Seiler C, Harshavardhan VT, Rajesh K, Reddy PS, Strickert M, Rolletschek H, Scholz U, Wobus U, Sreenivasulu N (2011) ABA biosynthesis and degradation contributing to ABA homeostasis during barley seed development under control and terminal drought-stress conditions. J Exp Bot 62:2615–2632

Sharp RE (2002) Interaction with ethylene: changing views on the role of abscisic acid in root and shoot growth responses to water stress. Plant Cell Environ 25:211–222

Sirichandra C, Wasilewska A, Vlad F, Valon C, Leung J (2009) The guard cell as a single-cell model towards understanding drought tolerance and abscisic acid action. J Exp Bot 60:1439–1463

Swarbreck D, Wilks C, Lamesch P, Berardini TZ, Garcia-Hernandez M, Foerster H, Li D, Meyer T, Muller R, Ploetz L, Radenbaugh A, Singh S, Swing V, Tissier C, Zhang P, Huala E (2008) The Arabidopsis Information Resource (TAIR): gene structure and function annotation. Nucleic Acid Res 36:D1009–D1014

Tanaka Y, Sano T, Tamaoki M, Nakajima N, Kondo N, Hasezawa S (2005) Ethylene inhibits abscisic acid-induced stomatal closure in *Arabidopsis*. Plant Physiol 138:2337–2343

Telewski FW (2006) A unified hypothesis of mechanoperception in plants. Am J Bot 93:1466–1476

Tester M, Bacic A (2005) Abiotic stress tolerance in grasses. From model plants to crop plants. Plant Physiol 137:791–793

Tran LSP, Urao T, Qin F, Maruyama K, Kakimoto T, Shinozaki K, Yamaguchi-Shinozaki K (2007) Functional analysis of AHK1/ATHK1 and cytokinin receptor histidine kinases in response to abscisic acid, drought, and salt stress in *Arabidopsis*. Proc Natl Acad Sci USA 104:20623–20628

Tsuchisaka A, Theologis A (2004) Unique and overlapping expression patterns among the *Arabidopsis* 1-amino-cyclopropane-1-carboxylate synthase gene family members. Plant Physiol 136:2982–3000

Umezawa T, Fujita M, Fujita Y, Yamaguchi-Shinozaki K, Shinozaki K (2006) Engineering drought tolerance in plants: discovering and tailoring genes to unlock the future. Curr Opin Biotechnol 17:113–122

Umezawa T, Nakashima K, Miyakawa T, Kuromori T, Tanokura M, Shinozaki K, Yamaguchi-Shinozaki K (2010) Molecular basis of the core regulatory network in ABA responses: sensing, signaling and transport. Plant Cell Physiol 51:1821–1839

Urao T, Yakubov B, Satoh R et al (1999) A transmembrane hybrid-type histidine kinase in *Arabidopsis* functions as an osmosensor. Plant Cell 11:1743–1754. doi:10.1105/tpc.11.9.1743

Walley JW, Dehesh K (2010) Molecular mechanisms regulating rapid stress signaling networks in *Arabidopsis*. J Integr Plant Biol 52:354–359

Walley JW, Coughlan S, Hudson ME, Covington MF, Kaspi R, Banu G, Harmer SL, Dehesh K (2007) Mechanical stress induces biotic and abiotic stress responses via a novel *cis*-element. PLoS Genet 3:1800–1812

Wasilewska A, Vlad F, Sirichandra C, Redko Y, Jammes F, Valon C, Frey NF, Leung J (2008) An update on abscisic acid signaling in plants and more. Mol Plant 1:198–217

Wasternack C (2007) Jasmonates: An update on biosynthesis, signal transduction and action in plant stress response, growth and development. Ann Bot 100:681–697

Werner T, Nehnevajovaa E, Köllmera I, Novákb O, Strnadb M, Krämerc U, Schmüllinga T (2010) Root-specific reduction of cytokinin causes enhanced root growth, drought tolerance, and leaf mineral enrichment in *Arabidopsis* and Tobacco. Plant Cell 22:3905–3920

Wilkinson S, Davies WJ (2009) Ozone suppresses soil drying- and abscisic acid (ABA)-induced stomatal closure via an ethylene-dependent mechanism. Plant Cell Environ 32:949–959

Winter D, Vinegar B, Nahal H, Ammar R, Wilson GV, Provart NJ (2007) An electronic fluorescent pictograph browser for exploring and analyzing large-scale biological data sets. PLoS One 2: e718

Wohlbach DJ, Quirino BF, Sussman MR (2008) Analysis of the *Arabidopsis* histidine kinase athk1 reveals a connection between vegetative osmotic stress sensing and seed maturation. Plant Cell 20:1101–1117

Wright AJ, Knight H, Knight MR (2002) Mechanically stimulated *TCH3* gene expression in *Arabidopsis* involves protein phosphorylation and EIN6 downstream of calcium. Plant Physiol 128:1402–1409

Yamaguchi M, Sharp RE (2010) Complexity and coordination of root growth at low water potentials: recent advances from transcriptomic and proteomic analyses. Plant Cell Environ 33:590–603

Yang Z, Tian L, Latoszek-Green M, Brown D, Wu K (2005) *Arabidopsis* ERF4 is a transcriptional repressor capable of modulating ethylene and abscisic acid responses. Plant Mol Biol 58:585–596

Zhang J, Jia W, Yang J, Ismail AM (2006) Role of ABA in integrating plant responses to drought and salt stresses. Field Crops Res 97:111–119

Zhu SY, Yu XC, Wang XJ, Zhao R, Li Y, Fan RC, Shang Y, Du SY, Wang XF, Wu FQ et al (2007) Two calcium-dependent protein kinases, CPK4 and CPK11, regulate abscisic acid signal transduction in *Arabidopsis*. Plant Cell 19:3019–3036

Zou JJ, Wei FJ, Wang C, Wu JJ, Ratnasekera D, Liu WX, Wu WH (2010) *Arabidopsis* calcium-dependent protein kinase CPK10 functions in abscisic acid- and Ca^{2+}-mediated stomatal regulation in response to drought stress. Plant Physiol 154:1232–12431

Chapter 3
Phytohormones in Salinity Tolerance: Ethylene and Gibberellins Cross Talk

Noushina Iqbal, Asim Masood, and Nafees A. Khan

Abstract Plants are severely affected by salinity due to its high magnitude of adverse impacts and worldwide distribution. Phytohormones are thought to be the most important endogenous substances involved in the mechanisms of tolerance or susceptibility of plants to salinity stress. The role of phytohormones under salinity stress is critical in modulating physiological responses that will eventually lead to adaptation to an unfavorable environment. Ethylene and gibberellins (GAs) are involved in mitigating the adverse effects of salinity stress by initiating a set of defense response or increasing plants' growth. However, both these phytohormones influence each other's action. On the one hand, GA is known to increase ethylene synthesis, and on the other hand, its signaling is itself affected by ethylene, and therefore, this interaction opens a cross talk between them. The present study focuses on both individual and interactive effect of the two in salinity tolerance to find out whether they have independent action or their action is dependent on each other.

3.1 Introduction

Agricultural crops face different types of biotic and abiotic stresses. Among abiotic stresses, salinity is very harmful and adversely affects the agricultural production (Syeed et al. 2010; Khan et al. 2010; Nazar et al. 2011a, b). It has been estimated that soil salinization affects about 7% of the global total land area and 20–50% of the global irrigated farmland (Koyro et al. 2008). It is expected to result in the loss of up to 50% fertile land by the middle of the twenty-first century (Manchanda and Garg 2008). Maintenance of ionic and water homeostasis is necessary for plant survival; however, salinity decreases crop productivity both by reducing leaf

N. Iqbal (✉) • A. Masood • N.A. Khan
Department of Botany, Aligarh Muslim University, Aligarh 202002, India
e-mail: naushina.iqbal@gmail.com

N.A. Khan et al. (eds.), *Phytohormones and Abiotic Stress Tolerance in Plants*,
DOI 10.1007/978-3-642-25829-9_3,

growth and inducing leaf senescence. This lowers the total photosynthetic capacity of the plant, thus limiting its ability to generate further growth or harvestable biomass and also to maintain defense mechanisms against the stress (Yeo 2007). Salinity stress leads to the production of reactive oxygen species (ROS), such as the superoxide anion radical (O_2^-), hydroxyl radical ($OH^•$), and hydrogen peroxide (H_2O_2) (Liu et al. 2007). Singha and Choudhuri (1990) reported that H_2O_2 and O_2^- were mainly responsible for NaCl-induced injury in *Vigna catjang* and *Oryza sativa* leaves. To minimize the ROS damage, plants have evolved the antioxidant defense system, comprised of enzymes that are responsible for scavenging excessively accumulated ROS under stress conditions such as superoxide dismutase (SOD), catalase (CAT), peroxidase (POD), and ascorbate peroxidase (APX) (Jung et al. 2000). Other strategies employed to overcome the adverse effects of salinity stress involve the application of phytohormones (Siddiqui et al. 2008). Reduced plant growth under stress conditions could result from an altered hormonal balance, and phytohormone application provides an attractive approach to cope with stress. The response of plants to salinity stress depends on multiple factors, but phytohormones are thought to be the most important endogenous substances involved in the mechanisms of tolerance or susceptibility of plants (Velitcukova and Fedina 1998).

The root, by sensing environmental constraints of the soil, may influence root-to-shoot signaling to control shoot growth and physiology, and ultimately agricultural productivity. Hormonal regulation of source–sink relations during the osmotic phase of salinity (independent of specific ions) affects whole-plant energy availability to prolong the maintenance of growth, root function, and ion homeostasis and could be critical to delay the accumulation of Na^+ or any other ion to toxic levels (Pérez-Alfocea et al. 2010). Plant hormones, including abscisic acid (ABA; Mauch-Mani and Mauch 2005), brassinosteroids (Nakashita et al. 2003), and auxin (Navarro et al. 2006; Wang et al. 2007), have been implicated in plant defense, but their significance is less well studied.

Ethylene is involved in various stress-related processes in plant. It plays a distinct role in response to biotic and abiotic forms of stress and can therefore be viewed as a stress hormone and mandatory participant in the development and expression of oxidative damage (Abeles et al. 1992; Kendrick and Chang 2008). In plants exposed to various types of abiotic and biotic stresses, increased ethylene levels correspond to increased damage, implying that stress ethylene is deleterious to plants. However, it was also reported that transcriptional activation of ERF (ethylene response factor) in ethylene-signaling process enhances stress tolerance in tobacco seedlings by decreasing ROS accumulation in response to salt, drought, and freezing (Wu et al. 2008). It may be that these discrepancies are due to differences in the amount of endogenous ethylene production, and in the period of stress treatment, in addition to the plant tissue studied (Wi et al. 2010). Therefore, ethylene, regarded as a ripening or senescence hormone, needs further consideration. Although stress leads to the production of stress ethylene, it interacts to send signals to initiate a set of defense response for stress tolerance.

Similarly, gibberellic acid (GA_3) is also involved in tolerance to salinity stress. An intimate relationship has been suggested to exist between gibberellin

(GA) levels and the acquisition of stress protection in barley (*Hordeum vulgare*) (Vettakkorumakankav 1999). Achard et al. (2006) reported that salt-treated *Arabidopsis* plants contain reduced levels of bioactive GAs, supporting the idea that salt slows down the growth by modulating the GA metabolism pathway. Exogenous application of GAs overcomes the effect of salinity stress and improves seed germination under saline conditions (Kaya et al. 2009). GA_3 has also been shown to alleviate the effects of salt stress on pigment content, Hill activity (Aldesuquy and Gaber 1993), and water-use efficiency (Shah 2007).

Further, the previously isolated abiotic signaling network that is controlled by ABA and the biotic network that is controlled by salicylic acid (SA), jasmonic acid (JA), and ethylene are interconnected at various levels. Whether all of the potential connections and shared nodes are actually used for cross talk remains to be determined. However, the present study deals with the role of ethylene in salinity tolerance and its cross talk with GA in bringing about tolerance.

3.2 Brief Overview of Plant Response to Salinity Stress and Methods of Its Alleviation

Salinity is one of the most deleterious abiotic stresses that affect plant growth by adversely affecting its photosynthetic potential. Salinity of soil and water is caused by the presence of excessive amounts of salts, and most common among them are high amounts of Na^+ and Cl^-. The deleterious effects of salinity on plant growth are associated with (1) low water potential of soil solution (water stress), (2) nutritional imbalance, (3) specific ion effect (salt stress), or (4) a combination of these factors (Ashraf et al. 1994; Marschner 1995). All of these factors cause adverse pleiotropic effects on plant growth and development at cellular (Gorham et al. 1985; Munns 2002) and at the molecular level (Mansour 2000; Tester and Davenport 2003). The toxic effect of salinity is through oxidative stress caused by enhanced production of ROS (Zhu et al. 2007; Giraud et al. 2008). The presence of high concentration of ROS can damage photosynthetic pigments, proteins, lipids, and nucleic acids by oxidation (Halliwell and Gutteridge 1985). Surviving such stresses led plants to acquire mechanisms by which they can sensitively perceive incoming stresses and regulate their physiology accordingly over a long evolutionary scale. Deciphering the mechanism by which plants perceive environmental signal is of critical importance for the development of rational breeding and transgenic strategies.

One of the methods by which plants respond directly and specifically to Na^+ within seconds is by increasing cytosolic (Ca^{2+}) (Munns and Tester 2008). One of the best-characterized signaling pathways specific to salinity involves sensing calcium by the calcineurin B-like protein (CBL) CBL4/SOS3 and its interacting protein kinase CIPK24/SOS2 (Zhu et al. 2007; Munns and Tester 2008; Luan et al. 2009). Lima-Costa et al. (2010) reported that proline may act on osmotic adjustment, as a free radical scavenger, protecting enzymes and avoiding DNA damages. The addition of external proline to salt-sensitive *Citrus sinensis* "Valencia late" cell line which has a smaller growth rate and accumulates proline in the presence of NaCl (>200 mM) was studied, and a positive influence on the relief of salt stress symptoms due to the presence of exogenous proline 5 and 100 mM NaCl was obtained, with increased growth of this salt-sensitive citrus cell line. Abbas et al. (2010) reported that the foliar-applied pure glycine betaine (GB) or GB-enriched sugar beet extract effectively minimized salt-induced harmful effects on growth of eggplant.

Ling et al. (2009) suggested that carbon monoxide might be involved in plant tolerance against salinity stress, and its alleviation of programmed cell death and inhibition of root growth was related to the decrease of superoxide anion overproduction partially via upregulation of SOD and downregulation of NADPH oxidase expression. Xu et al. (2011) reported that hemin, an inducer of heme oxygenase-1 (HO-1), could alleviate salinity damage during wheat seed germination in comparison with the pretreatment of a well-known nitric oxide (NO) donor sodium nitroprusside (SNP). Hemin and SNP could increase antioxidant enzyme activities, thus resulting in the alleviation of oxidative damage, as indicated by the decrease of thiobarbituric acid reactive substances content.

Manipulation of small RNA-guided gene regulation represents a novel and feasible approach to improve plant stress tolerance (Sunkar et al. 2006). Türkan and Demiral (2009) have studied the involvement of numerous small RNAs in salinity tolerance. High salt stress causes accumulation of H_2O_2, and both salt and H_2O_2 induce the expression of SRO5 protein. Work on the founding member of nat-siRNAs, which is derived from a *cis*-NAT gene pair of SRO5 and P5CDH genes, demonstrated an important role of nat-siRNAs in osmoprotection and oxidative stress management under salt stress in *Arabidopsis* (Borsani et al. 2005).

Nazar et al. (2011a) reported that S nutrition may provide a novel strategy to reduce the adverse effect of salinity through increased N utilization and synthesis of reduced S compounds such as Cys and GSH. Adaptation of sulfate uptake and assimilation is assumed to be a crucial determinant for plant survival in a wide range of adverse environmental conditions since different sulfur-containing compounds are involved in plant responses to both biotic and abiotic stresses (Rausch and Wachter 2005).

Besides, phytohormones are also involved in salinity tolerance, and the next section deals with their role in salinity tolerance.

3.3 Phytohormones in Salinity Tolerance

Plants have evolved a number of adaptive strategies to cope with salt stresses, and key to their operation is the ability to perceive stress and then to relay this information to allow the triggering of appropriate physiological and cellular responses. Many of the proteins produced by the plant under abiotic stress are induced by phytohormones (Hamayun et al. 2010). Salinity reduces the root/shoot ratio (an important adaptive response) due to the rapid inhibition of shoot growth (which limits plant productivity), while root growth is maintained and both processes may be regulated by changes in plant hormone concentrations (Albacete et al. 2008). The major physiological processes believed to be involved in the control of plant growth under salinity have been water relations, hormonal balance, and carbon supply, with their respective importance depending on the timescale of the response (Munns 2002). It is thought that the repressive effect of salinity on germination could be related to a decline in endogenous levels of plant growth hormones or phytohormones (Jackson 1997). Indeed, the exogenous application of plant growth regulators, for example, gibberellins (Prakash and Prathapasenan 1990; Afzal et al. 2005), auxins (Gul et al. 2000), and cytokinins (Dhingra and Varghese 1985; Gul et al. 2000) produced some benefit in alleviating the adverse effects of salt stress and they also improved germination, growth, fruit setting, fresh vegetable and seed yields, and yield quality (Saimbhi 1993). Similarly, presowing wheat seeds with plant growth regulators like IAA and GA alleviated the growth inhibiting effect of salt stress (Sastry and Shekhawa 2001; Afzal et al. 2005).

Salinity increased plant ABA concentration in all plant compartments (Wolf et al. 1990; Kefu et al. 1991); however, its role in growth regulation has been equivocal as different studies have suggested it can inhibit (Dodd and Davies 1996) or maintain growth by restricting the evolution of ethylene, another potential growth inhibitor (Sharp and LeNoble 2002). ABA acts as a major internal signal, enabling plants to survive adverse environmental conditions such as salt stress (Keskin et al. 2010). It has been reported that exogenous application of ABA reduces ethylene release and leaf abscission under salt stress in citrus, probably by decreasing the accumulation of toxic Cl^- ions in leaves (Gomez et al. 2002). In addition, Cabot et al. (2009) reported that salt-induced ABA mediated the inhibition of leaf expansion and limited the accumulation of Na^+ and Cl^- in leaves. Mutation analysis has demonstrated that the salt overly sensitive (SOS) pathway is important for the maintenance of intracellular ionic homeostasis and salt tolerance in *Arabidopsis thaliana*. Qiu et al. (2002) described three of the genetic components of this pathway, namely, SOS1 (a plasma Na^+/H^+ antiporter), SOS2 (a protein kinase), and SOS3 (an EF and Ca^{2+} sensor). SOS3 transduces the Ca^{2+} signal via the activation of SOS2, whereupon the SOS3–SOS2 complex activates SOS1 (Mahajan et al. 2008). SOS2 appears to interact with ABI2 (Ohta et al. 2003), suggesting the existence of some cross talk between the SOS and ABA pathways. Farhoudi and Saeedipour (2011) reported that the foliar application of ABA (15 μmol/l) under salinity condition improved shoot dry matter, photosynthesis rate, peroxidase, catalase activity, and shoot K^+ concentration, while decreased shoot Na^+ concentration in Okamer, thereby increasing salinity tolerance, but the high level of ABA treatments (30 μmol/l) reduced growth in this cultivar.

Cytokinins (CKs) are often considered as ABA antagonists and auxin antagonists/synergists in various processes in plants (Pospíšilová 2003) and could increase salt tolerance in wheat plants by interacting with other plant hormones, especially auxins and ABA (Iqbal et al. 2006). It was suggested that the decrease in CK content was an early response to salt stress, but that the effects of NaCl on salt-sensitive varieties are not mediated by CKs since the reduction in growth rate preceded any decline in CK levels (Walker and Dumbroff 1981).

It has been reported that jasmonate treatments (or endogenous of these compounds) are accompanied by the synthesis of abundant proteins in response to abiotic stress, called JIPs (Sembdner and Parthier 1993). JA levels in tomato cultivars changed in response to salt stress, and JA increase was observed in salt-tolerant cultivar HF (Hellfrucht Fruhstamm) from the beginning of salinization, while in salt-sensitive cultivar Pera, JA level decreased after 24 h of salt treatment (Pedranzani et al. 2003). Kramell et al. (2000) found a rapid increase in endogenous JA content in barley leaf segments subjected to osmotic stress with sorbitol or mannitol; however, endogenous jasmonates did not increase when they were treated with a high NaCl concentration (Kramell et al. 1995). The changes of endogenous JA levels in rice plants under various salt stresses were investigated. Kang et al. (2005) reported that the concentrations of JA in salt-sensitive cultivar plants were lower than in salt-tolerant cultivar plants. In addition, MeJA levels in rice roots increased significantly in 200 mM NaCl (Moons et al. 1997). Therefore, high levels of JA in salt-tolerant plants accumulated after salt treatments can be an effective protection against high salinity.

Similar role of brassinosteroids in salt tolerance was reported by Kagale et al. (2007) in *Brassica napus*. They reported that epibrassinlide (EBR) helped to overcome inhibition of seed germination by salt. Divi et al. (2010) reported that 24-EBR increased thermotolerance and salt tolerance of *Arabidopsis* mutants. Using a collection of *Arabidopsis* mutants that are either deficient in or insensitive to ABA, ethylene, JA, and SA, they studied the effects of 24-EBR on thermotolerance and salt tolerance of these mutants. The positive impact of EBR on thermotolerance in mutant proportion to wild type was evident in all mutants studied, with the exception of the SA-insensitive npr1-1 mutant. EBR could rescue the ET-insensitive ein2 mutant from its hypersensitivity to salt stress-induced inhibition of seed germination, but remained ineffective in increasing the survival of eto1-1 (ET-overproducer) and npr1-1 seedlings on salt. The positive effect of EBR was significantly greater in the ABA-deficient aba1-1 mutant as compared to wild type, indicating that ABA masks BR effects in plant stress responses. Treatment with EBR increased expression of various hormone marker genes in both wild-type and mutant seedlings, although to different levels. They concluded that BR shares transcriptional targets with other hormones.

Another phytohormone involved in salt tolerance is salicylic acid (Zahra et al. 2010; Hussain et al. 2010; Nazar et al. 2011). This alleviation effect could be due to SA-induced increase in proline or glycine betaine content (Misra and Saxena 2009; Palma et al. 2009), or SA may upregulate the antioxidant system (Syeed et al. 2010; Khan et al. 2010; Nazar et al. 2011b).

Amor and Cuadra-Crespo (2011) studied the foliar application of urea or methyl-jasmonate (MeJA) on the salinity tolerance of broccoli plants (*Brassisca oleracea* L. var. italica). Plant dry weight, leaf CO_2 assimilation, and root respiration were reduced significantly under moderate saline stress (40 mM NaCl), but application of either urea or MeJA maintained growth, gas exchange parameters, and leaf $N\text{–}NO_3^-$ concentrations at values similar to those of nonsalinized plants.

Besides the role of these phytohormones in salinity tolerance, the role of ethylene and gibberellins in stress tolerance is gaining increasing concern. Although ethylene is known as a senescing hormone, its definition needs elaboration. It is being recognized as a signal for inducing stress tolerance, and its high concentration under stress is probably a means to initiate a set of defense response. Similarly, much work is available on the role of the gibberellins in salinity tolerance. However, since phytohormone interaction is not independent and the network is interconnected, the interaction between ethylene and gibberellins in salinity tolerance needs further investigation (Fig. 3.1).

3.4 Ethylene in Salt Tolerance

Ethylene has long been regarded as a stress-related hormone, but only recently, the link between ethylene-signaling pathway and salt stress was primarily established. The production of ethylene is induced by many stresses (Morgan and Drew 1997), and variation in the ability of rice genotypes to generate it is regarded as a diagnostic for salt tolerance (Khan et al. 1987). Overexpression of the ethylene receptor ETR2 is associated with sensitivity to salinity stress, and similarly, reduced function mutants of a positive ethylene regulator were less tolerant than the wild type from which the mutants were derived (Cao et al. 2006). Cao et al. (2007) reported that ethylene signaling might be required for plant salt tolerance, as evidenced by the fact that ethylene-insensitive mutants etr1-1 and ein2-1 were more sensitive to salt stress. The salt-resistant cultivar Pokkali produced higher amounts of ethylene than the salt-sensitive cultivar IKP, and exogenous putrescine increased ethylene synthesis in both cultivars, suggesting no direct antagonism between polyamine and ethylene pathways in rice. Li et al. (2010) studied the expression of TaDi19A (a salt-responsive gene) and reported that this gene not only responded transcriptionally to exogenous ethylene but also reduced ethylene sensitivity in transgenic *A. thaliana*. This indicates that it is involved in ethylene-signaling pathway and may further affect the salinity tolerance. Zhu et al. (2005) reported that *A. thaliana* plants transformed with an ethylene-responsive factor (ERF)-like regulation factor hos10-1 gene accumulated more Na^+ than wild plants but had lower sensitivity to NaCl, implying that salt sensitivity was not directly associated with the change in Na^+ accumulation. Li et al. (2008) reported that expression of tomato jasmonic ethylene-responsive factor (JERFs) gene that encodes the ethylene-responsive factor (ERF)-like transcription factor to the genome of a hybrid poplar (*Populus alba* × *Populus berolinensis*) enhanced their salt tolerance.

Ethylene signaling modulates salt response at different levels, including membrane receptors, components in cytoplasm, and nuclear transcription factors in the pathway (Cao et al. 2008). It is involved not only in plant growth and development but also in plant responses to biotic stress such as pathogen attack and abiotic

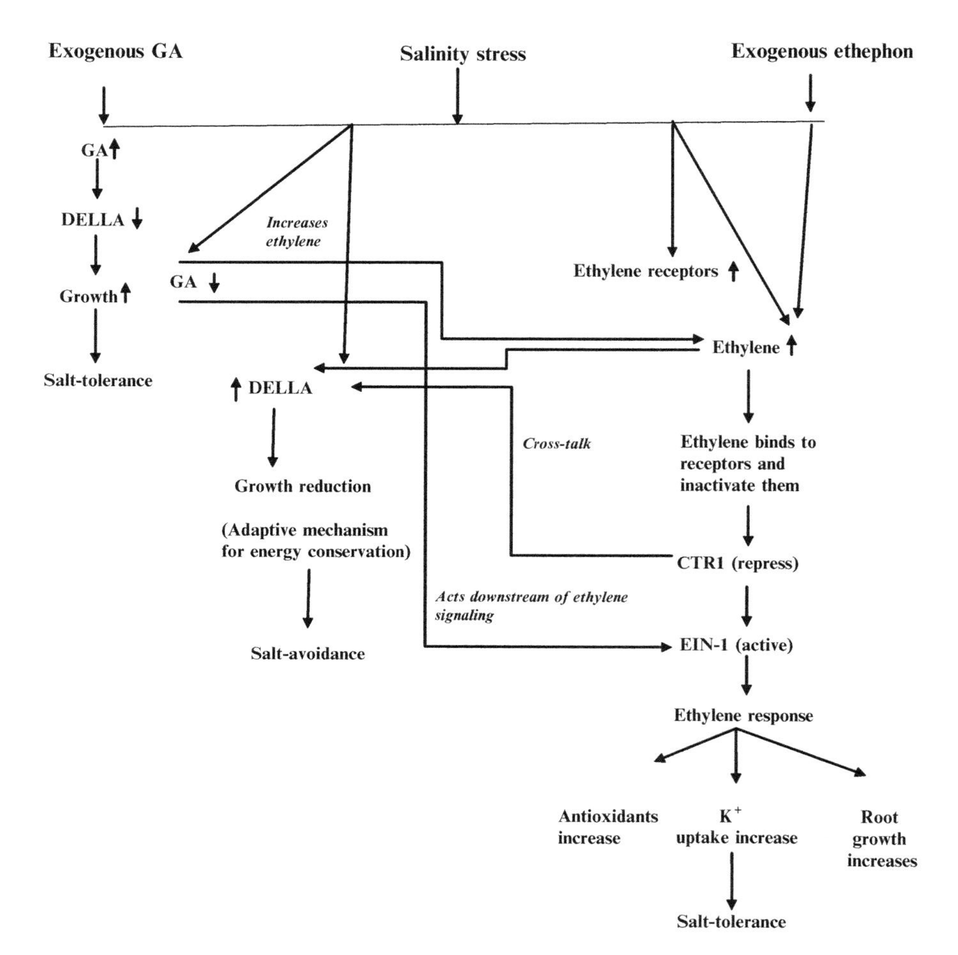

Fig. 3.1 Diagrammatic representation of the role of ethylene and gibberellins (GA) in salinity tolerance. Salinity stress increases GA which increases growth by decreasing DELLA and thus salt tolerance. Similarly, ethylene released under salinity stress increases salinity tolerance by enhancing antioxidants and root growth for nutrient uptake. Ethylene and GA also interact for salinity tolerance. Under salinity stress increased ethylene might act independently to bring salinity tolerance or might decrease GA or increase DELLA to cause growth reduction and this act as an adaptive mechanism for energy conservation and salt-avoidance. GA might act downstream of ethylene signaling to bring about salt tolerance via ethylene

stresses such as wounding, ozone, and salt (Abeles et al. 1992; Cao et al. 2007; Wang et al. 2009). ETR1 and EIN2 are two well-characterized ethylene-signaling molecules (Buer et al. 2006). It has been reported that plant responses to salt stress are modulated by changes in the expression level of ethylene receptor ETR1 (Zhao and Schaller 2004). Mutations in the ETR1 gene resulted in reduced ethylene responses (O'Malley et al. 2005). Besides, it has been reported that the salt-resistant rice cultivars produce higher amounts of ethylene than salt-sensitive cultivars (Khan et al. 1987; Lutts et al. 1996). It is interesting to find that 1-amino cyclopropane 1-carboxylic acid synthase (ACC, an ethylene precursor) suppresses the salt sensitivity conferred by NTHK1 in transgenic *Arabidopsis* plants, suggesting that ethylene is required for counteraction of receptor function to improve salt tolerance. Ethylene receptors may function in stress response through regulation of salt-responsive gene expressions. Salt stress induces ethylene-biosynthesis genes and ethylene production (Cao et al. 2007; Achard et al. 2006); ethylene may then inhibit its receptors, suppress salt sensitivity conferred by ethylene receptors, and promote ethylene responses including salt tolerance. Negative regulation between ethylene and its receptors has been established (Hua and Meyerowita 1998).

Ethylene-responsive factors, within a subgroup of the AP2/ERF transcription factor family, are involved in diverse plant reactions to biotic or abiotic stresses. Seo et al. (2010) reported that overexpression of an ERF gene from *Brassica rapa* species pekinensis (BrERF4) led to improved tolerance to salt and drought stress in *Arabidopsis*. The role of BrERF4 in responses to salt and drought by those plants could result from cross talk between different pathways for activating multiple stress responses, and that various stress pathways can be linked by certain transcriptional regulators, for example, Tsi1 acting in ethylene-signaling and salt stress-signaling pathways (Park et al. 2001).

Recently, ethylene was reported to alter *Arabidopsis* salt tolerance by interfering with other hormone pathways and NO signals (Achard et al. 2006; Cao et al. 2007; Wang et al. 2009). Ethylene-signaling pathways have been well characterized especially in *Arabidopsis*, and the pathway involves ethylene receptors, CTR1 (Constitutive Triple Response), EIN2 and EIN3, and other components (Bleecker and Kende 2000; Chen et al. 2005). The receptors activate a Raf-like protein kinase, CTR1, which negatively regulates downstream ethylene signaling. The loss of function of receptor CTR1 generates *Arabidopsis* mutant ctr1-1, which shows constitutive ethylene responses (Guo and Ecker 2004). Downstream of EIN2, EIN3 transcription factor stimulates the expression of ethylene response target genes, and the loss-of-function ethylene mutants ein2-5 and ein3-1 show entirely or partly ethylene insensitivity (Alonso et al. 2003; Chen et al. 2005). These ethylene mutants and the wild-type plant Col-0 provide useful genetic tools to dissect the role of ethylene in plant salt responses. Yang et al. (2010) studied *Arabidopsis* ein2-5, ein3-1, and ctr1-1 mutants and Col-0 plants to analyze the regulation of the salt response by altered ethylene signaling. The ethylene entirely or partly insensitive mutant plant (ein2-5 or ein3-1) increased *Arabidopsis* salt sensitivity than the wild-type plant (Col-0). On the contrary, the ethylene constitutively sensitive mutant (ctr1-1) largely improves plant salt tolerance with respect to biomass accumulation, photosystem function. The exogenous supply of ethephon (an ethylene precursor) to salt-stressed Col-0 plants also improved *Arabidopsis* salt tolerance. The analysis on the Na^+ and K^+ homeostasis found that ethylene could help plants to retain higher K^+ nutrition in the short- or long-term salt-stressed plants and improve salt tolerance. The ROS redox, sucrose, and antioxidant system were found to be regulated by ethylene to improve salt tolerance. Additionally, ethylene was reported to stimulate PM H^+-ATPase activity to modulate ion homeostasis and salt tolerance in *Arabidopsis* (Wang et al. 2009).

3.5 Gibberellins in Salinity Tolerance

The biosynthesis of GAs is regulated by both developmental and environmental stimuli (Yamaguchi and Kamiya 2000; Olszewski et al. 2002). Gibberellic acid accumulates rapidly when plants are exposed to both biotic (McConn et al. 1997) and abiotic stresses (Lehmann et al. 1995). In order to alleviate deleterious effects of salinity, different types of phytohormones have been used. Among them, gibberellins have been the main focus of some plant scientists (Basalah and Mohammad 1999; Hisamatsu et al. 2000). For instance, GA_3 has been reported to be helpful in enhancing wheat and rice growth under saline conditions (Parasher and Varma 1988; Prakash and Prathapasenan 1990). Plant growth of wheat decreased with increasing salinity levels but was increased by seed treatment with GA_3 (Kumar and Singh 1996). Gibberellic acid signaling is involved in adjustment of plants under limiting environmental conditions and maintains source–sink relation (Iqbal et al. 2011) since salinity causes a reduction in sink enzyme activities, leading to an increase in sucrose in source leaves, with a decrease in photosynthesis rate by feedback inhibition (Poljakoff-Mayber and Lerner 1994) responses. However, the mechanisms by which GA_3-priming could induce salt tolerance in plants are not yet clear. Salinity perturbs the hormonal balance in plants; therefore, hormonal homeostasis under salt stress might be the possible mechanism of GA_3-induced plant salt tolerance (Iqbal and Ashraf 2010).

Hamayun et al. (2010) reported that GA_3 application significantly promoted plant length and plant fresh/dry biomass which was markedly hindered by NaCl-induced salt stress in soybean plants. Phytohormonal analysis of soybean showed that the level of bioactive gibberellins (GA_1 and GA_4) and jasmonic acid increased in GA_3-treated plants, while the endogenous ABA and SA contents declined under the same treatment. GA_3 mitigated the adverse effects of salt stress by regulating the level of phytohormones, thus aiding the plant in resuming its normal growth and development.

Maggio et al. (2010) analyzed the functional role of GA in stress adaptation and reported that exogenous GA applications may benefit plant growth and yield at low to moderate salinity, whereas it may enhance stress sensitivity at moderate to high salinity levels. Few studies have previously demonstrated the ability of GA_3 to mitigate the adverse effects of salt stress (Chakraborti and Mukherji 2003). GA_3 has been shown to alleviate the effects of salt stress on pigment content, Hill activity (Aldesuquy and Gaber 1993), and water-use efficiency (Aldesuquy and Ibrahim 2001). Exogenous application of GA alleviates the effect of salinity stress on the growth of Sorghum (Amzallag et al. 1992), wheat (Naqvi et al. 1982), and rice seedlings (Prakash and Prathapasenan 1990). The GA_3-mediated increase in grain yield in the salt-intolerant cultivar of wheat was due to the increased number of fertile tillers per plant and grain weight rather than number of grains per year particularly when under salt stress. The beneficial effects of GA_3-priming can be attributed to its effect on hormonal homeostasis and ionic uptake and partitioning (within shoots and roots) in the salt-stressed wheat plants (Iqbal et al. 2011).

Gibberellic acid may increase growth under salinity stress by increasing both nitrogen and magnesium contents in the leaves and roots (Tuna et al. 2008). Overexpression of dwarf and delayed flowering 1 (DDF1) caused a reduction in GA_4 content and dwarfism in *Arabidopsis* (Magome et al. 2004). There is a correlation among the survival of salt toxicity and the function of DELLA proteins (Achard et al. 2006). These results suggest that the salt-inducible DDF1 gene is involved in growth responses under high salinity conditions in part through altering GA levels and improves seed germination (Kaya et al. 2009). Kumar and Singh (1996) reported increased growth and grain yield under saline condition on GA treatment. Achard et al. (2006) reported that salt-treated *Arabidopsis* plants contain reduced levels of bioactive GAs, supporting the idea that salt slows down the growth by modulating the GA metabolism pathway. However, exogenous GA treatment of salt-stressed wheat plants resulted in an increased photosynthetic capacity, which was discussed as a major factor for greater dry matter production (Iqbal and Ashraf 2010). The increase in pigment content, photosynthetic capacity, and growth through GA treatment under salinity stress indicates its potential in the regulation of source–sink metabolism. Mohammed (2007) suggested that the multiple effects of GA_3 which could be involved in alleviating the adverse effect of salinity on mung-bean seedlings include the stimulation of growth parameters; increase in photosynthetic pigments concurrent with the marked increase in reducing sugars and sucrose; increase in protein synthesis, including de novo synthesis of proteins and accumulation of certain existing proteins; increase in the activities of catalase and peroxidase; and decreases in the activities of the ribonuclease and polyphenol oxidase.

Thus, GA_3 counteracts with salinity stress by improving membrane permeability and nutrient levels in leaves which ultimately leads to better seedling growth and shoot, root, and total biomass. Further, it may also improve salinity tolerance by maintaining antioxidant enzyme activities. It is, therefore, possible that foliar GA_3 could be a useful tool in promoting good seedling growth and establishment under saline soil conditions. Exogenous application of growth hormones may be useful to return metabolic activities to their normal levels. At a certain concentration, GA_3 has been shown to be beneficial for the physiology and metabolism of many plants under abiotic stress since it may provide a mechanism to regulate the metabolic process as a function of sugar signaling and antioxidative enzymes (Iqbal et al. 2011).

3.6 Ethylene and Gibberellin Acid in Salt Tolerance: Cross Talk

Both ethylene and gibberellins are involved in salinity tolerance individually, but their interaction to bring about salinity tolerance is still not detailed out. Transcript meta-analysis showed that GA and ethylene metabolism genes are expressed in the majority of plant organs. Both GAs and the ethylene precursor ACC may thus be synthesized ubiquitously (Dugardeyn et al. 2008). Bialecka and Kepczynski (2009) clearly demonstrate the ability of exogenous ethephon and GA_3 to alleviate inhibition of *Amaranthus caudatus* seed germination under salinity. The stimulatory effect of ethephon in the presence of NaCl was seen earlier than that of GA_3, and ethephon was found to be more effective than GA_3. Alleviation of the detrimental effects of salinity on seed germination by ethylene and/or GA_3 has been shown (Khan and Huang 1988; Kumar and Singh 1996; Mohammed 2007). Interaction between the two was reported by Foo et al. (2006) for pea where the phytochromes negatively controlled ethylene production, which in turn lowered GA synthesis. In the *Arabidopsis* hypocotyl, ethylene and GA have a synergistic effect on elongation in the light (De Grauwe et al. 2007).

De Grauwe et al. (2008) reported that a functional GA response pathway is required for the increased ethylene biosynthesis in eto2-1 (*ethylene overproducing* mutant) since the gai eto2-1 (*gibberellins insensitive; ethylene overproducing*) double mutant does not overproduce ethylene, suggesting that the stability of the ACS5 protein is dependent on GA. More recently, it was demonstrated that active ethylene signaling results in decreased GA content, thus stabilizing DELLA proteins (Vandenbussche et al. 2007). Proteins from the DELLA family, which comprises five members in *Arabidopsis* (GAI, repressor of ga1-3 (RGA), RGA-like (RGL1/2/3)), are rapidly destabilized after GA treatment through degradation by the 26S proteasome (Dill and Sun 2001; Fu et al. 2002). In addition, ethylene affects DELLA stability primarily via changes in GA concentration, supposedly by post-transcriptional control of GA20ox/GA3ox/GA2ox genes (Achard et al. 2007; Vandenbussche et al. 2007). Furthermore, analysis of microarray data using Genevestigator has shown that in 28-day-old plants, none of the genes encoding DELLA proteins, neither the F-box protein SLEEPY (SLY), are transcriptionally affected by ethylene. The only gene from the GA-signaling cascade which appears to be regulated by ethylene is GIBBERELLIN INSENSITIVE DWARF1 (GID1c) (more than twofold induction after ethylene treatment) (Dugardeyn et al. 2008; Zimmermann et al. 2004). GA-promoted destabilization of DELLA proteins is modulated by environmental signals (such as salt and light) and other plant hormone signaling (such as auxin and ethylene), which reveals the mechanisms of this cross-talking at the molecular level (Achard et al. 2006). Cross talk between ethylene and GA during the transition from seed dormancy to germination has been postulated from the hormonal regulation of a GA 20-oxidase in *Fagus sylvatica* (Calvo et al. 2004). GA treatment induces ACO expression in imbibed ga1-3 seeds (Ogawa et al. 2003). An excess of ethylene can bypass the GA requirement and induce germination of *Arabidopsis* ga1 mutant seeds imbibed in the light, the effect being much weaker in darkness (Karssen et al. 1989; Koornneef and Karssen 1994). The GA–ethylene interaction seems reciprocal because high GA concentrations restore the germination of etr1 mutant seeds to wild-type levels (Bleecker et al. 1988).

Steffens et al. (2006) reported that GA is ineffective on its own but acts in a synergistic manner together with ethylene to promote the number of penetrating roots and the growth rate of emerged roots. The effects of GA and ethylene are not additive but synergistic, and therefore, sharing signaling components must be predicted. GA by itself was nearly ineffective in promoting root penetration or growth, and when roots were treated with NBD, an inhibitor of ethylene perception, neither ethylene (Lorbiecke and Sauter 1999) nor GA was able to promote root penetration or growth. Therefore, it appeared that GA activity on adventitious roots required ethylene signaling, GA acts downstream of the ethylene receptor, and GA activity requires activated ethylene signaling achieved through ethylene binding to its receptor.

Additional examples come from studies of the relationship between GA–GID1–DELLA, the phytohormones ethylene, and plant stress responses. The molecular characterization of various GA-response mutants led to the discovery of the GID1 and DELLA proteins, key components of the molecular GA–GID1–DELLA mechanism that enables plants to respond to GA (Harberd et al. 2009). GA–GID1–DELLA mechanism enables plants to maintain transient growth arrest and thus to survive periods of adversity. Initial investigations of the relationship between ethylene signaling and the GA–GID1–DELLA mechanism showed that ethylene inhibits DELLA-deficient mutant *Arabidopsis* seedling root growth less than that of the wild type, and that ethylene inhibits the GA-induced disappearance of GFP-RGA via CTR1-dependent signaling (Achard et al. 2003). In addition, the maintenance of the exaggerated apical hook structure typical of dark-grown ethylene-treated seedlings was shown to be dependent on loss of DELLA-mediated growth inhibition (Achard et al. 2003; Vriezen et al. 2004). Thus, there is a connection between ethylene response and the GA–GID1–DELLA mechanism and a correlation between ethylene-mediated growth inhibition and DELLA accumulation. The conclusion from these studies is that ethylene inhibits growth (at least in part) via a DELLA-dependent mechanism. However, there are circumstances in which ethylene promotes (rather than inhibits) the growth of light-grown hypocotyls, and in this case, promotion interestingly is also accompanied by an accumulation (rather than a depletion) of GFP-RGA in hypocotyl nuclei (Vandenbussche et al. 2007). It seems that in this case ethylene still promotes GFP-RGA accumulation, but this accumulation is not translated into growth inhibition. Thus, the frequently observed negative correlation between growth and DELLA accumulation is not absolute and can be broken in specific circumstances.

Salt-activated DELLA-dependent growth inhibition presumably has adaptive significance because plants lacking DELLAs are more susceptible, whereas plants in which DELLAs accumulate (e.g., ga1-3) are more resistant to the lethal effects of extreme salt (Achard et al. 2006). These observations indicate that the GA–GID1–DELLA mechanism provides plants with a means of regulating growth appropriate to environmental conditions, enabling a slowing of growth and reduced energetic commitment during periods of environmental adversity. The interaction between GA and the stress-related gaseous hormone ethylene is rather complex, as both negative and positive reciprocal effects have been demonstrated. Ethylene inhibits growth in a GA-antagonistic manner. Achard et al. (2003) have shown that at least part of the inhibitory effect of ethylene on growth and its interaction with GA in this regard are mediated by the DELLA proteins. GA promotes seedling root elongation in *Arabidopsis*, and this effect is inhibited by ethylene. However, in gai-rga double mutants, GA stimulated root elongation also in the presence of ethylene, suggesting that ethylene acts through these DELLA proteins in this process. A synergistic promotive effect of GA and ethylene was also shown in light-grown *Arabidopsis* seedlings (Saibo et al. 2003).

During submergence, SUB1A dampens ethylene production and enhances mRNA and protein accumulation of two negative regulators of GA signaling, SLENDER RICE1 (SLR1) and SLR1-like 1 (SLRL1), resulting in the suppression of the energy-consuming escape response (Fukao et al. 2006; Fukao and Bailey-Serres 2008). Consistently, global-scale transcriptome analysis revealed that SUB1A regulates the abundance of mRNAs associated with ethylene and GA production and signaling during submergence (Jung et al. 2010). SUB1A, an ERF transcription factor found in limited rice accessions, dampens ethylene production and GA responsiveness during submergence, economizing carbohydrate reserves and significantly prolonging endurance (Fukao et al. 2011). Induction of PCD is dependent on ethylene signaling and is further promoted by GA. Ethylene and GA act in a synergistic manner, indicating converging signaling pathways. Treatment of plants with GA alone did not promote PCD. Treatment with the GA biosynthesis inhibitor paclobutrazol resulted in increased PCD in response to ethylene and GA presumably due to an increased sensitivity of epidermal cells to GA (Steffens and Sauter 2005).

Previous studies have suggested that ethylene may increase the level of GAs during internode elongation of deepwater rice upon submergence (Hoffmann-Benning and Kende 1992). Ethylene was shown to delay *Arabidopsis* flowering by reducing bioactive GA levels (Achard et al. 2007). In addition, the ctr1-1 loss-of-function mutation, which confers constitutive ethylene responses, causes later flowering, especially in short-day photoperiods. All these ethylene-stimulated late-flowering phenotypes are rescued by exogenous GA treatment. The levels of GA1 and GA4 are substantially reduced in the ctr1-1 mutant, suggesting that the ethylene signal is initially targeted to the GA metabolism pathway and then to DELLA proteins as a consequence of altered GA content. In the ctr1-1 mutant, transcript abundance of AtGA3ox1 and AtGA20ox1 genes is elevated, presumably through the negative feedback mechanism caused by the decreased bioactive GA levels. Further studies are required to clarify how the ctr1-1 mutation alters bioactive GA levels (Yamaguchi 2008).

Perhaps the two hormones interact with each another; however, their interaction is not clear. The available literature suggests that both these hormones may interact synergistically or antagonistically. However, under stress, the decreased GA is due to ethylene action on DELLA. This has been explained through a model which shows that ethylene increases DELLA protein and reduces GA action, whereas GA might work downstream of ethylene signaling and does not influence ethylene action or may increase ethylene synthesis. However, the exact mechanism of interplay of both hormones under salt stress remains to be elucidated.

3.7 Conclusion

Plants are continuously exposed to various biotic and abiotic stresses that adversely affect their growth, and among these stresses predominant is salinity stress. In order to survive under these abiotic stress conditions, plants activate their defense mechanism. Phytohormone signaling is one such defense response that is activated in response to stress. Whether there is increase or decrease in the level of these hormones, it is an adaptive mechanism to tolerate stress. Ethylene increases under stress, whereas GA decreases; however, both these are involved in salinity tolerance or avoidance. Exogenous application of both ethylene and GA is involved in inducing salt tolerance in plants, but whether they act independently or are dependent on each other is debatable. In fact, phytohormones are integrated at several levels during plant growth and development, and cross talks between hormonal and defense signaling pathways should reveal new potential targets for the development mechanisms against stress.

References

Abbas W, Ashraf M, Akram NA (2010) Alleviation of salt-induced adverse effects in eggplant (*Solanum melongena* L.) by glycinebetaine and sugarbeet extracts. Sci Hortic 125:188–195

Abeles FB, Morgan PW, Saltveit ME Jr (1992) Ethylene in plant biology, 2nd edn. Academic, San Diego, USA

Achard P, Vriezen WH, van der Straeten D, Harberd NP (2003) Ethylene regulates *Arabidopsis* development via the modulation of DELLA protein growth repressor function. Plant Cell 15:612816–612825

Achard P, Cheng H, De Grauwe L, Decat J, Schoutteten H, Moritz T, Van Der Straeten D, Peng J, Harberd NP (2006) Integration of plant responses to environmentally activated phytohormonal signals. Science 311:91–94

Achard P, Baghour M, Chapple A, Hedden P, Van Der Straeten D et al (2007) The plant stress hormone ethylene controls floral transition via DELLA-dependent regulation of floral meristem-identity genes. Proc Natl Acad Sci USA 104:6484–6489

Afzal I, Basra SMA, Iqbal A (2005) The effects of seed soaking with plant growth regulators on seedling vigor of wheat under salinity stress. J Stress Physiol Biochem 1:6–14

Albacete A, Ghanem ME, Martínez-Andújar C, Acosta M, Sánchez-Bravo J, Martínez V, Lutts S, Dodd IC, Pérez-Alfocea F (2008) Hormonal changes in relation to biomass partitioning and

shoot growth impairment in salinized tomato (*Solanum lycopersicum* L.) plants. J Exp Bot 19:4119–4131
Aldesuquy HS, Gaber AM (1993) Effect of growth regulators on *Vicia faba* plants irrigated by seawater, leaf area, pigment content and photosynthetic activity. Biol Plant 35:519–527
Aldesuquy HS, Ibrahim AH (2001) Interactive effect of seawater and growth bio-regulators on water relations, absicisic acid concentration, and yield of wheat plants. J Agron Crop Sci 187:185–193
Alonso JM, Stepanova AN, Solano R, Wisman E, Ferrari S, Ausubel FM, Ecker JR (2003) Five components of the ethylene-response pathway identified in a screen for weak ethylene-insensitive mutants in *Arabidopsis*. Proc Natl Acad Sci USA 100:2992–2997
del Amor FM, Cuadra-Crespo P (2011) Alleviation of salinity stress in broccoli using foliar urea or methyl-jasmonate: analysis of growth, gas exchange, and isotope composition. Plant Growth Regul 63:55–62
Amzallag GN, Lerner H, Poljakoff-Mayber A (1992) Interaction between mineral nutrients, cytokinins and gibberellic acid during growth of sorghum at higher NaCl salinity. J Exp Bot 43:81–87
Ashraf MY, Azmi AR, Khan AH, Ala SA (1994) Effect of water stress on total phenol, peroxidase activity and chlorophyll contents in wheat (*Triticum aestivum* L.). Acta Physiol Plant 16:185–191
Basalah MO, Mohammad S (1999) Effect of salinity and plant growth regulators on seed germination of *Medicago sativa* L. Pak J Biol Sci 2:651–653
Bialecka B, Kepczynski J (2009) Effect of ethephon and gibberellin A_3 on *Amaranthus caudatus* seed germination and α- and β-amylase activity under salinity stress. Acta Biol Cracov Ser Bot 51:119–125
Bleecker AB, Kende H (2000) Ethylene: a gaseous signal molecule in plants. Annu Rev Cell Dev Biol 16:1–18
Bleecker AB, Estelle MA, Somerville C, Kende H (1988) Insensitivity to ethylene conferred by a dominant mutation in *Arabidopsis thaliana*. Science 241:1086–1089
Borsani O, Zhu J, Verslues PE, Sunkar R, Zhu JK (2005) Endogenous siRNAs derived from a pair of natural cis-antisense transcripts regulate salt tolerance in *Arabidopsis*. Cell 123: 1279–1291
Buer CS, Sukumar P, Muday GK (2006) Ethylene modulates flavonoid accumulation and gravitropic responses in roots of *Arabidopsis*. Plant Physiol 140:1384–1396
Cabot C, Sibole JV, Barcelo J, Poschenrieder C (2009) Abscisic acid decreases leaf Na^+ exclusion in salt treated *Phaseolus vulgaris* L. J Plant Growth Regul 28:187–192
Calvo AP, Nicolás C, Nicolás G, Rodríguez D (2004) Evidence of a cross-talk regulation of a GA 20-oxidase (FsGA20ox1) by gibberellins and ethylene during the breaking of dormancy in *Fagus sylvatica* seeds. Physiol Plant 120:623–630
Cao WH, Liu J, Zhou QY, Cao YR, Zheng SF, Du BX, Zhang JS, Chen SY (2006) Expression of tobacco ethylene receptor NTHK1 alters plant responses to salt stress. Plant Cell Environ 29:1210–1219
Cao WH, Liu J, He XJ, Mu RL, Zhou HL, Chen SY, Zhang JS (2007) Modulation of ethylene responses affects plant salt-stress responses. Plant Physiol 143:707–719
Cao YR, Chen SY, Zhang JS (2008) Ethylene signaling regulates salt stress response. Plant Signal Behav 3:761–763
Chakraborti N, Mukherji S (2003) Effect of phytohormone pretreatment on nitrogen metabolism in *Vigna radiata* under salt stress. Biol Plant 46:63–66
Chen YF, Etheridge N, Schaller E (2005) Ethylene signal transduction. Ann Bot (Lond) 95:901–915
De Grauwe L, Dugardeyn J, Van Der Straeten D (2008) Novel mechanisms of ethylene-gibberellin crosstalk revealed by the gai eto2-1 double mutant. Plant Signal Behav 3:1113–1115
De Grauwe L, Vriezen WH, Bertrand S, Phillips A, Vidal AM, Hedden P, Van Der Straeten D (2007) Reciprocal influence of ethylene and gibberellins on response-gene expression in *Arabidopsis thaliana*. Planta 226:485–498

Dhingra HR, Varghese TM (1985) Effect of growth regulators on the *in vitro* germination and tube growth of maize (*Zea mays* L.) pollen from plants raised under sodium chloride salinity. New Phytol 100:563–569

Dill A, Sun T (2001) Synergistic derepression of gibberellin signaling by removing RGA and GAI function in *Arabidopsis thaliana*. Genetics 159:777–785

Divi UK, Rahman T, Krishna P (2010) Brassinosteroid-mediated stress tolerance in *Arabidopsis* shows interactions with abscisic acid, ethylene and salicylic acid pathways. BMC Plant Biol 10:151

Dodd IC, Davies WJ (1996) The relationship between leaf growth and ABA accumulation in the grass leaf elongation zone. Plant Cell Environ 19:1047–1056

Dugardeyn J, Vandenbussche F, Van Der Straeten D (2008) To grow or not to grow: what can we learn on ethylene–gibberellin cross-talk by in silico gene expression analysis? J Exp Bot 59:1–16

Farhoudi R, Saeedipour S (2011) Effect of exogenous abscisic acid on antioxidant activity and salt tolerance in rapeseed (*Brassica napus*) cultivars. Res Crops 12:122–130

Foo E, Ross JJ, Davies NW, Reid JB, Weller JL (2006) A role for ethylene in the phytochrome-mediated control of vegetative development. Plant J 46:911–921

Fu X, Richards DE, Ait-Ali T, Hynes LW, Ougham H, Peng J, Harberd NP (2002) Gibberellin-mediated proteasome-dependent degradation of the barley DELLA protein SLN1 repressor. Plant Cell 14:3191–3200

Fukao T, Bailey-Serres J (2008) Submergence tolerance conferred by Sub1A is mediated by SLR1 and SLRL1 restriction of gibberellin responses in rice. Proc Natl Acad Sci USA 105:16814–16819

Fukao T, Yeung E, Bailey-Serres J (2011) The submergence tolerance regulator SUB1A mediates crosstalk between submergence and drought tolerance in rice. Plant Cell 23:412–442

Fukao T, Xu K, Ronald PC, Bailey-Serres J (2006) A variable cluster of ethylene response factor-like genes regulates metabolic and developmental acclimation responses to submergence in rice. Plant Cell 18:2021–2034

Giraud E, Ho LH, Clifton R et al (2008) The absence of alternative oxidase 1a in *Arabidopsis* results in acute sensitivity to combined light and drought stress. Plant Physiol 147:595–610

Gomez CA, Arbona V, Jacas J, PrimoMillo E, Talon M (2002) Abscisic acid reduces leaf abscission and increases salt tolerance in citrus plants. J Plant Growth Regul 21:234–240

Gorham JE, McDonnel E, Budrewicz JRGW (1985) Salt tolerance in the Triticeae: growth and solute accumulation in leaves of *Thinopyrum bessarabicum*. J Exp Bot 36:1021–1031

Gul B, Khan MA, Weber DJ (2000) Alleviation salinity and darken forced dormancy in *Allenrolfea occidentalis* seeds under various thermo periods. Aust J Bot 48:745–752

Guo H, Ecker JR (2004) The ethylene signaling pathway: new insights. Curr Opin Plant Biol 7:40–49

Halliwell B, Gutteridge JMC (1985) Free radicals in biology and medicine. Clarendon, Oxford

Harberd NP, Belfield E, Yasumura Y (2009) The angiosperm gibberellin-GID1-DELLA growth regulatory mechanism: how an "inhibitor of an inhibitor" enables flexible response to fluctuating environments. Plant Cell 21:1328–1339

Hisamatsu T, Koshioka M, Kubota S, Fujime Y, King RW, Mander LN (2000) The role of gibberellin in the control of growth and flowering in *Matthiola incana*. Physiol Plantarium 109:97–105

Hoffmann-Benning S, Kende H (1992) On the role of abscisic acid and gibberellin in the regulation of growth in rice. Plant Physiol 99:1156–1161

Hua J, Meyerowita EM (1998) Ethylene responses are negatively regulated by a receptor gene family in *Arabidopsis thaliana*. Cell 94:261–271

Hamayun M, Khan SA, Khan AL, Shin JH, Ahmad B, Shin DH, Lee IJ (2010) Exogenous gibberellic acid reprograms soybean to higher growth and salt stress tolerance. J Agric Food Chem 58:7226–7232

Hussain K, Nawaz K, Majeed A, Khan F, Lin F, Ghani A, Raza G, Afghan S, Zia-ul-Hussnain S, Ali K, Shahazad A (2010) Alleviation of salinity effects by exogenous applications of salicylic acid in pearl millet (*Pennisetum glaucum* (L.) R. Br.) seedlings. Afr J Biotechnol 9:8602–8607

Iqbal M, Ashraf M (2010) Gibberellic acid mediated induction of salt tolerance in wheat plants: growth, ionic partitioning, photosynthesis, yield and hormonal homeostasis. Env Exp Bot http://dx.doi.org/10.1016/j.envexpbot.2010.06.002

Iqbal N, Nazar R, Khan MIR, Masood A, Khan NA (2011) Role of gibberellins in regulation of source–sink relations under optimal and limiting environmental conditions. Curr Sci 100:110

Iqbal M, Ashraf M, Jamil A, Ur-Rehman S (2006) Does seed priming induce changes in the levels of some endogenous plant hormones in hexaploid wheat plants under salt stress? J Integr Plant Biol 48:81–189

Jackson M (1997) Hormones from roots as signals for the shoots of stressed plants. Elsevier Trends J 2:22–28

Jung S, Kim JS, Cho KY, Tae GS, Kang BG (2000) Antioxidant responses of cucumber (*Cucumis sativus*) to photoinhibition and oxidative stress induced by norflurazon under high and low PPFDs. Plant Sci 153:145–154

Jung KH, Seo YS, Walia H, Cao P, Fukao T, Canlas PE, Amonpant F, Bailey-Serres J, Ronald PC (2010) The submergence tolerance regulator Sub1A mediates stress-responsive expression of AP2/ERF transcription factors. Plant Physiol 152:1674–1692

Kagale S, Divi UK, Krochko JE, Keller WA, Krishna P (2007) Brassinosteroids confers tolerance in *Arabidopsis thaliana and Brassica napus* to a range of abiotic stresses. Planta 225:353–364

Kang DJ, Seo YJ, Lee JD, Ishii R, Kim KU, Shin DH, Park SK, Jang SW, Lee IJ (2005) Jasmonic acid differentially affects growth, ion uptake and abscisic acid concentration in salt-tolerant and salt-sensitive rice cultivars. J Agron Crop Sci 191:273–282

Karssen CM, Zagórsky S, Kepczynski J, Groot SPC (1989) Key role for endogenous gibberellins in the control of seed germination. Ann Bot 63:71–80

Kaya C, Tuna AL, Yokas I (2009) The role of plant hormones in plants under salinity stress. Book Salinity Water Stress 44:45–50

Kefu Z, Munns R, King RW (1991) Abscisic-acid levels in NaCl-treated barley, cotton and saltbush. Aust J Plant Physiol 18:17–24

Kendrick MD, Chang C (2008) Ethylene signaling: new levels of complexity and regulation. Curr Opin Plant Biol 11:479–485

Keskin BC, Sarikaya AT, Yuksel B, Memon AR (2010) Abscisic acid regulated gene expression in bread wheat. Aust J Crop Sci 4:617–625

Khan AA, Huang XL (1988) Synergistic enhancement of ethylene production and germination with kinetin and 1-aminocyclopropane-1-carboxylic acid in lettuce seeds exposed to salinity stress. Plant Physiol 87:847–852

Khan MN, Siddiqui MH, Mohammad F, Naeem M, Khan MMA (2010) Calcium chloride and gibberellic acid protect Linseed (*Linum usitatissimum* L.) from NaCl stress by inducing antioxidative defence system and osmoprotectant accumulation. Acta Physiol Plant 32:121–132

Khan AA, Akbar M, Seshu DV (1987) Ethylene as an indicator of salt tolerance in rice. Crop Sci 27:1242–1248

Koornneef M, Karssen CM (1994) Seed dormancy and germination. In: Meyerowitz EM, Somerville CR (eds) Arabidopsis. Cold Spring Harbor Laboratory, New York, pp 313–334

Koyro HW, Geissler N, Hussin S, Debez A, Huchzermeyer B (2008) Strategies of halophytes to survive in a salty environment. In: Khan NA, Singh S (eds) Abiotic stress and plant responses. I.K. International Publishing House, New Delhi, pp 83–104

Kramell R, Atzorn R, Schneider G, Miersch O, Bruckner C, Schmidt J, Sembdner G, Parthier B (1995) Occurrence and identification of jasmonic acid and its amino acid conjugates induced by osmotic stress in barley leaf tissue. J Plant Growth Regul 14:29–36

Kramell R, Miersch O, Atzorn R, Parthier B, Wasternack C (2000) Octadecanoid-derived alteration of gene expression and the 'oxylipin signature' in stressed barley leaves. Implications for different signaling pathways. Plant Physiol 123:177–187

Kumar B, Singh B (1996) Effect of plant hormones on growth and yield of wheat irrigated with saline water. Ann Agric Res 17:209–212

Lehmann J, Atzorn R, Bruckner C, Reinbothe S, Leopold J, Wasternack C, Parthier B (1995) Accumulation of jasmonate, abscisic acid, specific transcripts and proteins in osmotically stressed barley leaf segments. Planta 197:156–162
Li Y, Su X, Zhang B, Huang Q, Zhang X, Huang R (2008) Expression of jasmonic ethylene responsive factor gene in transgenic poplar tree leads to increased salt tolerance. Tree Physiol 29:273–279
Li S, Xu C, Yang Y, Xia G (2010) Functional analysis of TaDi19A, a salt-responsive gene in wheat. Plant Cell Environ 33:117–129
Lima-Costa ME, Ferreira S, Duarte A, Ferreira AL (2010) Alleviation of salt stress using exogenous proline on a citrus cell line. Acta Hortic 868:109–112
Ling T, Zhang B, Cui W, Wu M, Lin J, Zhou W, Huang J, Shen WB (2009) Carbon monoxide mitigates salt-induced inhibition of root growth and suppresses programmed cell death in wheat primary roots by inhibiting superoxide anion overproduction. Plant Sci 177:331–340
Liu YG, Wu RR, Wan Q, Xie GQ, Bi YR (2007) Glucose-6-phosphate dehydrogenase plays a pivotal role in nitric oxide-invomagomelved defense against oxidative stress under salt stress in red kidney bean roots. Plant Cell Physiol 48:511–522
Lorbiecke R, Sauter M (1999) Adventitious root growth and cell cycle induction in deepwater rice. Plant Physiol 119:21–29
Luan S, Lana W, Lee SC (2009) Potassium nutrition, sodium toxicity, and calcium signaling: connections through the CBL–CIPK network. Curr Opin Plant Biol 12:339–346
Lutts S, Kinet JM, Bouharmont J (1996) Ethylene production by leaves of rice (*Oryza sativa* L.) in relation to salinity tolerance and exogenous putrescine application. Plant Sci 116:15–25
Maggio A, Barbieri G, Raimondi G, De Pascale S (2010) Contrasting effects of GA3 treatments on tomato plants exposed to increasing salinity. J Plant Growth Regul 29:63–72
Magome H, Yamaguchi S, Hanada A, Kamiya Y, Oda K (2004) Dwarf and delayed-flowering 1, a novel *Arabidopsis* mutant deficient in gibberellin biosynthesis because of overexpression of a putative AP2 transcription factor. Plant J 37:720–729
Mahajan S, Pandey GK, Tuteja N (2008) Calcium and salt stress signaling in plants: shedding light on SOS pathway. Arch Biochem Biophys 471:146–158
Manchanda G, Garg N (2008) Salinity and its effect on the functional biology of legumes. Acta Physiol Plant 30:595–618
Mansour MMF (2000) Nitrogen containing compounds and adaptation of plants to salinity stress. Biol Plant 43:491–500
Marschner H (1995) Mineral nutrition of higher plants. Academic, London
Mauch-Mani B, Mauch F (2005) The role of abscisic acid in plant-pathogen interactions. Curr Opin Plant Biol 8:409–414
McConn M, Creelman RA, Bell F, Mullet JE, Browse J (1997) Jasmonate is essential for insect defense in *Arabidopsis*. Proc Natl Acad Sci USA 94:5473–5477
Misra N, Saxena P (2009) Effect of salicylic acid on proline metabolism in lentil grown under salinity stress. Plant Sci 177:181–189
Mohammed AHMA (2007) Physiological aspects of mungbean plant (*Vigna radiata* L. Wilczek) in response to salt stress and gibberellic acid treatment. Res J Agr Biol Sci 3:200–213
Moons A, Prisen E, Bauw G, Montagu MV (1997) Antagonistic effects of abscisic acid and jasmonates on salt-inducible transcripts in rice roots. Plant Cell 92:243–259
Morgan PW, Drew MC (1997) Ethylene and plant responses to stress. Physiol Plant 100:620–630
Munns R (2002) Comparative physiology of salt and water stress. Plant Cell Environ 25:239–250
Munns R, Tester M (2008) Mechanisms of salinity tolerance. Annu Rev Plant Biol 59:651–681
Nakashita H, Yasuda M, Nitta T, Asami T, Fujioka S, Arai Y, Sekimata K, Takatsuto S, Yamaguchi I, Yoshida S (2003) Brassinosteroid functions in a broad range of disease resistance in tobacco and rice. Plant J 33:887–898
Naqvi SSM, Ansari R, Kuawada AN (1982) Responses of salt stressed wheat seedlings to kinetin. Plant Sci Lett 26:279–283

Navarro L, Dunoyer P, Jay F, Arnold B, Dharmasiri N, Estelle M, Voinnet O, Jones JDG (2006) A plant miRNA contributes to antibacterial resistance by repressing auxin signaling. Science 312:436–439

Nazar R, Iqbal N, Masood A, Syeed S, Khan NA (2011a) Understanding the significance of sulfur in improving salinity tolerance in plants. Environ Exp Bot 70:80–87

Nazar R, Iqbal N, Syeed S, Khan NA (2011b) Salicylic acid alleviates decreases in photosynthesis under salt stress by enhancing nitrogen and sulfur assimilation and antioxidant metabolism differentially in two mungbean cultivars. J Plant Physiol 168:807–815

O'Malley RC, Rodriguez FI, Esch JJ, Binder BM, O'Donnell P, Klee HJ et al (2005) Ethylene-binding activity, gene expression levels, and receptor system output for ethylene receptor family members from *Arabidopsis* and tomato. Plant J 41:651–659

Ogawa M, Hanada A, Yamauchi Y, Kuwahara A, Kamiya Y, Yamaguchi S (2003) Gibberellin biosynthesis and response during *Arabidopsis* seed germination. Plant Cell 15:1591–1604

Ohta M, Guo Y, Halfter U, Zhu JK (2003) A novel domain in the protein kinase SOS2 mediates interaction with the protein phosphatase 2C ABI2. Proc Natl Acad Sci USA 100:11771–11776

Olszewski N, Sun TP, Gubler F (2002) Gibberellin signalling biosynthesis, catabolism, and response pathways. Plant Cell 14(suppl):S61–S80

Palma F, Lluch C, Iribarne C, Garcia-Garrida JM, Garcia NAT (2009) Combined effect of salicylic acid and salinity on some antioxidant activities, oxidative stress and metabolite accumulation in *Phaseolus vulgaris*. Plant Growth Regul 58:307–331

Parasher A, Varma SK (1988) Effect of pre-sowing seed soking in gibberellic acid on growth of wheat (*Triticum aestivum* L.) under saline conditions. Indian J Biol Sci 26:473–475

Park JM, Park CJ, Lee SB, Ham BK, Shin R, Paek KH (2001) Overexpression of the tobacco Tsi1 gene encoding an EREBP/AP2-type transcription factor enhances resistance against pathogen attack and osmotic stress in tobacco. Plant Cell 13:1035–1046

Pedranzani H, Racagni G, Alemano S, Miersch O, Ramírez I, Peña-Cortés H, Taleisnik E, Domenech EM, Abdala G (2003) Salt tolerant tomato plants show increased levels of jasmonic acid. Plant Growth Regul 41:149–158

Pérez-Alfocea F, Albacete A, Ghanem ME, Dodd IC (2010) Hormonal regulation of source–sink relations to maintain crop productivity under salinity: a case study of root-to-shoot signalling in tomato. Funct Plant Biol 37:592–603

Poljakoff-Mayber A, Lerner HR (1994) Plants in saline environments. In: Pessarakli M (ed) Handbook of plant and crop stress. Dekker, New York, pp 65–96

Pospíšilová J (2003) Interaction of cytokinins and abscisic acid during regulation of stomatal opening in bean leaves. Photosynthetica 41:49–56

Prakash L, Prathapasenan G (1990) NaCl and gibberellic acid-induced changes in the content of auxin, the activity of cellulose and pectin lyase during leaf growth in rice (*Oryza sativa*). Ann Bot 365:251–257

Qiu QS, Guo Y, Dietrich MA, Schumaker KS, Zhu JK (2002) Regulation of SOS1, a plasma membrane Na^+/H^+ exchanger in *Arabidopsis thaliana*, by SOS2 and SOS3. Proc Natl Acad Sci USA 99:8436–8441

Rausch T, Wachter A (2005) Sulfur metabolism: a versatile platform for launching defence operations. Trends Plant Sci 10:503–509

Saibo NJM, Vriezen WH, Beemster GTS, Van der Straeten D (2003) Growth and stomata development of *Arabidopsis* hypocotyls are controlled by gibberellins and modulated by ethylene and auxins. Plant J 33:989–1000

Saimbhi MS (1993) Growth regulators on vegetable crops. In: Chadha KL, Kallo G (eds) Advances in horticulture. Malhotra, New Delhi, pp 619–642

Sastry EVD, Shekhawa KS (2001) Alleviatory effect of GA3 on the effect of salt at seedling stage in wheat (*Triticum aestivum*). Indian J Agr Res 35:226–231

Sembdner G, Parthier B (1993) The biochemistry and physiology and molecular actions of jasmonates. Ann Rev Plant Physiol Plant Mol Biol 44:569–586

Seo YJ, Park JB, Cho YJ, Jung C, Seo HS, Park SK, Nahm BH, Song JT (2010) Overexpression of the ethylene-responsive factor gene BrERF4 from *Brassica rapa* increases tolerance to salt and drought in *Arabidopsis*. Plants Mol Cells 30:271–277

Shah SH (2007) Effects of salt stress on mustard as affected by gibberellic acid application. Gen Appl Plant Physiol 33:97–106

Sharp R, LeNoble ME (2002) ABA, ethylene and the control of shoot and root growth under water stress. J Exp Bot 53:33–37

Siddiqui MH, Khan MN, Mohammad F, Khan MMA (2008) Role of nitrogen and gibberellin (GA_3) in the regulation of enzyme activities and in osmoprotectant accumulation in *Brassica juncea* L. under salt stress. J Agron Crop Sci 194:214–224

Singha S, Choudhuri MA (1990) Effect of salinity (NaCl) stress on H_2O_2 metabolism in *Vigna* and *Oryza* seedlings. Biochem Physiol Pflan 186:69–74

Steffens B, Wang J, Sauter M (2006) Interactions between ethylene, gibberellin and abscisic acid regulate emergence and growth rate of adventitious roots in deepwater rice. Planta 223:604–612

Steffens B, Sauter M (2005) Epidermal cell death in rice (*Oryza sativa* L.) is regulated by ethylene, gibberellin and abscisic acid. Plant Physiol 139:1–9

Sunkar R, Kapoor A, Zhu JK (2006) Posttranscriptional induction of two Cu/Zn superoxide dismutase genes in Arabidopsis is mediated by downregulation of miR398 and important for oxidative stress tolerance. Plant Cell 18:2051–2065

Syeed S, Anjum NA, Nazar R, Iqbal N, Masood A, Khan NA (2010) Salicylic acid-mediated changes in photosynthesis, nutrients content and antioxidant metabolism in two mustard (*Brassica juncea* L.) cultivars differing in salt tolerance. Acta Physiol Plant 33(877):886

Tester M, Davenport R (2003) Na^+ tolerance and Na^+ transport in higher plants. Ann Bot 91:503–507

Türkan I, Demiral T (2009) Recent developments in understanding salinity tolerance. Environ Exp Bot 1(Special issue):2–9

Tuna AL, Kaya C, Dikilitas M, Higgs D (2008) The combined effects of gibberellic acid and salinity on some antioxidant enzyme activities, plant growth parameters and nutritional status in maize plants. Environ Exp Bot 62:1–9

Vandenbussche F, Vancompernolle B, Rieu I, Ahmad M, Phillips A, Moritz T, Hedden P, Van Der Straeten D (2007) Ethylene induced *Arabidopsis* hypocotyl elongation is dependent on but not mediated by gibberellins. J Exp Bot 58:4269–4281

Velitcukova M, Fedina I (1998) Response of photosynthesis of *Pisum sativum* to salt stress as affected by methyl jasmonate. Photosynthetica 35:89–97

Vettakkorumakankav NA (1999) Crucial role for gibberellin in stress protecting of plants. Plant Cell Physiol 40:542–548

Vriezen WH, Achard P, Harberd NP, Van Der Straeten D (2004) Ethylene-mediated enhancement of apical hook formation in etiolated *Arabidopsis thaliana* seedlings is gibberellin dependent. Plant J 37:505–516

Walker MA, Dumbroff EB (1981) Effects of salt stress on abscisic acid and cytokinin levels in tomato. ZPfl anzenphysiol 101:461–470

Wang D, Pajerowska-Mukhtar K, Hendrickson Culler A, Dong X (2007) Salicylic acid inhibits pathogen growth in plants through repression of the auxin signaling pathway. Curr Biol 17:1784–1790

Wang HH, Liang XL, Wan Q, Wang XM, Bi YR (2009) Ethylene and nitric oxide are involved in maintaining ion homeostasis in *Arabidopsis* callus under salt stress. Planta 230:293–307

Wi SJ, Jang SJ, Park KY (2010) Inhibition of biphasic ethylene production enhances tolerance to abiotic stress by reducing the accumulation of reactive oxygen species in *Nicotiana tabacum*. Mol Cells 30:37–39

Wolf O, Jeschke WD, Hartung W (1990) Long-distance transport of abscisic-acid in NaCl-treated intact plants of *Lupinus albus*. J Exp Bot 41:593–600

Wu L, Zhang Z, Zhang H, Wang XC, Huang R (2008) Transcriptional modulation of ethylene response factor protein JERF3 in the oxidative stress response enhances tolerance of tobacco seedlings to salt, drought, and freezing. Plant Physiol 148:1953–1963

Xu S, Lou T, Zhao N, Gao Y, Dong L, Jiang D, Shen W, Huang L, Wang R (2011) Presoaking with hemin improves salinity tolerance during wheat seed germination. Acta Physiol Plant 33:1173–1183

Yamaguchi S (2008) Gibberellin metabolism and its regulation. Annu Rev Plant Biol 59:225–251

Yamaguchi S, Kamiya Y (2000) Gibberellin biosynthesis: its regulation by endogenous and environmental signals. Plant Cell Physiol 41:251–257

Yang L, Zua YG, Tang ZH (2010) Ethylene improves *Arabidopsis* salt tolerance mainly via retaining K^+ in shoots and roots rather than decreasing tissue Na^+ content. Environ Exp Bot. doi:10.1016/j.envexpbot.2010.08.006

Yeo AR (2007) Salinity. In: Yeo AR, Flowers TJ (eds) Plant solute transport. Blackwell, Oxford, pp 340–365

Zahra S, Amin B, Mohamad Ali VS, Mehdi Y (2010) The salicylic acid effect on the tomato (*Lycopersicum esculentum* Mill.) sugar, protein and praline contents under salinity stress (NaCl). J Biophys Struct Biol 2:35–41

Zhao XC, Schaller GE (2004) Effect of salt and osmotic stress upon expression of the ethylene receptor ETR1 in *Arabidopsis thaliana*. FEBS Lett 562:189–192

Zhu JH, Verslues PE, Zheng XW et al (2005) HOS10 encodes an R2R3-type MYB transcription factor essential for cold acclimation in plant. Proc Natl Acad Sci USA 102:9966–9971

Zhu J, Fu X, Koo YD et al (2007) An enhancer mutant of *Arabidopsis* salt overly sensitive 3 mediates both ion homeostasis and the oxidative stress response. Mol Cell Biol 27:5214–5224

Zimmermann P, Hirsch-Hoffmann M, Hennig L, Gruissem W (2004) Genevestigator. Arabidopsis microarray database and analysis toolbox. Plant Physiol 136:2621–2632

Chapter 4
Function of Nitric Oxide Under Environmental Stress Conditions

Marina Leterrier, Raquel Valderrama, Mounira Chaki, Morak Airaki, José M. Palma, Juan B. Barroso, and Francisco J. Corpas

Abstract Nitric oxide (NO) is a key signaling molecule in different physiological processes of plants. However, under adverse stress conditions, plants can undergo a deregulation in its production which can provoke a process of nitrosative stress. In addition, the exogenous application of NO seems to alleviate or even prevent cellular damage under some specific environmental stresses, suggesting the involvement of this molecule in the mechanism of defense against abiotic stresses. In this article, the current knowledge of the implication of NO under environmental stresses is briefly reviewed with a special emphasis in its interaction with some phytohormones.

4.1 Introduction

Nitric oxide is one of the molecules which have received much attention during the last decade from plant researchers. The main reason is that this free radical is involved as signal molecule in many physiological processes during plant growth and development including seed germination, primary and lateral root growth, flowering, pollen tube growth regulation, fruit ripening, and senescence, among others (Wojtaszek 2000; Corpas et al. 2001, 2004; Lamattina et al. 2003; Magalhaes et al. 2005; Shapiro 2005; Besson-Bard et al. 2008a, b), as well under different environmental stress conditions including heavy metal, salinity, wounding, extreme temperature, etc. (Corpas et al. 2011). However, the real significance of NO and

M. Leterrier • M. Chaki • M. Airaki • J.M. Palma • F.J. Corpas (✉)
Departamento de Bioquímica, Biología Celular y Molecular de Plantas, Estación Experimental del Zaidín (EEZ), CSIC, Granada, Spain
e-mail: javier.corpas@eez.csic.es

R. Valderrama • J.B. Barroso
Grupo de Señalización Molecular y Sistemas Antioxidantes en Plantas, Unidad Asociada al CSIC (EEZ), Área de Bioquímica y Biología Molecular, Universidad de Jaén, Jaen, Spain

N.A. Khan et al. (eds.), *Phytohormones and Abiotic Stress Tolerance in Plants*,
DOI 10.1007/978-3-642-25829-9_4,

related molecules designated as reactive nitrogen species (RNS) is only at the beginning of this attractive research area of plant physiology since many aspects of its biochemistry and physiology are still to be elucidated. Here, it will give a general overview of the basic biochemistry of NO and its interaction with some classical phytohormones. Then, it will summarize briefly the implication of NO under environmental stress conditions taking in consideration that NO depending of its cellular concentration could be a plant regulator (low concentration) or be part of the mechanism of defense as toxic molecule (high concentration).

4.2 Basic Biochemistry of NO

Nitric oxide is a free radical because the nitrogen has an unpaired electron in its π orbital (˙NO) which determinates its biochemistry. Moreover, NO has a family of related molecules designated as reactive nitrogen species (RNS). Table 4.1 summarizes the main RNS including radical and nonradical molecules. Among the different RNS, peroxynitrite ($ONOO^-$) which is produced by the reaction between NO and superoxide radical ($O_2^{\cdot -}$) has a relevant significance because it is a powerful oxidant that can mediate nitration process and provokes cellular injury (Szabó et al. 2007; Corpas et al. 2009a; Arasimowicz-Jelonek and Floryszak-Wieczorek 2011).

Nitric oxide can interact with different biomolecules including lipids, nucleic acids, and proteins affecting its functions. However, the interaction with proteins has been the most studied. In this sense, NO directly or indirectly can react with proteins in different ways: (1) with transition metals present in the protein to give complexes called metal nitrosyls, (2) with sulfhydryl groups to render a process of *S*-nitrosylation, and (3) by adding a nitro ($-NO_2$) group in a process of nitration. So far, the analysis of NO binding to plant metal-containing protein has been done mainly with plant hemoglobins (Besson-Bard et al. 2008a, b); however, there are some experimental data showing that certain enzyme activities such as cytochrome *c* oxidase, catalase, or ascorbate peroxidase can be modulated by this mechanism (Millar and Day 1996; Clark et al. 2000). Protein *S*-nitrosylation is a posttranslational modification of cysteine residues produced by NO which can modify the

Table 4.1 Reactive nitrogen species (RNS) including radicals and nonradicals molecules

Radicals	Nonradicals
Nitric oxide (˙NO)	Nitroxyl anion (NO^-)
Nitrogen dioxide (NO_2)	Nitrosonium cation (NO^+)
	Nitrous acid (HNO_2)
	Dinitrogen trioxide (N_2O_3)
	Dinitrogen tetroxide (N_2O_4)
	Peroxynitrite ($ONOO^-$)
	Peroxynitrous acid (ONOOH)
	Alkyl peroxynitrite (RNOONO)

function of a broad spectrum of proteins (Stamler et al. 2001; Lindermayr et al. 2005; Wang et al. 2006; Lindermayr and Durner 2009). Special attention must be given in the process of *S*-nitrosylation of the tripeptide glutathione (GSH) to form the *S*-nitrosoglutathione (GSNO) since this molecule can function as mobile reservoir of NO (Durner and Klessig 1999; Barroso et al. 2006) and it can regulate the equilibrium between GSNO and *S*-nitrosylated proteins by a process of transnitrosylation. In this sense, the enzyme GSNO reductase seems to be a key element because it catalyzes the NADH-dependent reduction of (GSNO) to GSSG and NH_3. Consequently, this enzyme controls the intracellular level of GSNO and, as a result, the effects of NO in cells (Leterrier et al. 2011). Protein nitration is another process that introduces a nitro group, ($-NO_2$) and there are several amino acids which are preferentially nitrated, such as tyrosine(Y), tryptophan (W), cysteine (C), and methionine (M). However, in plants, most studies are focused in tyrosine nitration (Corpas et al. 2009a, b; Chaki et al. 2009). Figure 4.1 shows a straightforward model of NO metabolism in plant cells under environmental stress conditions.

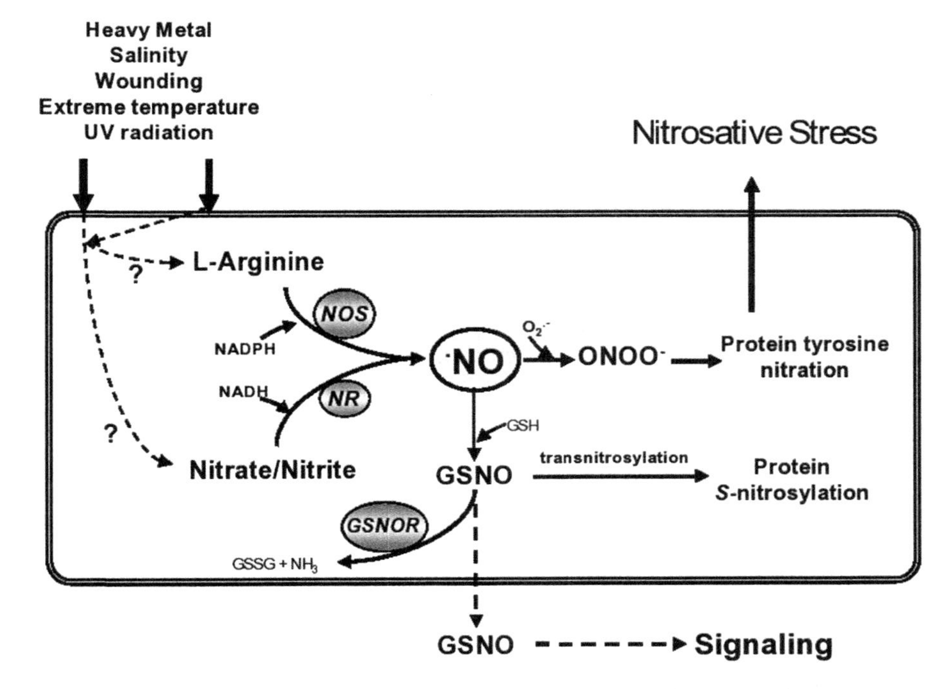

Fig. 4.1 Schematic model of nitric oxide (NO) metabolism in plant cells under environment stress conditions. Under a specific environmental stress, L-arginine-dependent nitric oxide synthase (NOS) and/or nitrate reductase (NR) can generate NO which can react with reduced glutathione (GSH) in the presence of O_2 to form *S*-nitrosoglutathione (GSNO). This metabolite can be converted by the enzyme GSNO reductase (GSNOR) into oxidized glutathione (GSSG) and NH_3. Alternatively, GSNO by a process of transnitrosylation can mediate protein *S*-nitrosylation or diffuse out to the cell where it can act a signal molecule with adjacent cells. On the other hand, NO can react with superoxide radicals (O_2^-) to generate peroxynitrite ($ONOO^-$), a powerful oxidant molecule that can mediate the rise of protein tyrosine nitration which can be considered as a marker of nitrosative stress

4.3 Nitric Oxide: A New Hormone?

Independent of the name, plant hormones, phytohormones, or plant growth regulators are molecules produced inside of plant cells that at low concentrations, in the range nM to pM, promote and influence the growth, development, and differentiation of cells and tissues (Davies 1995; Beligni and Lamattina 2001). Thus, the biosynthesis of plant hormones within plant tissues is often diffuse and not always localized. Classically, there are five groups of plant hormones including auxin (indole-3-acetic acid), cytokinin, gibberellins, ethylene, and abscisic acid (ABA). However, there are also other compounds that have important growth-regulating activities including jasmonic acid (JA), salicylic acid (SA), and brassinosteroids which are considered to function as phytohormones. In this framework, NO could be considered a new member of this group of molecules. However, even when arguments in both directions can be presented, the relevance of NO in physiological processes such as germination, growth, senescence, photosynthesis, stomatal movement, etc. is clear, even though more research must be done in this direction.

4.4 Cross Talk Between NO and Other Hormones Under Stress Conditions

Such as it has been mentioned, NO is implied in numerous plant functions which are also regulated by others phytohormones. So, interactions between NO and other hormones through synergy or antagonism permit a finest level of control depending on the situation of environmental stress.

4.4.1 Abscisic Acid

One of the main roles of ABA is to regulate plant response to drought by inducing stomatal closure, therefore reducing water loss via transpiration (Davies and Zhang 1991; Zhu 2002). It is well known that NO is involved in ABA-induced stomatal closure (Garcia-Mata and Lamattina 2002; Desikan et al. 2004; Bright et al. 2006). Moreover, cross talk between ABA and NO also enhances activities of antioxidant enzymes such as catalase and superoxide dismutase. Thus, ABA induces H_2O_2 accumulation, leading to NO generation which in turn activates mitogen-activated protein kinase (MAPK) and upregulates gene expression of antioxidant enzymes (Zhang et al. 2007; Lu et al. 2009). Recently, it has also been shown that cGMP acts downstream of H_2O_2 and NO in the signaling pathway by which ABA induces stomatal closure (Dubovskaya et al. 2011).

NO is also part of the signaling cascade mediating the brassinosteroid-induced ABA biosynthesis in maize leaves under water stress (Zhang et al. 2011b). In maize leaves under UV-B, stress induces ABA production which triggers NO synthesis (Tossi et al. 2009). So, there are some clear interactions between ABA and NO, but the exact nature of this interaction appears to depend on the system studied (plant species and stress).

4.4.2 Ethylene

Ethylene is involved not only in plant growth and senescence but also in plant response to abiotic stress such as salinity, iron deficiency, ozone, and UV-B (Romera and Alcantara 1994; Mackerness et al. 1999; Romera et al. 1999; Vahala et al. 2003; Cao et al. 2007; Ahlfors et al. 2009; Wang et al. 2009).

NO is known to have a protective effect on salt stress (Zhao et al. 2004; Zhang et al. 2006). In fact, in Arabidopsis callus under 100 mM NaCl, NO accumulation greatly stimulates ethylene emission, which in turn induces expression of the plasma membrane H^+-ATPase genes (Wang et al. 2009). The higher activity of the plasma membrane H^+-ATPase modulates ion homeostasis for a better salt tolerance. Under salt stress, NO-induced ethylene also stimulates the alternative respiratory pathway (Wang et al. 2010a, b).

Ethylene and NO also cooperate in iron homeostasis as it was determined that Fe-related genes upregulated by ethylene were also responsive to nitric oxide (García et al. 2010). Moreover, under iron deficiency, NO induces expression of genes involved in ethylene synthesis, and conversely, ethylene enhances NO production. Hence, both NO and ethylene are necessary for the upregulation of Fe-acquisition genes, and each one influences the production of the other (García et al. 2011). Ethylene and NO also cooperate in ozone stress: both NO- and ethylene-dependent pathways are required for O_3-induced upregulation of alternative oxidase (AOX) in ozone-treated tobacco plants (Ederli et al. 2006). In *Vicia faba*, ethylene participates to UV-B-induced stomatal closure and acts upstream of NO generation (He et al. 2011a, b).

4.4.3 Salicylic Acid

Salicylic acid (SA) induces tolerance to various abiotic stresses such as salinity, heavy metals, and drought (through regulation of stomatal closure) (Manthe et al. 1992; Metwally et al. 2003; Horvath et al. 2007; Szepesi et al. 2009).

NO is involved in the mechanism of salt tolerance generated by SA pretreatment in tomato. Thus, salt stress increases NO content in tomato roots, but pretreatment of the plants with SA changes that response and prevents NO accumulation (Gémes et al. 2011). In Arabidopsis roots, SA triggers the NO production thorough

NOS-dependent pathways where calcium and protein phosphorylation are essential components (Zottini et al. 2007).

SA and NO are both known to reduce heavy metal toxicity separately (Metwally et al. 2003; Arasimowicz and Floryszak-Wieczorek 2007). However, a synergic effect occurs when exogenous applications of both products are combined in canola plants under nickel stress (Kazemi et al. 2010). SA can also induce stomatal closure (Manthe et al. 1992). NO and ROS are both required in SA-induced stomatal closure; in consequence, a model has been proposed where SA activates a peroxidase (sensitive to the inhibitor salicylhydroxamic acid) to produce extracellular ROS, leading to ROS accumulation and NO production in guard cells, and inactivates K_{in}^{+} channels, causing stomatal closure (Khokon et al. 2011).

4.5 Nitric Oxide and Abiotic Stresses

Such it has been mentioned, under stress conditions, there are many reports demonstrating the participation of different phytohormones. For example, under wounding, the participation of several hormones including abscisic acid, ethylene, and jasmonic acid is well recognized (Stratmann 2003), but they are also involved under other stresses such as cadmium or aluminum stress where the production of jasmonic acid, ethylene, gibberellins (GA), or ABA is affected (Sanitá di Toppi and Gabbrielli 1999; Rodríguez-Serrano et al. 2009; He et al. 2011a, b). In this context, there are also accumulating data showing the participation of NO in different types of environmental stresses.

4.5.1 Heavy Metal

In plant biology, the term heavy metal is used to design a series of metals and also metalloids that can be toxic to plants even at very low concentrations being for that reason these phytotoxic elements (Corpas et al. 2010; Xiong et al. 2010; Rascio and Navari-Izzo 2011). Thus, heavy metals can also distinguish two categories: (1) essential elements required for normal growth and metabolism such as Co, Cu, Fe, Mn, Mo, Ni, and Zn (micronutrients); and (2) not essential element since they do not perform any known physiological such as Cd, Hg, Se, Pb, or As. Although during the last 10 years, certain numbers of studies have been done in how NO could be involved in the mechanism of response against heavy metals. It must be mentioned that the basic biochemical mechanism is very rudimentary.

4.5.1.1 Cadmium

This is one of the most commonly found heavy metal in soil (Arasimowicz-Jelonek et al. 2011). There are some reports showing the participation of NO under cadmium stress. For example, in roots of *Pisum sativum* and *Brassica juncea* in the presence of 100 μM Cd, there was a time-dependent endogenous NO production (Bartha et al. 2005). These data contrast with the results observed in leaves and roots of *P. sativum* grown with 50 μM $CdCl_2$, which produced growth inhibition and oxidative damages (Romero-Puertas et al. 2004), and a drastic reduction of the NO content (Barroso et al. 2006; Rodríguez-Serrano et al. 2006). In addition, the lower NO content in leaves was accompanied by a reduction of GSH, GSNO, and GSNO reductase activity and transcript (Barroso et al. 2006). In contrast, pre-treatment of sunflower seedlings with 100 μM sodium nitroprusside (SNP) protected leaves against Cd-induced oxidative stress (Laspina et al. 2005). In *Lupinus* roots grown with 50 μM Cd, a similar behavior has been observed, and it was proposed that the protective effect of NO could involve the stimulation of superoxide dismutase activity to counteract the overproduction of superoxide radicals (Kopyra and Gwózdz 2003). In the case of rice (*Oryza sativa*) plantlets subjected to 0.2 mM $CdCl_2$ and treated with 0.1 mM SNP, the toxicity of Cd is alleviated. Accordingly, the authors proposed that the NO-induced increase of Cd tolerance is produced by a rise in pectin and hemicellulose content in the cell wall of rice roots, and this provoked a diminished distribution of Cd in the soluble fraction of leaves and roots but with the concomitant increases of Cd in the cell walls of roots (Xiong et al. 2009). Additionally, the treatment with SNP induces the antioxidant system including ascorbate, catalase, glutathione reductase, and peroxidase which counteracts the oxidative stress provoked by Cd (Hsu and Kao 2004; Panda et al. 2011).

4.5.1.2 Arsenic

It is a metalloid constituent of a wide range of minerals which has become in an important environmental contaminant that can provoke health problems to humans by its accumulation in food crops or in drinking water (Tripathi et al. 2007; Zhao et al. 2010). Under arsenic stress conditions, plants suffer alteration at different levels including elements' uptake and transport, metabolism, and gene expression (Abercrombie et al. 2008; Verbruggen et al. 2009; Zhao et al. 2009). Accumulating data indicate that metabolism of reactive oxygen species (ROS) can be involved and can cause an oxidative stress (Dwivedi et al. 2010). However, there are only few reports analyzing the NO function under arsenic toxicity in higher plants, and they have been done by the exogenous application of NO with sodium nitroprusside (SNP). Thus, in roots of *O. sativa*, the application of SNP provides resistance against As toxicity and has an ameliorating effect against As-induced stress (Singh et al. 2009). Similar observation has been reported in tall fescue where the application of 100 μM SNP reduced arsenic-induced oxidative damage in leaves (Jin et al. 2010). In Arabidopsis seedlings exposed to 0.5 mM arsenate, a significant reduction in growth parameters such as length of roots affecting its membrane integrity and provoking an increase of lipid oxidation has been observed. These changes were accompanied by an alteration of antioxidative enzymes (catalase and glutathione reductase) and nitric oxide metabolism with a significant increase of

NO content, *S*-nitrosoglutathione reductase (GSNOR) activity, and protein tyrosine nitration, which the concomitant reduction of GH and GSNO content. Thus, in this case, arsenate seems to provoke both oxidative and nitrosative stress being the glutathione reductase and GSNOR activities key components in the mechanism of response (Leterrier et al. 2010).

4.5.1.3 Aluminum

Aluminum (Al^{3+}) is not a heavy metal, but it makes up about 8% of the surface of the earth and is the third most abundant element being a major factor limiting crop growth and yield in acid soils since it inhibits cell division, cell extension, and transport (Ma et al. 2001). Thus, *Hibiscus moscheutos* exposed to100 μM $AlCl_3$ induced inhibition of root growth. This is accompanied by inhibition of nitric oxide synthase activity and reduced NO content (Tian et al. 2007). In *Arabidopsis thaliana* roots, the cells of the distal portion of the transition zone emitted large amounts of NO, but the treatment with 90 μM aluminum blocked this NO emission (Illés et al. 2006).

In the same way with that of cadmium and arsenic, the treatment with NO donors seems to alleviate the aluminum toxicity. In *Cassia tora* plants pretreated for 12 h with 0.4 mM SNP and subsequently exposed to 10 μM aluminum treatment for 24 h exhibited a significantly greater root elongation and a decrease in Al^{3+} accumulation in root apexes as compared with plants without SNP treatment (Wang and Yang 2005). More recently, it has been also shown that SNP treatment in rice (*O. sativa*) provoked an enhancement of Al tolerance in roots. Thus, the mechanism of tolerance is because NO induced a decrease in the contents of pectin and hemicellulose, an increase in the degree of methylation of pectin, and a decrease in the Al accumulation in root cell walls (Zhang et al. 2011a). In the cases of rye (*Secale cereale* L) and wheat(*Triticum aestivum*), Al treatment provoked an inhibition in root growth that was accompanied by a reduction in gibberellin (GA) content and an increase in the values of IAA/GA and ABA/GA. However, treatment with SNP reversed Al toxicity due to an alteration of endogenous hormones in the roots. Thus, SNP reduced the inhibition of root elongation by increasing GA content and decreasing the values of IAA/GA and IAA/zeatin riboside under Al stress (He et al. 2011a, b).

4.5.2 Wounding

Different types of stresses, for example, herbivores, wind, or rain, can produce mechanical injury in plants. To avoid the potential infection by opportunistic microorganisms in the damage zone, plants respond with a cascade of signal that provokes the induction of numerous genes (Schilmiller and Howe 2005). In this sense, there are also data indicating that NO metabolism is also involved.

For example, in *A. thaliana*, mechanical wounding induced a rapid accumulation of NO that could be involved in jasmonic-acid-associated defense responses and adjustments (Huang et al. 2004). In the case of pea (*P. sativum* L) seedlings, mechanical wounding in leaves provoked an accumulation of NO content after 4 h, and this was accompanied by an increase in the content of *S*-nitrosothiols (SNOs) and a general induction of nitric oxide synthase (NOS) and GSNO reductase activities, although the pattern of proteins that undergo tyrosine nitration did not appear to be affected (Corpas et al. 2008). In sunflower (*Helianthus annuus*) hypocotyls, mechanical wounding apparently did not affect the NO content, but it triggered the accumulation of SNOs, specifically GSNO, due to a downregulation of GSNOR activity, while protein tyrosine nitration increases. Consequently, a process of nitrosative stress is induced, and *S*-nitrosothiols seem to be a new wound signal in plants (Chaki et al. 2011). In other cases, the application of exogenous NO has been reported to modulate the response against wounding. Thus, in tomato (*Lycopersicon esculentum*) plants, the application of NO donors such as SNP or SNAP inhibited the expression of wound-inducible proteinase inhibitors (Orozco-Cardenas and Ryan 2002).

4.5.3 Salinity

Salinity stress takes place when soluble salts (usually NaCl) are elevated in soil, and this affects plant productivity due to its negative effects on plant growth, ion balance, and water relations (Munns and Tester 2008).

One more time, the application of NO donors has been used as tool to study the involvement of NO in plants exposed to salinity stress. For example, in the calluses of reed (*Phragmites communis*) exposed to 200 mM NaCl, the addition of SNP stimulated the expression of the plasma membrane H^+-ATPase, indicating that NO serves as a signal-inducing salt resistance by increasing the K^+-to-Na^+ ratio (Zhao et al. 2004). Similar results have been reported to be found in maize, where the addition of exogenous NO also booted the salt-stress tolerance by elevating the activities of the proton pump and the Na^+/H^+ antiport of the tonoplast (Zhang et al. 2006). An 8-day-old rice (*O. sativa*) plant treated with 1 μM SNP or 10 μM H_2O_2 and then exposed to salinity stress has been shown to present an increased tolerance since it is induced by both antioxidant enzymes and some stress-related genes (Uchida et al. 2002). In the case of orange (*Citrus aurantium* L.) trees, similar behavior has been observed suggesting that the induction of antioxidant enzymes as consequence of SNP pretreatment provided a major resistance to salinity (Tanou et al. 2009).

However, there are also reports indicating that salinity affects the endogenous metabolism of NO. Hence, in olive (*Olea europaea* L.) plants grown under *in vivo* salt stress (200 mM NaCl), biochemical analyses demonstrated a general increase in the production of NO, *S*-nitrosothiols, and protein nitration. These data seem to indicate that salinity induced a nitrosative stress (Valderrama et al. 2007). Similar data have been reported in *A. thaliana* with 100 mM NaCl, where using genetic strategies, it was reported that peroxisomes are responsible for the NO accumulation observed in the cytosol of root cells under this salinity stress conditions (Corpas et al. 2009b).

4.5.4 Atmospheric Pollutants (Ozone and Ultraviolet Radiation)

Ozone (O_3) layer located in the upper atmosphere is a natural component that protects Earth against potential cellular damage by ultraviolet radiation. However, air pollutants resulted in industrial and vehicle emissions such as hydrocarbons and nitrogen oxides found in the troposphere (the lowest layer of the atmosphere) can produce ozone by photochemical reactions, and this ozone negatively affects plants and animals. In plants, the effects of ozone depend of the concentration and exposure time. Low level of ozone reduces photosynthesis and growth and triggers premature leaf senescence in sensitive plant species and cultivars. On the other hand, high concentration of ozone induces cell death with visible injuries in the leaves.

Thus, the interaction of NO with some phytohormones in response to O_3 treatments has been reported. In Arabidopsis, ozone induces the production of NO which is preceded by an accumulation of salicylic acid and then cell death. Interestingly, the application of exogenous NO increased the levels of ozone-induced ethylene production and leaf injury (Rao and Davis 2001). In addition, the accumulation of NO observed after ozone treatments provoked also the induction of genes involved in salicylic acid biosynthesis in Arabidopsis (Ahlfors et al. 2009) and ethylene in tobacco plants (Ederli et al. 2006).

As mentioned before, the increase of atmospheric pollution by compounds such as chlorofluorocarbon used as refrigerants, propellants (in aerosol applications), and solvents contributes also to the destruction of the O_3 layer located in the upper atmosphere which protects against UV radiation. Consequently, its destruction provokes an increase of UV-B radiation (280–320 nm) which affects plant growth and usually induces oxidative stress (reduced photosynthesis, increased damage to DNA). The involvement of NO has also been studied under UV-B radiation. For example, in maize leaves, the treatment with UV-B induced a rise in the content of NO, H_2O_2, and ABA, being this ABA required for the NO-mediated attenuation of deleterious effect of this stress (Tossi et al. 2009). On the other hand, the application of NO donors in bean seedlings subjected to UV-B radiation reduced the UV-B effect characterized by a decrease in chlorophyll contents and oxidative damage to the thylakoid membrane (Shi et al. 2005).

4.6 Perspectives

Many environmental stresses as excess salts, extreme temperatures, toxic metals, air pollutants, etc. constitute a major limitation to agricultural production. To palliate this negative effect is important to progress and integrate different disciples such as plant physiology, plant breeding, biochemistry, genetics, molecular biology, agricultural engineering, among others. Considering that NO is involved in a plethora of plant functions under physiological and stress conditions, basic research in plant NO metabolism can be a new piece to contribute and make progress in this direction. So far, there are some promising experimental data which support the relevance of NO in plants under stress conditions. For example, the application of exogenous NO to plants seems to active different biochemical pathways that provide some resistance against several types of stresses (salinity, heavy metal, ozone, etc.). Therefore, to elucidate NO metabolism in plants can contribute, in coordination with other disciplines, to establish biotechnological strategies against abiotic stresses, which are responsible for important losses in plant yield and crop productivity.

Acknowledgments This work was supported by ERDF-cofinanced grants from the Ministry of Science and Innovation (ACI2009-0860, BIO2009-12003-C02-01 and BIO2009-12003-C02-02).

References

Abercrombie JM, Halfhill MD, Ranjan P, Rao MR, Saxton AM, Yuan JS, Stewart CN Jr (2008) Transcriptional responses of *Arabidopsis thaliana* plants to As (V) stress. BMC Plant Biol 8:87

Ahlfors R, Brosche M, Kollist H, Kangasjarvi J (2009) Nitric oxide modulates ozone-induced cell death, hormone biosynthesis and gene expression in *Arabidopsis thaliana*. Plant J 58:1–12

Arasimowicz M, Floryszak-Wieczorek J (2007) Nitric oxide as a bioactive signalling molecule in plant stress responses. Plant Sci 172(5):876–887

Arasimowicz-Jelonek M, Floryszak-Wieczorek J (2011) Understanding the fate of peroxynitrite in plant cells from physiology to pathophysiology. Phytochemistry 72(8):681–688

Arasimowicz-Jelonek M, Floryszak-Wieczorek J, Gwózdz EA (2011) The message of nitric oxide in cadmium challenged plants. Plant Sci 181(5):612–620. doi:10.1016/j.plantsci.2011.03.019

Bartha B, Kolbert Z, Erdei L (2005) Nitric oxide production induced by heavy metals in *Brassica juncea* L. Czern. and *Pisum sativum* L. Acta Biologica Szegediensis 49:9–12

Barroso JB, Corpas FJ, Carreras A, Rodríguez-Serrano M, Esteban FJ, Fernández-Ocaña A, Chaki M, Romero-Puertas MC, Valderrama R, Sandalio LM, del Río LA (2006) Localization of *S*-nitrosoglutathione and expression of *S*-nitrosoglutathione reductase in pea plants under cadmium stress. J Exp Bot 57:1785–1793

Beligni MV, Lamattina L (2001) Nitric oxide: a nontraditional regulator of plant growth. Trends Plant Sci 6:508–509

Besson-Bard A, Pugin A, Wendehenne D (2008a) New insights into nitric oxide signaling in plants. Annu Rev Plant Biol 59:21–39

Besson-Bard A, Courtois C, Gauthier A, Dahan J, Dobrowolska G, Jeandroz S, Pugin A, Wendehenne D (2008b) Nitric oxide in plants: production and cross-talk with Ca^{2+} signalling. Mol Plant 1:218–228

Bright J, Desikan R, Hancock JT, Weir IS, Neill SJ (2006) ABA-induced NO generation and stomatal closure in Arabidopsis are dependent on H_2O_2 synthesis. Plant J 45:113–122

Cao WH, Liu J, He XJ, Mu RL, Zhou HL, Chen SY, Zhang JS (2007) Modulation of ethylene responses affects plant salt-stress responses. Plant Physiol 143:707–719

Chaki M, Valderrama R, Fernández-Ocaña AM, Carreras A, López-Jaramillo J, Luque F, Palma JM, Pedrajas JR, Begara-Morales JC, Sánchez-Calvo B, Gómez-Rodríguez MV, Corpas FJ, Barroso

JB (2009) Protein targets of tyrosine nitration in sunflower (*Helianthus annuus* L.) hypocotyls. J Exp Bot 60:4221–4234

Chaki M, Valderrama R, Fernández-Ocaña AM, Carreras A, Gómez-Rodríguez MV, Pedradas JR, Begara-Morales JC, Sánchez-Calvo B, Luque F, Leterrier M, Corpas FJ, Barroso JB (2011) Mechanical wounding induces a nitrosative stress by downregulation of GSNO reductase and a rise of *S*-nitrosothiols in sunflower (*Helianthus annuus*) seedlings. J Exp Bot 62:1803–1813

Clark D, Durner J, Navarre DA, Klessig DF (2000) Nitric oxide inhibition of tobacco catalase and ascorbate peroxidase. Mol Plant Microbe Interact 13:1380–1384

Corpas FJ, Barroso JB, del Río LA (2001) Peroxisomes as a source of reactive oxygen species and nitric oxide signal molecules in plant cells. Trends Plant Sci 6:145–150

Corpas FJ, Barroso JB, Carreras A, Quirós M, León AM, Romero-Puertas MC, Esteban FJ, Valderrama R, Palma JM, Sandalio LM, Gómez M, del Río LA (2004) Cellular and subcellular localization of endogenous nitric oxide in young and senescent pea plants. Plant Physiol 136:2722–2733

Corpas FJ, Chaki M, Fernández-Ocaña A, Valderrama R, Palma JM, Carreras A, Begara-Morales JC, Airaki M, del Río LA, Barroso JB (2008) Metabolism of reactive nitrogen species in pea plants under abiotic stress conditions. Plant Cell Physiol 49:1711–1722

Corpas FJ, Chaki M, Leterrier M, Barroso JB (2009a) Protein tyrosine nitration: a new challenge in plants. Plant Signal Behav 4:920–923

Corpas FJ, Hayashi M, Mano S, Nishimura M, Barroso JB (2009b) Peroxisomes are required for in vivo nitric oxide accumulation in the cytosol following salinity stress of Arabidopsis plants. Plant Physiol 151(4):2083–2094

Corpas FJ, Palma JM, Leterrier M, del Río LA, Barroso JB (2010) Nitric oxide and abiotic stress in higher plants. In: Hayat S, Mori M, Pichtel J, Ahmad A (eds) Nitric oxide in plant physiology. Wiley-VCH, Germany, pp 51–63. ISBN 978-3-527-32519-1

Corpas FJ, Leterrier M, Valderrama R, Airaki M, Chaki M, Palma JM, Barroso JB (2011) Nitric oxide imbalance provokes a nitrosative response in plants under abiotic stress. Plant Sci 181:604–611

Davies PJ (1995) The plant hormone concept: concentration, sensitivity and transport. In: Davies PJ (ed) Plant hormones: physiology, biochemistry and molecular biology. Kluwer, Dordrecht, pp 13–18

Davies W, Zhang J (1991) Root signals and the regulation of growth and development of plants in drying soil. Annu Rev Plant Physiol Plant Mol Biol 42:55–76

Desikan R, Cheung MK, Bright J, Henson D, Hancock JT, Neill SJ (2004) ABA, hydrogen peroxide and nitric oxide signalling in stomatal guard cells. J Exp Bot 55:205–212

Dubovskaya LV, Bakakina YS, Kolesneva EV, Sodel DL, McAinsh MR, Hetherington AM, Volotovski ID (2011) cGMP-dependent ABA-induced stomatal closure in the ABA-insensitive Arabidopsis mutant abi1-1. New Phytol 191(1):57–69

Durner J, Klessig DF (1999) Nitric oxide as a signal in plants. Curr Opin Plant Biol 2:369–374

Dwivedi S, Tripathi RD, Tripathi P, Kumar A, Dave R, Mishra S, Singh R, Sharma D, Rai UN, Chakrabarty D, Trivedi PK, Adhikari B, Bag MK, Dhankher OP, Tuli R (2010) Arsenate exposure affects amino acids, mineral nutrient status and antioxidants in rice (*Oryza sativa* L.) genotypes. Environ Sci Technol 44:9542–9549

Ederli L, Morettini R, Borgogni A, Wasternack C, Miersch O, Reale L, Ferranti F, Tosti N, Pasqualini S (2006) Interaction between nitric oxide and ethylene in the induction of alternative oxidase in ozone-treated tobacco plants. Plant Physiol 142:595–608

García MJ, Lucena C, Romera FJ, Alcántara E, Pérez-Vicente R (2010) Ethylene and nitric oxide involvement in the up-regulation of key genes related to iron acquisition and homeostasis in Arabidopsis. J Exp Bot 61:3885–3899

García MJ, Suárez V, Romera FJ, Alcántara E, Pérez-Vicente R (2011) A new model involving ethylene, nitric oxide and Fe to explain the regulation of Fe-acquisition genes in strategy I plants. Plant Physiol Biochem 49:537–544

Garcia-Mata C, Lamattina L (2002) Nitric oxide and abscisic acid cross talk in guard cells. Plant Physiol 128:790–792

Gémes K, Poór P, Horváth E, Kolbert Z, Szopkó D, Szepesi A, Tari I (2011) Cross-talk between salicylic acid and NaCl-generated reactive oxygen species and nitric oxide in tomato during acclimation to high salinity. Physiol Plant 142(2):179–192

He JM, Zhang Z, Wang RB, Chen YP (2011b) UV-B-induced stomatal closure occurs via ethylene-dependent NO generation in *Vicia faba*. Funct Plant Biol 38:293–302

Horvath E, Szalai G, Janda T (2007) Induction of abiotic stress tolerance by salicylic acid signaling. J Plant Growth Regul 26:290–300

Hsu YT, Kao CH (2004) Cadmium toxicity is reduced by nitric oxide in rice leaves. J Plant Growth Regul 42:227–238

Huang X, Stettmaier K, Michel C, Hutzler P, Mueller MJ, Durner J (2004) Nitric oxide is induced by wounding and influences jasmonic acid signaling in *Arabidopsis thaliana*. Planta 218:938–946

Illés P, Schlicht M, Pavlovkin J, Lichtscheidl I, Baluska F, Ovecka M (2006) Aluminium toxicity in plants: internalization of aluminium into cells of the transition zone in Arabidopsis root apices related to changes in plasma membrane potential, endosomal behaviour, and nitric oxide production. J Exp Bot 57:4201–4213

Jin JW, Xu YF, Huang YF (2010) Protective effect of nitric oxide against arsenic-induced oxidative damage in tall fescue leaves. Afr J Biotechnol 9:1619–1627

Kazemi N, Khavari-Nejad RA, Fahimi H, Saadatmand S, Nejad-Sattari T (2010) Effects of exogenous salicylic acid and nitric oxide on lipid peroxidation and antioxidant enzyme activities in leaves of *Brassica napus* L. under nickel stress. Sci Hortic 126:402–407

Khokon AR, Okuma E, Hossain MA, Munemasa S, Uraji M, Nakamura Y, Mori IC, Murata Y (2011) Involvement of extracellular oxidative burst in salicylic acid-induced stomatal closure in Arabidopsis. Plant Cell Environ 34:434–443

Kopyra M, Gwóźdź EA (2003) Nitric oxide stimulates seed germination and counteracts the inhibitory effect of heavy metals and salinity on root growth of *Lupinus luteus*. Plant Physiol Biochem 41:1011–1017

Laspina NV, Groppa MD, Tomaro ML, Benavides MP (2005) Nitric oxide protects sunflower leaves against Cd-induced oxidative stress. Plant Sci 169:323–330

Lamattina L, Garcia-Mata C, Graziano M, Pagnussat G (2003) Nitric oxide: the versatility of an extensive signal molecule. Annu Rev Plant Biol 54:109–136

Leterrier M, Airaki M, Barroso JB, Palma JM, del Río LA, Corpas FJ (2010) Arsenic impairs the metabolism of RNS and ROS in Arabidopsis plant. In: international symposium on the pathophysiology of reactive oxygen and nitrogen species, Salamanca, Spain, p 220 (ISBN: 978-84-692-9284-6)

Leterrier M, Chaki M, Airaki M, Valderrama R, Palma JM, Barroso JB, Corpas FJ (2011) Function of *S*-nitrosoglutathione reductase (GSNOR) in plant development and under biotic/abiotic stress. Plant Signal Behav 6:789–793

Lindermayr C, Durner J (2009) *S*-Nitrosylation in plants: pattern and function. J Proteomics 73(1): 1–9

Lindermayr C, Saalbach G, Durner J (2005) Proteomic identification of *S*-nitrosylated proteins in Arabidopsis. Plant Physiol 137(3):921–930

Lu S, Su W, Li H, Guo Z (2009) Abscisic acid improves drought tolerance of triploid bermudagrass and involves H_2O_2- and NO-induced antioxidant enzyme activities. Plant Physiol Biochem 47(2):132–138

Ma JF, Ryan PR, Delhaize E (2001) Aluminium tolerance in plants and the complexing role of organic acids. Trends Plant Sci 6:273–278

Mackerness SAH, Surplus SL, Blake P, John CF, Buchanan-Wollaston V, Jordan BR, Thomas B (1999) Ultraviolet-B-induced stress and changes in gene expression in *Arabidopsis thaliana*: role of signalling pathways controlled by jasmonic acid, ethylene and reactive oxygen species. Plant Cell Environ 22:1413–1423

Magalhaes JR, Singh RN, Passos LP (2005) Nitric oxide signaling in higher plants. Studium Press LLC, Houston, pp 1–347

Manthe B, Schulz M, Schnabl H (1992) Effects of salicylic acid on growth and stomatal movements of *Vicia faba* L: evidence for salicylic acid metabolization. J Chem Ecol 18:1525–1539

Metwally A, Finkemeier I, Georgi M, Dietz KJ (2003) Salicylic acid alleviates the cadmium toxicity in barley seedlings. Plant Physiol 132:272–281

Millar AH, Day DA (1996) Nitric oxide inhibits the cytochrome oxidase but not the alternative oxidase of plant mitochondria. FEBS Lett 398:155–158

Munns R, Tester M (2008) Mechanisms of salinity tolerance. Annu Rev Plant Biol 59:651–681

Orozco-Cardenas ML, Ryan CA (2002) Nitric oxide negatively modulates wound signaling in tomato plants. Plant Physiol 130:487–493

Panda P, Nath S, Chanu TT, Sharma GD, Panda SK (2011) Cadmium stress-induced oxidative stress and role of nitric oxide in rice (*Oryza sativa* L.). Acta Physiol Plant 33:1737–1747

Rao MV, Davis KR (2001) The physiology of ozone induced cell death. Planta 213:682–690

Rascio N, Navari-Izzo F (2011) Heavy metal hyperaccumulating plants: how and why do they do it? And what makes them so interesting? Plant Sci 180:169–181

Rodríguez-Serrano M, Romero-Puertas MC, Zabalza A, Corpas FJ, Gómez M, del Río LA, Sandalio LM (2006) Cadmium effect on the oxidative metabolism of pea (*Pisum sativum* L.) roots. Imaging of ROS and NO accumulation in vivo. Plant Cell Environ 29:1532–1544

Rodríguez-Serrano M, Romero-Puertas MC, Pazmiño DM, Testillano PS, Risueño MC, del Río LA, Sandalio LM (2009) Cellular response of pea plants to cadmium toxicity: cross talk between reactive oxygen species, nitric oxide, and calcium. Plant Physiol 150:229–243

Romera FJ, Alcantara E (1994) Iron deficiency stress response in cucumber (*Cucumis sativus* L) roots: a possible role for ethylene. Plant Physiol 105:1133–1138

Romera FJ, Alcantara E, de la Guardia M (1999) Ethylene production by Fe-deficient roots and its involvement in the regulation of Fe-deficiency stress responses by strategy I plants. Ann Bot 83(1):51–55

Romero-Puertas MC, Rodríguez-Serrano M, Corpas FJ, Gómez M, del Río LA, Sandalio LM (2004) Cadmium-induced subcellular accumulation of O_2^- and H_2O_2 in pea leaves. Plant Cell Environ 27:1122–1134

Sanitá di Toppi L, Gabbrielli R (1999) Response to cadmium in higher plants. Environ Exp Bot 41:105–130

Schilmiller AL, Howe GA (2005) Systemic signaling in the wound response. Curr Opin Plant Biol 8:369–377

Shapiro AD (2005) Nitric oxide signaling in plants. Vitam Horm 72:339–398

Shi SY, Wang G, Wang YD, Zhang LG, Zhang LX (2005) Protective effect of nitric oxide against oxidative stress under ultraviolet-B radiation. Nitric Oxide-Biol Chem 13:1–9

Singh HP, Kaur S, Batish DR, Sharma VP, Sharma N, Kohli RK (2009) Nitric oxide alleviates arsenic toxicity by reducing oxidative damage in the roots of *Oryza sativa* (rice). Nitric Oxide 20(4):289–297

Stamler JS, Lamas S, Fang FC (2001) Nitrosylation the prototypic redox based signaling mechanism. Cell 106:675–683

Stratmann JW (2003) Long distance run in the wound response-jasmonic acid is pulling ahead. Trends Plant Sci 8:247–250

Szabó C, Ischiropoulos H, Radi R (2007) Peroxynitrite: biochemistry, pathophysiology and development of therapeutics. Nat Rev Drug Discov 6(8):662–680

Szepesi A, Csiszár J, Gémes K, Horváth E, Horváth F, Simon ML, Tari I (2009) Salicylic acid improves acclimation to salt stress by stimulating abscisic aldehyde oxidase activity and abscisic acid accumulation, and increases Na^+ content in leaves without toxicity symptoms in *Solanum lycopersicum* L. J Plant Physiol 166:914–925

Tanou G, Molassiotis A, Diamantidis G (2009) Hydrogen peroxide- and nitric oxide-induced systemic antioxidant prime-like activity under NaCl-stress and stress-free conditions in citrus plants. J Plant Physiol 166:1904–1913

Tian QY, Sun DH, Zhao MG, Zhang WH (2007) Inhibition of nitric oxide synthase (NOS) underlies aluminum-induced inhibition of root elongation in *Hibiscus moscheutos*. New Phytol 174:322–331
Tossi V, Lamattina L, Cassia R (2009) An increase in the concentration of abscisic acid is critical for nitric oxide-mediated plant adaptive responses to UV-B irradiation. New Phytol 181:871–879
Tripathi RD, Srivastava S, Mishra S, Singh N, Tuli R, Gupta DK, Maathuis FJ (2007) Arsenic hazards: strategies for tolerance and remediation by plants. Trends Biotechnol 25(4):158–165
Uchida A, Jagendorf AT, Hibino T, Takabe T (2002) Effects of hydrogen peroxide and nitric oxide on both salt and heat stress tolerance in rice. Plant Sci 163:515–523
Vahala J, Ruonala R, Keinänen M, Tuominen H, Kangasjärvi J (2003) Ethylene insensitivity modulates ozone-induced cell death in birch. Plant Physiol 132(1):185–195
Valderrama R, Corpas FJ, Carreras A, Fernández-Ocaña A, Chaki M, Luque F, Gómez-Rodríguez MV, Colmenero-Varea P, del Río LA, Barroso JB (2007) Nitrosative stress in plants. FEBS Lett 581:453–461
Verbruggen N, Hermans C, Schat H (2009) Mechanisms to cope with arsenic or cadmium excess in plants. Curr Opin Plant Biol 12:364–372
Wang YS, Yang ZM (2005) Nitric oxide reduces aluminum toxicity by preventing oxidative stress in the roots of *Cassia tora* L. Plant Cell Physiol 46:1915–1923
Wang Y, Yun BW, Kwon E, Hong JK, Yoon J, Loake GJ (2006) *S*-nitrosylation: an emerging redox-based post-translational modification in plants. J Exp Bot 57:1777–1784
Wang H, Liang X, Wan Q, Wang X, Bi Y (2009) Ethylene and nitric oxide are involved in maintaining ion homeostasis in Arabidopsis callus under salt stress. Planta 230(2):293–307
Wang H, Huang J, Bi Y (2010a) Induction of alternative respiratory pathway involves nitric oxide, hydrogen peroxide and ethylene under salt stress. Plant Signal Behav 5:1636–1637
Wang H, Liang X, Huang J, Zhang D, Lu H, Liu Z, Bi Y (2010b) Involvement of ethylene and hydrogen peroxide in induction of alternative respiratory pathway in salt-treated Arabidopsis calluses. Plant Cell Physiol 51(10):1754–1765
Wojtaszek P (2000) Nitric oxide in plants: to NO or not to NO. Phytochemistry 54:1–4
Xiong J, An L, Lu H, Zhu C (2009) Exogenous nitric oxide enhances cadmium tolerance of rice by increasing pectin and hemicellulose contents in root cell wall. Planta 230:755–765
Xiong J, Fu G, Tao L, Zhu C (2010) Roles of nitric oxide in alleviating heavy metal toxicity in plants. Arch Biochem Biophys 497:13–20
Zhang YY, Wang LL, Liu YL, Zhang Q, Wei QP, Zhang WH (2006) Nitric oxide enhances salt tolerance in maize seedlings through increasing activities of proton-pump and Na^+/H^+ antiport in the tonoplast. Planta 224:545–555
Zhang A, Jiang M, Zhang J, Ding H, Xu S, Hu X, Tan M (2007) Nitric oxide induced by hydrogen peroxide mediates abscisic acid-induced activation of the mitogen-activated protein kinase cascade involved in antioxidant defense in maize leaves. New Phytol 175:36–50
Zhang Z, Wang H, Wang X, Bi Y (2011a) Nitric oxide enhances aluminium tolerance by affecting cell wall polysaccharides in rice roots. Plant Cell Rep 30(9):1701–1711
Zhang A, Zhang J, Zhang J, Ye N, Zhang H, Tan M, Jiang M (2011b) Nitric oxide mediates brassinosteroid-induced ABA biosynthesis involved in oxidative stress tolerance in maize leaves. Plant Cell Physiol 52:181–192
Zhao LQ, Zhang F, Guo JK, Yang YL, Li BB, Zhang LX (2004) Nitric oxide functions as a signal in salt resistance in the calluses from two ecotypes of reed. Plant Physiol 134:849–857
Zhao FJ, Ma JF, Meharg AA, McGrath SP (2009) Arsenic uptake and metabolism in plants. New Phytol 181:777–794
Zhao FJ, McGrath SP, Meharg AA (2010) Arsenic as a food chain contaminant: mechanisms of plant uptake and metabolism and mitigation strategies. Annu Rev Plant Biol 61:535–559
Zhu JK (2002) Salt and drought stress signal transduction in plants. Annu Rev Plant Biol 53:247–273
Zottini M, Costa A, De Michele R, Ruzzene M, Carimi F, Lo Schiavo F (2007) Salicylic acid activates nitric oxide synthesis in Arabidopsis. J Exp Bot 58:1397–1405

Chapter 5
Auxin as Part of the Wounding Response in Plants

Claudia A. Casalongué, Diego F. Fiol, Ramiro París, Andrea V. Godoy, Sebastián D'Ippólito, and María C. Terrile

Abstract In plants, different types of injury and physical damage are commonly referred as wounding. Some organs such as leaves and shoots have cutin as a protective barrier, but once a wound occurs, putative pathogens may gain entrance into the plant through the injured tissue. Consequently, plants have developed orchestrated responses to wounding at the histological, genetic, and biochemical levels resulting in a complex defense mechanism. Therefore, the response to wounding is aimed at restoring the physiological status of the damaged tissue and is critical to prevent further lesions.

Interestingly, the classical growth regulator auxin has been implicated in the wounding response. Even though initial reports showed an apparent antagonism between auxin and wounding, novel findings suggest a more intricate relationship between auxin, stress, and other plant defense pathways. Transcriptomic studies carried out in *Arabidopsis* and solanaceous have offered a wider comprehensive picture on the regulation of auxin-related genes by wounding.

In this chapter, we reviewed the participation of auxin-related genes as part of the complex mechanism that takes place during wounding in plants particularly in *Arabidopsis thaliana* and solanaceous. In addition, we also raised a discussion, about the participation of small molecules downstream wound signal such as NO, ROS, and eATP.

C.A. Casalongué (✉) • D.F. Fiol • R. París • A.V. Godoy • S. D'Ippólito • M.C. Terrile
Facultad de Ciencias Exactas y Naturales, Instituto de Investigaciones Biológicas, UE-CONICET-UNMDP, Universidad Nacional de Mar del Plata, Funes 3250, 7600 Mar del Plata, Argentina
e-mail: casalong@mdp.edu.ar

N.A. Khan et al. (eds.), *Phytohormones and Abiotic Stress Tolerance in Plants*,
DOI 10.1007/978-3-642-25829-9_5,

5.1 How Do Plants Integrate Wounding and Auxin Signals?

Wounding commonly refers to different types of injury and physical damage caused by herbivory, wind, hail and other environmental stresses. The plant response to wounding has been studied since the early 1970s when Green and Ryan (1972) discovered that a chymotrypsin inhibitor (Pin2) was accumulated in tomato leaves in response to wounding. Since chymotrypsin-like proteins are common in insect digestive tracts, the expression of this protein was considered as part of a defense system. To this date, a very high number of genes and proteins have been identified as being wound inducible giving a more comprehensive view of the functional response to such mechanical injury. The types of genes and their activation timings have led to the recognition of different phases, including a mechanical barrier formation, wounded tissue sealing and the defensive compound activation. Thus, all these events are orchestrated to avoid pathogen invasion and to keep metabolic integrity at the wounded site. Hence, wound healing is aimed at restoring the physiological status of damaged tissue and is critical to prevent further lesions.

In plants, it had been reported the activation of several signal transduction pathways upon mechanical stress. For example, jasmonic acid (JA), an endogenous chemical produced in wounded tissue has been proposed as a long-distance signal of wounding (Glauser et al. 2009). The phytohormones ethylene, salicylic acid (SA), and abscisic acid (ABA) were all largely associated to the induction of plant gene expression upon wounding (Hildmann et al. 1992; O'Donnell et al. 1996; Pena-Cortes et al. 1991). More recently, auxin widely recognized as a key growth regulator hormone has also emerged as a player in the defense response against biotic and abiotic stresses (Iglesias et al. 2010; Kazan and Manners 2009; Park et al. 2007).

Briefly, we intended to provide a discussion about the participation of auxin and other small related molecules such as nitric oxide (NO), reactive oxygen species (ROS), and extracellular ATP (eATP) associated with wounding, rather to present a comprehensive compilation of the extensive literature on wounding responses.

5.2 Auxin-Related Gene Expression Is Modulated by Wounding

Many studies showed that exogenous auxin application can downregulate a number of wound-induced responses, including the well-characterized chymotrypsin inhibitor, Pin2 (Kernan and Thornburg 1989), and two methyl jasmonate-induced genes such as S-*adenosylmethionine synthase* and *putrescine* N-*methyltransferase* (Imanishi et al. 1998a, b). More recently, a transcriptional analysis has suggested an antagonism between wounding and auxin responses (Cheong et al. 2002).

Table 5.1 Summary of auxin-related genes regulated in *A. thaliana* and *S. tuberosum* by wounding

Description	Accession no.	Regulation	Time after wounding	References
Auxin biosynthesis				
Nitrilase (A.t.)	n.i	Up	n.i	Cheong et al. (2002)
Auxin homeostasis				
IAA glucosyltranferase (A.t.)	n.i	Up	n.i	Cheong et al. (2002)
IAA-Ala hydrolase (IAR3) (S.t.)	GT888414	Up	24 h	Unpublished data
Regulation of auxin signaling				
Similar to NPK1 protein kinase (A.t.)	At2g30040	Up	0.5 h	Cheong et al. (2002)
Auxin-responsive genes				
Auxin-responsive GH3-like protein (JAR1) (A.t.)	At2g46370	Up	6 h	Cheong et al. (2002)
StIAA1 protein (S.t.)	AY098938	Up	8 h	Zanetti et al. (2003)
SAUR-like auxin-responsive protein (A.t.)	At2g45210	Up	6 h	Cheong et al. (2002)
Similar to auxin-induced protein SAUR-AC1 (S.t.)	GT888928	Up	24 h	Unpublished data
Auxin-regulated protein (S.t.)	GT888976	Up	24 h	Unpublished data
Putative GH3-like protein (DFL2) (A.t.)	At4g03400	Down	0.5 h	Cheong et al. (2002)
Auxin-responsive protein (IAA2) (A.t.)	At3g23030	Down	6 h	Cheong et al. (2002)
Indole-3-acetic acid inducible 3 (IAA3) (A.t.)	At1g04240	Down	6 h	Cheong et al. (2002)
Auxin-induced protein (IAA20) (A.t.)	At2g46990	Down	6 h	Cheong et al. (2002)
SAUR-AC1 (A.t.)	At4g38850	Down	0.5 h	Cheong et al. (2002)
SAUR-like auxin-responsive protein (A.t.)	At2g46690	Down	0.5 h	Cheong et al. (2002)
SAUR-like auxin-responsive protein (A.t.)	At4g38840	Down	0.5 h	Cheong et al. (2002)
SAUR-like auxin-responsive protein (A.t.)	At4g38860	Down	0.5 h	Cheong et al. (2002)
SAUR-like auxin-responsive protein (A.t.)	At4g00880	Down	6 h	Cheong et al. (2002)
Putative auxin-regulated protein (A.t.)	At2g21210	Down	6 h	Cheong et al. (2002)

n.i not identified, *A.t. A. thaliana*, *S.t. S. tuberosum*

The authors found that wounding downregulates a number of genes positively associated with the response to auxin, including synthesis, availability, signaling and transport (Table 5.1). Among them, a nitrilase gene which is involved in auxin biosynthesis, a number of transcripts encoding auxin-responsive genes, such as *IAA2*, *IAA3*, *IAA20*, and some other transcripts encoding putative auxin-induced proteins are downregulated by wounding. In addition, two genes encoding IAA glucosyltransferases, which reduce endogenous active IAA levels through the formation of IAA–sugar conjugates are also highly induced by wounding (Cheong et al. 2002). Interestingly, in a study aimed to identify genes upregulated upon wounding and *Fusarium solani* f. sp. *eumartii* infection in potato tubers, we identified several auxin-related genes (Godoy et al. 2000)

Table 5.1 shows an IAA-amino acid hydrolase (*IAR3*), an Aux/IAA protein (*StIAA1*), a gene belonging to SAUR family, and an auxin-regulated protein upregulated in wounded potato tubers.

Apparently, other aspect of wounding and auxin regulation is articulated through a negative crosstalk of the hormone signaling pathway. This hypothesis may be demonstrated by the upregulation of *NPK1*-like gene, which is proposed to negatively regulates IAA-responsive genes (Kovtun et al. 1998; Cheong et al. 2002). Thus, this crosstalk between wounding and auxin was also evidenced by the study of a wound-induced protein kinase (WIPK) from tobacco plants (Chung and Sano 2007). *WIPK* is a MAPK which plays a central role in stress responses being activated within a few minutes after wounding (Seo et al. 1995). Then, WIPK itself activates a transcription factor, NtWIF homolog to the auxin-responsive factors ARF1 and ARF9, which is able to recognize and activate genes containing the auxin-responsive element (TGTCTC) in their promoter regions (Yap et al. 2005).

Taking advantage of the gene expression databases and tools available at Genevestigator (http://www.genevestigator.com) (Zimmermann et al. 2004), we analyzed the expression of the full amount of genes interlinked with auxin regulation and wounding. A total of 181 transcripts with available probesets were examined, including the auxin-responsive gene families *Aux/IAA, SAUR,* and *GH3* and other gene involved in auxin synthesis, auxin-amino acid conjugation and hydrolysis, transport and sensing. From this set, 22 transcripts resulted upregulated and 10 downregulated by wounding in experiments carried out either in green tissues or in root samples. Interestingly, Fig. 5.1 shows that induced and repressed transcripts include members of ARF, Aux/IAA, GH3, and SAUR families. On the other hand, other families were only represented in the set of upregulated genes, as IAA-amido synthases and IAA-amino acid conjugate hydrolases. In addition, genes identified as wounding responsive may be modulated in diverse ways by auxin. This particular fact occurs even into the same gene family, describing a complex picture for the relationships between auxin and wounding (Fig. 5.1). In general, many components of the auxin regulation network belong to multiprotein families, opening the possibility for a combinatorial interaction with their coregulator partners, e.g. ARFs and Aux/IAAs or Aux/IAAs and TIR1/AFBs. Hence, taking into account all these facts and the putative crosstalk with other different hormone signaling pathways, the analysis of auxin-related genes upon wounding seems to be indeed very complex.

In addition, evidences on several auxin-related genes that are regulated by wounding and a variety of stresses suggest that their gene products may be part of a general stress regulator node and/or may influence the action of detoxifying or repairing enzymes in damaged tissue (Pickett and Lu 1989). In conclusion, the

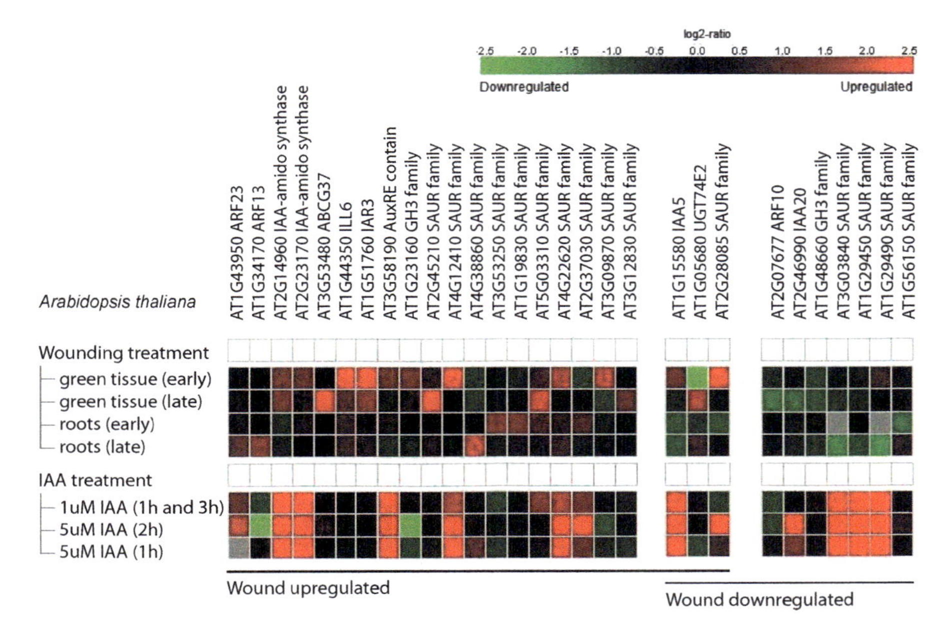

Fig. 5.1 Analysis of auxin-related genes. Transcripts up- or downregulated by wounding treatment were included in the figure. A threshold of twofold induction/repression was set. Experiments included the following: for wounding treatments, AT-00120 (early 0.5, 1, and 3 h; late 6, 12, and 24 h after treatment) and for auxin treatments, AT-00110, AT-00164, and AT-00167. ABCG37 is an auxin transporter; UGT74E2 is a UDP-glucosyltransferase (http://www.genevestigator.com)

molecular basis for such versatility seems to be the highly complex nature of the networks regulating auxin metabolism, transport, and response (Del Bianco and Kepinski 2011).

5.3 Auxin Involvement on Wounding Response Is Spatially Limited

As it is expected, the metabolic response to wounding is also spatially modulated. An early report showed that apically applied auxin stimulated xylem regeneration in okra and pea wounded areas. However, basal application did not have effect, suggesting that auxin transport is important for tissue regeneration upon wounding (Thompson 1970). Similarly, the polar movement of IAA in the cut surface of sweet potato roots was described by Tanaka and Uritani (1979). In support of these old evidences, new reports advert that many aspects of plant development, such as organogenesis or tissue regeneration, require the coordinated polarization of individual cells, which at the end leads to the establishment of a new axis of polarity for the whole tissue or organ. Fabbri et al. (2000) showed an increase of the mitotic activity limited to procambium and the cells of the axillary bud domes upon wounding. It has been assumed that the coordination between polarizing cells occurs via a feedback mechanism by which auxin influences the directionality of its own transport (Sauer et al. 2006).

Therefore, we propose that several regulatory networks and nodes in which auxin takes place may operate during wounding and wound-healing response in plant tissues. Indeed, this fact may depend on plant genetic background, plant organ or tissue, combinatorial environmental conditions, and time window upon wounding.

5.4 Other Downstream Signaling Molecules in the Wound Signal Transduction Pathway

Among the earliest responses associated to wounding, it is noteworthy to mention a transient NO burst. NO production increases from minutes to a few hours after wounding, returning to basal levels afterwards (Arasimowicz et al. 2009; Garcês et al. 2001; Huang et al. 2004; Paris et al. 2007). The source of NO at the wounded site has not been defined yet. However, NO production can be attributed to nonenzymatic or enzymatic sources (Chaki et al. 2011; Cooney et al. 1994). For this former, the conversion of nitrate by nitrate reductase (Rockel et al. 2002; Yamasaki et al. 1999) or the oxidation of L-arginine by NO synthase as in mammals was described (Corpas et al. 2004, 2008a). Certainly, the NO production is accompanied by a ROS burst, and both, together, could act as modulators of the cell-redox status (Beligni and Lamattina 1999; Corpas et al. 2008b; de Pinto et al. 2002), and a proper cell-redox balance is of great importance to establish an efficient wound-healing response in plants (Arasimowicz et al. 2009; Chaki et al. 2011; Corpas et al. 2008b). The action of NO after wounding could follow diverse paths, among them we summarize the following: (1) hormone-like signaling through second messengers like Ca^{2+} and cyclic GMP, MAPK, and phospholipase D (Lanteri et al. 2008; Pagnussat et al. 2004); (2) specific posttranslational protein modifications as metal nitrosylation of metalloproteins, S-nitrosylation, and nitration (reviewed by Besson-Bard et al. 2008); (3) gene expression regulation, as demonstrated by high-throughput expression experiments (Parani et al. 2004; Polverari et al. 2003); and (4) crosstalk with hormones and endogenous signals (Lamattina et al. 2003; Wendehenne et al. 2004). Initial evidences on NO action indicated that it might share common steps with auxin because both signals could elicit a similar physiological response (Gouvea et al. 1997). More recently, NO dependence was demonstrated for several auxin-associated processes such as adventitious root promotion (Pagnussat et al. 2002), lateral root formation (Correa-Aragunde et al. 2004), and root hair development (Lombardo et al. 2006). Despite a positive association between NO and auxin, in tobacco, the endogenous levels of IAA decline two- to threefold within 6 h after wounding (Kernan and Thornburg 1989; Thornburg and Li 1991). However, we found that endogenous IAA did not drastically change when we analyzed IAA level after 24 h of wounding in potato tubers (Terrile et al. 2006). Our interpretation is that at earlier times after wounding, when transient accumulation of ROS and NO is maximal, auxin level may decrease (Orozco-Cardenas and Ryan 1999). Alternatively, when ROS and NO levels decline, the auxin homeostasis is gradually reestablished. Moreover, the recovery of the initial levels of active auxin was proposed as a mechanism to limit the duration of the response to wounding (León et al. 2001). In addition, auxin and NO were demonstrated to be transiently modulated after wounding and eventually to take part during the adequacy of wound-healing response in plants (Arasimowicz et al. 2009; Paris et al. 2007).

Thereafter, as a consequence of wounding, NO could exert its action throughout any of the above mentioned mechanisms. Otherwise, we cannot discard that other unknown NO-mediated mechanism may operate upon wounding in plant tissues. Recently, eATP was described as a wound signal. Extracellular ATP mediates H_2O_2 accumulation (Jeter et al. 2004; Song et al. 2006) and NO production (Foresi et al. 2007; Sueldo et al. 2010). Moreover, eATP was described as an inhibitor of auxin transport in roots (Tang et al. 2003). In addition, we have demonstrated that eATP-mediated NO production counteracts the redox imbalance in etiolated seedlings (Tonón et al. 2010). All these findings allowed us to hypothesize that NO may control cell-redox status, as a signal itself and/or as a second messenger of signal transduction pathways that operate during development and plant defense against stress. However, it still remains elusive how eATP, NO, and ROS interplay with auxin to coordinate plant development and wound signal transduction pathway.

5.5 Conclusions

In summary, we have a glimpse of the components that interplay in the auxin and wounding pathways. Future understandings of the still unraveled network involving wounding, auxin, and the small related molecules such as NO, ROS, and eATP will shed light to new and interesting questions. The postgenomic era and new technologies will offer decisive progress in this subject. Thus, it is an exciting time for hormone research and its crosstalk with other signaling in plants.

References

Arasimowicz M, Floryszak-Wieczorek J, Milczarek G, Jelonek T (2009) Nitric oxide, induced by wounding, mediates redox regulation in pelargonium leaves. Plant Biol (Stuttg) 11:650–663
Beligni M, Lamattina L (1999) Is nitric oxide toxic or protective? Trends Plant Sci 4:299–300
Besson-Bard A, Pugin A, Wendehenne D (2008) New insights into nitric oxide signaling in plants. Annu Rev Plant Biol 59:21–39

Chaki M, Valderrama R, Fernandez-Ocana AM, Carreras A, Gomez-Rodriguez MV, Pedrajas JR, Begara-Morales JC, Sanchez-Calvo B, Luque F, Leterrier M, Corpas FJ, Barroso JB (2011) Mechanical wounding induces a nitrosative stress by down-regulation of GSNO reductase and an increase in S-nitrosothiols in sunflower (*Helianthus annuus*) seedlings. J Exp Bot 62:1803–1813

Cheong YH, Chang HS, Gupta R, Wang X, Zhu T, Luan S (2002) Transcriptional profiling reveals novel interactions between wounding, pathogen, abiotic stress, and hormonal responses in Arabidopsis. Plant Physiol 129:661–677

Chung KM, Sano H (2007) Transactivation of wound-responsive genes containing the core sequence of the auxin-responsive element by a wound-induced protein kinase-activated transcription factor in tobacco plants. Plant Mol Biol 65:763–773

Cooney RV, Harwood PJ, Custer LJ, Franke AA (1994) Light-mediated conversion of nitrogen dioxide to nitric oxide by carotenoids. Environ Health Perspect 102:460–462

Corpas FJ, Barroso JB, Del Río LA (2004) Enzymatic sources of nitric oxide in plant cells—beyond one protein-one function. New Phytol 162:246–248

Corpas FJ, Carreras A, Esteban FJ, Chaki M, Valderrama R, del Rio LA, Barroso JB (2008a) Localization of S-nitrosothiols and assay of nitric oxide synthase and S-nitrosoglutathione reductase activity in plants. Methods Enzymol 437:561–574

Corpas FJ, Chaki M, Fernandez-Ocana A, Valderrama R, Palma JM, Carreras A, Begara-Morales JC, Airaki M, del Rio LA, Barroso JB (2008b) Metabolism of reactive nitrogen species in pea plants under abiotic stress conditions. Plant Cell Physiol 49:1711–1722

Correa-Aragunde N, Graziano M, Lamattina L (2004) Nitric oxide plays a central role in determining lateral root development in tomato. Planta 218:900–905

de Pinto MC, Tommasi F, De Gara L (2002) Changes in the antioxidant systems as part of the signaling pathway responsible for the programmed cell death activated by nitric oxide and reactive oxygen species in tobacco Bright-Yellow 2 cells. Plant Physiol 130:698–708

Del Bianco M, Kepinski S (2011) Context, specificity, and self-organization in auxin response. Cold Spring Harb Perspect Biol 3:a001578

Fabbri AA, Fanelli C, Reverberi M, Ricelli A, Camera E, Urbanelli S, Rossini A, Picardo M, Altamura MM (2000) Early physiological and cytological events induced by wounding in potato tuber. J Exp Bot 51:1267–1275

Foresi N, Laxalt AM, Tonón C, Casalongue C, Lamattina L (2007) Extracellular ATP induces nitric oxide production in tomato cell suspensions. Plant Physiol 145:589–592

Garcês H, Durzan D, Pedroso MC (2001) Mechanical stress elicits nitric oxide formation and DNA fragmentation in *Arabidopsis thaliana*. Ann Bot 87:567–574

Glauser G, Dubugnon L, Mousavi SAR, Serge Rudaz S, Wolfender J-L, Farmer EE (2009) Velocity estimates for signal propagation leading to systemic jasmonic acid accumulation in wounded *Arabidopsis*. J Biol Chem 284:34506–34513

Godoy AV, Lazzaro AS, Casalongue CA, San Segundo B (2000) Expression of a Solanum tuberosum cyclophilin gene is regulated by fungal infection and abiotic stress conditions. Plant Sci 152:123–134

Gouvea CMCP, Souza JF, Magalhaes ACN, Martins IS (1997) NO-releasing substances that induce growth elongation in maize root segments. Plant Growth Regul 21:183–187

Green TR, Ryan CA (1972) Wound-induced proteinase inhibitor in plant leaves: a possible defense mechanism against insects. Science 175:776–777

Hildmann T, Ebneth M, Pena-Cortes H, Sanchez-Serrano JJ, Willmitzer L, Prat S (1992) General roles of abscisic and jasmonic acids in gene activation as a result of mechanical wounding. Plant Cell 4:1157–1170

Huang X, Stettmaier K, Michel C, Hutzler P, Mueller M, Durner J (2004) Nitric oxide is induced by wounding and influences jasmonic acid signaling in *Arabidopsis thaliana*. Planta 218:938–946

Iglesias MJ, Terrile MC, Bartoli CG, D'Ippolito S, Casalongue CA (2010) Auxin signaling participates in the adaptative response against oxidative stress and salinity by interacting with redox metabolism in Arabidopsis. Plant Mol Biol 74:215–222

Imanishi S, Hashizume K, Kojima H, Ichihara A, Nakamura K (1998a) An mRNA of tobacco cell, which is rapidly inducible by methyl jasmonate in the presence of cycloheximide, codes for a putative glycosyltransferase. Plant Cell Physiol 39:202–211

Imanishi S, Hashizume K, Nakakita M, Kojima H, Matsubayashi Y, Hashimoto T, Sakagami Y, Yamada Y, Nakamura K (1998b) Differential induction by methyl jasmonate of genes encoding ornithine decarboxylase and other enzymes involved in nicotine biosynthesis in tobacco cell cultures. Plant Mol Biol 38:1101–1111

Jeter CR, Tang W, Henaff E, Butterfield T, Roux SJ (2004) Evidence of a novel cell signaling role for extracellular adenosine triphosphates and diphosphates in *Arabidopsis*. Plant Cell 16:2652–2664

Kazan K, Manners JM (2009) Linking development to defense: auxin in plant-pathogen interactions. Trends Plant Sci 14:373–382

Kernan A, Thornburg RW (1989) Auxin levels regulate the expression of a wound-inducible proteinase inhibitor ii-chloramphenicol acetyl transferase gene fusion in vitro and in vivo. Plant Physiol 91:73–78

Kovtun Y, Chiu WL, Zeng W, Sheen J (1998) Suppression of auxin signal transduction by a MAPK cascade in higher plants. Nature 395:716–720

Lamattina L, Garcia-Mata C, Graziano M, Pagnussat G (2003) Nitric oxide: the versatility of an extensive signal molecule. Annu Rev Plant Biol 54:109–136

Lanteri ML, Laxalt AM, Lamattina L (2008) Nitric oxide triggers phosphatidic acid accumulation via phospholipase D during auxin-induced adventitious root formation in cucumber. Plant Physiol 147:188–198

León J, Rojo E, Sánchez-Serrano JJ (2001) Wound signalling in plants. J Exp Bot 52:1–9

Lombardo MC, Graziano M, Polacco J, Lamattina L (2006) Nitric oxide functions as a positive regulator of root hair development. Plant Signal Behav 1:28–33

O'Donnell PJ, Calvert C, Atzorn R, Wasternack C, Leyser HMO, Bowles DJ (1996) Ethylene as a signal mediating the wound response of tomato plants. Science 274:1914–1917

Orozco-Cardenas M, Ryan CA (1999) Hydrogen peroxide is generated systemically in plant leaves by wounding and systemin via the octadecanoid pathway. Proc Natl Acad Sci USA 96:6553–6557

Pagnussat GC, Simontacchi M, Puntarulo S, Lamattina L (2002) Nitric oxide is required for root organogenesis. Plant Physiol 129:954–956

Pagnussat GC, Lanteri ML, Lombardo MC, Lamattina L (2004) Nitric oxide mediates the indole acetic acid induction activation of a mitogen-activated protein kinase cascade involved in adventitious root development. Plant Physiol 135:279–286

Parani M, Rudrabhatla S, Myers R, Weirich H, Smith B, Leaman DW, Goldman SL (2004) Microarray analysis of nitric oxide responsive transcripts in Arabidopsis. Plant Biotechnol J 2:359–366

Paris R, Lamattina L, Casalongue CA (2007) Nitric oxide promotes the wound-healing response of potato leaflets. Plant Physiol Biochem 45:80–86

Park JE, Park JY, Kim YS, Staswick PE, Jeon J, Yun J, Kim SY, Kim J, Lee YH, Park CM (2007) GH3-mediated auxin homeostasis links growth regulation with stress adaptation response in Arabidopsis. J Biol Chem 282:10036–10046

Pena-Cortes H, Willmitzer L, Sanchez-Serrano JJ (1991) Abscisic acid mediates wound induction but not developmental-specific expression of the proteinase inhibitor II gene family. Plant Cell 3:963–972

Pickett CB, Lu AY (1989) Glutathione S-transferases: gene structure, regulation, and biological function. Annu Rev Biochem 58:743–764

Polverari A, Molesini B, Pezzotti M, Buonaurio R, Marte M, Delledonne M (2003) Nitric oxide-mediated transcriptional changes in *Arabidopsis thaliana*. Mol Plant Microbe Interact 16:1094–1105

Rockel P, Strube F, Rockel A, Wildt J, Kaiser WM (2002) Regulation of nitric oxide (NO) production by plant nitrate reductase in vivo and in vitro. J Exp Bot 53:103–110

Sauer M, Balla J, Luschnig C, Wisniewska J, Reinohl V, Friml J, Benkova E (2006) Canalization of auxin flow by Aux/IAA-ARF-dependent feedback regulation of PIN polarity. Genes Dev 20:2902–2911

Seo S, Okamoto M, Seto H, Ishizuka K, Sano H, Ohashi Y (1995) Tobacco MAP kinase: a possible mediator in wound signal transduction pathways. Science 270:1988–1992

Song CJ, Steinebrunner I, Wang X, Stout SC, Roux SJ (2006) Extracellular ATP induces the accumulation of superoxide via NADPH oxidases in *Arabidopsis*. Plant Physiol 140:1222–1232

Sueldo DJ, Foresi NP, Casalongue CA, Lamattina L, Laxalt AM (2010) Phosphatidic acid formation is required for extracellular ATP-mediated nitric oxide production in suspension-cultured tomato cells. New Phytol 185:909–916

Tanaka Y, Uritani I (1979) Effect of auxin and other hormones on the metabolic response to wounding in sweet potato roots. Plant Cell Physiol 20:1557–1564

Tang W, Brady SR, Sun Y, Muday GK, Roux SJ (2003) Extracellular ATP inhibits root gravitropism at concentrations that inhibit polar auxin transport. Plant Physiol 131:147–154

Terrile MC, Olivieri FP, Bottini R, Casalongue C (2006) Indole-3-acetic acid attenuates the fungal lesions in infected potato tubers. Physiol Plant 127:205–211

Thompson NP (1970) The transport of auxin and regeneration of xylem in okra and pea stems. Am J Bot 57:390–393

Thornburg RW, Li X (1991) Wounding *Nicotiana tabacum* leaves causes a decline in endogenous indole-3-acetic acid. Plant Physiol 96:802–805

Tonón C, Terrile MC, Iglesias MJ, Lamattina L, Casalongué CA (2010) Extracellular ATP, nitric oxide and superoxide act coordinately to regulate hypocotyl growth in etiolated Arabidopsis seedlings. J Plant Physiol 167:540–546

Wendehenne D, Durner J, Klessig DF (2004) Nitric oxide: a new player in plant signalling and defense responses. Curr Opin Plant Biol 7:449–455

Yamasaki H, Sakihama Y, Takahashi S (1999) An alternative pathway for nitric oxide production in plants: new features of an old enzyme. Trends Plant Sci 4:128–129

Yap YK, Kodama Y, Waller F, Chung KM, Ueda H, Nakamura K, Oldsen M, Yoda H, Yamaguchi Y, Sano H (2005) Activation of a novel transcription factor through phosphorylation by WIPK, a wound-induced mitogen-activated protein kinase in tobacco plants. Plant Physiol 139:127–137

Zanetti ME, Terrile MC, Godoy AV, San Segundo B, Casalongue CA (2003) Molecular cloning and characterization of a potato cDNA encoding a stress regulated Aux/IAA protein. Plant Physiol Biochem 41:755–760

Zimmermann P, Hirsch-Hoffmann M, Hennig L, Gruissem W (2004) GENEVESTIGATOR. Arabidopsis microarray database and analysis toolbox. Plant Physiol 136:2621–2632

Chapter 6
How Do Lettuce Seedlings Adapt to Low-pH Stress Conditions? A Mechanism for Low-pH-Induced Root Hair Formation in Lettuce Seedlings

Hidenori Takahashi

Abstract Plants are always surrounded by various environmental factors that may act as stressors. Acid rain and acid soil are serious environmental problems that inhibit plant growth. Most studies on acid stress have focused on the toxicity of aluminum (Al) solubilized from the soil by low pH; studies on the effect of low pH alone, however, are limited.

Recently, the H^+-hypersensitive mutant *stop1* was identified. The *STOP1* gene is predicted to be involved in signal transduction of H^+ and Al tolerance. Analysis of the *stop1* mutant facilitated our understanding of the molecular basis of H^+ tolerance in plants and the linkage between the H^+ and Al toxicity signaling pathways.

Low-pH-induced root hair formation in lettuce seedlings is an excellent model for studying adaptation of plants to low-pH stress. Lettuce seedlings form many root hairs at pH 4.0, whereas no root hairs are formed at pH 6.0. Root hairs increase the absorption of water and nutrients from the growth-inhibited main root at pH 4.0. Various key factors in root hair formation have been identified: medium pH, auxin, ethylene, light, cortical microtubule (CMT) randomization, manganese (Mn), sugar, and chlorogenic acid (CGA), all of which interact within a complex network. Light signals are mediated by auxin and ethylene and induce CMT randomization and root hair elongation. Expression of the 1-aminocyclopropane-1-carboxylic acid (ACC) synthase (ACS), ACC oxidase (ACO), and ethylene receptor gene families is differentially regulated by pH, auxin, ethylene, and light. General opinions on microtubule reorganization and its protection against biotic/abiotic stresses are also reviewed.

H. Takahashi (✉)
Department of Biology, Faculty of Science, Toho University, 2-2-1 Miyama, Funabashi, Chiba 274-8510, Japan
e-mail: takahide@bio.sci.toho-u.ac.jp

N.A. Khan et al. (eds.), *Phytohormones and Abiotic Stress Tolerance in Plants*,
DOI 10.1007/978-3-642-25829-9_6,

Abbreviations

ACC	1-Aminocyclopropane-1-carboxylic acid
ACO	ACC oxidase
ACS	ACC synthase
AVG	Aminoethoxyvinylglycine
CGA	Chlorogenic acid
CMT	Cortical microtubule
IAA	Indole-3-acetic acid
MAPK	Mitogen-activated protein kinase
PA	Phosphatidic acid
PCIB	2-(*p*-chlorophenoxy)-2-methylpropionic acid
PLD	Phospholipase D
ROS	Reactive oxygen species
SAM	*S*-adenosylmethionine
SIMK	Stress-induced MAPK

6.1 Introduction

Unlike animals, plants cannot move if their surrounding environment is unsuitable. Although stress factors such as injury, disease, temperature extremes, drought, salinity, and low pH sometimes adversely affect plants, they cannot escape these unfavorable conditions. Instead, plants modify their physiological conditions and ultimately change their shape to adapt to environmental conditions.

Acid rain and acid soil are serious environmental problems that damage forests and agricultural crops around the globe; soil acidity is thought to limit plant growth in more than 40% of the world's arable land (von Uexküll and Mutert 1995). Acid rain is mainly a mixture of sulfuric and nitric acids. The interaction of these acids with other constituents of the atmosphere releases protons, which cause increased soil acidity. Lowering soil pH mobilizes and leaches away nutrient cations, causing increased availability of toxic heavy metals. Aluminum (Al) is the most well-known and well-studied heavy metal solubilized under low-pH conditions. When the soil pH is below 5.0, Al is solubilized in the form of the toxic trivalent cation Al^{3+} because of its pH-dependent solubility. The first Al toxicity symptom is inhibition of root growth, leading to a damaged and reduced radicular system that results in limited absorption of water and mineral nutrients (Zheng et al. 2005). Micromolar concentrations of Al^{3+} reportedly inhibit root growth in many important crop species within a very short time (Kochian 1995). Such changes in soil chemical characteristics due to soil acidification reduce soil fertility, which ultimately inhibits growth and productivity of forest and crop plants.

For more than 100 years, numerous genetic, physiological, biochemical, and proteomic research efforts have been devoted to understanding the mechanisms of Al toxicity and Al tolerance in plants (Miyake 1916 and literature cited therein).

Plants possess two distinct tolerance mechanisms against soil Al toxicity. One is external detoxification of Al, which protects the root apex against Al penetration. The other is intracellular compartmentalization of Al^{3+}. The former is well studied and is via exudation of organic acids from the root, such as citric, malic, and oxalic acids (Ma et al. 1998; Ma 2000; Ryan et al. 2001; Kochian et al. 2004). These form stable complexes with Al^{3+} in the rhizosphere, reducing or even annulling its toxic effects, since such complexes cannot pass through the plasmatic membrane (Kochian et al. 2004). Since there are many excellent reviews available on the toxicity of and tolerance to Al (e.g., Kochian et al. 2004; Jardim 2007; Poschenrieder et al. 2008; Panda et al. 2009; Horst et al. 2010), I do not intend to recompile a century of intense research here. For details on the mechanisms of Al toxicity and Al tolerance, refer to these references.

The main focus of this chapter is the effect of low pH itself and how plants adapt to acidic conditions. The effect of low pH can be classified into direct and indirect aspects. The former includes the toxicity of H^+ itself, and the latter refers to the toxicity of heavy metals solubilized from the soil by low pH, as mentioned above. Compared with the latter secondary stress factors in acid soils such as Al toxicity, our knowledge of the molecular basis of the effect of H^+ itself is limited, although it is also highly toxic to plants (Koyama et al. 1995; Yokota and Ojima 1995; Kinraide 2003; Watanabe and Okada 2005). Recently, Iuchi et al. (2007) isolated the *Arabidopsis stop1* mutant, which is sensitive to H^+ toxicity. *STOP1* is assumed to be involved in the signal transduction pathway of not only H^+ but also Al tolerance. Isolation of *stop1* mutant therefore facilitated our understanding of the molecular basis of H^+ tolerance in plants and linkage of the signaling pathways of H^+ and Al toxicity. In this chapter, I first summarize research on the *STOP1* gene.

Following the summary of *STOP1*, most of the chapter is dedicated to the adaptation response of lettuce seedlings to low-pH stress. About a decade ago, Inoue et al. (2000) found an interesting response to low pH in lettuce (*Lactuca sativa* L. cv. Grand Rapids) seedlings. When the medium pH of hydroponically cultured lettuce seedlings was lowered, it induced the formation of root hairs (Fig. 6.1). Main root growth was inhibited at low pH, and therefore, absorption

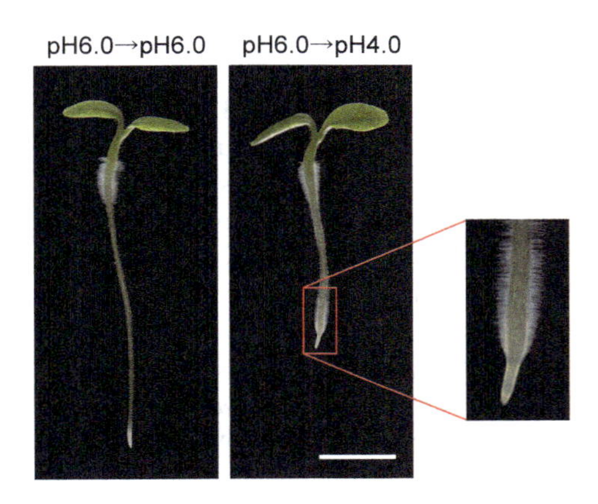

Fig. 6.1 Binocular microscopy images of seedling roots. Seedlings were precultured at pH 6.0 for 24 h, transferred to the indicated media, and harvested 24 h later. Seedlings were cultured throughout in the light. Scale bar = 5 mm

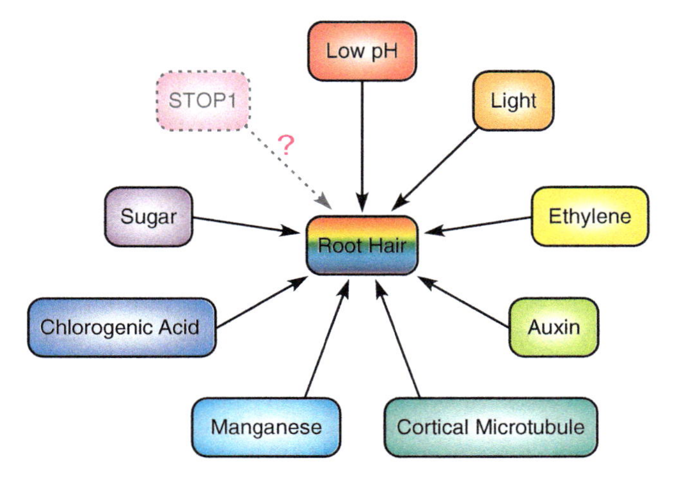

Fig. 6.2 Regulatory factors with known roles in low-pH-induced root hair formation in lettuce seedlings

through the root seemed to decrease; however, this decrease was compensated for by the development of root hairs, and ultimately, the shoot growth occurs. Thus, root hair formation is recognized as an adaptive response to low-pH stress (Inoue et al. 2000). Since the discovery of this response, various factors, including plant hormones, have been identified that regulate low-pH-induced root hair formation (Fig. 6.2). The latest information on these factors is introduced here, with emphasis on the crosstalk between them.

6.2 The *STOP1* Gene in H^+ and Al Tolerances

Under acid soil conditions, Al causes severe yield loss in food and biomass production via inhibition of root growth. Since the root contributes to water absorption, its growth inhibition further enhances the sensitivity to drought stress. Because of such widespread agricultural problems, the mechanism of Al tolerance has been studied, and a number of genes responsible for Al tolerance have been isolated from various plant species, such as genes encoding a malate transporter in *Arabidopsis* (*AtALMT1*; Hoekenga et al. 2006) and wheat (*Triticum aestivum*; *TaALMT1*; Sasaki et al. 2004), a citrate transporter in sorghum (*Sorghum bicolor*; Magalhaes et al. 2007), a type of multidrug and toxic compound exclusion protein in *Arabidopsis* (*AtMATE*; Liu et al. 2009), and an ABC-transporter-like protein in *Arabidopsis* (*ALS3*; Larsen et al. 2005). H^+ toxicity also causes severe root growth inhibition in wheat, *Arabidopsis*, and spinach (Kinraide 1998, 2003; Koyama et al. 2001; Yang et al. 2005). However, identification of genes that regulate H^+ tolerance awaits an improved understanding of the molecular basis of H^+ tolerance in plants.

In contrast to the accumulating information on Al tolerance, little is known about the genes controlling H^+ tolerance.

It has been postulated that H^+ and Al toxicities induce damage by different processes (Koyama et al. 1995; Kinraide 2003). In fact, the knockout mutant of *AtALMT1* is not sensitive to H^+ (Kobayashi et al. 2007). On the other hand, there are studies showing that H^+ and Al tolerances are genetically linked (Rangel et al. 2005; Yang et al. 2005). This discrepancy suggests that H^+ and Al tolerances are regulated in part by factors common to both pathways.

Recently, Iuchi et al. (2007) found an *Arabidopsis* mutant named *stop1* (sensitive to proton rhizotoxicity), whose root growth is hypersensitive to H^+. The *STOP1* gene encodes a Cys2/His2-type zinc-finger protein. When the gene was fused with GFP, fluorescence of the product was observed only in the nucleus, suggesting that *STOP1* may regulate transcription in the nucleus, either as a transcription factor or by affecting other members of the H^+ signal transduction pathway (Sawaki et al. 2009).

Interestingly, the *stop1* mutant showed hypersensitivity to Al, but normal sensitivity to other toxic ions, such as cadmium, copper, sodium, manganese, and lanthanum. The expression of genes for Al tolerance, namely, *AtALMT1* and *AtMATE*, is suppressed in the *stop1* mutant, even when the mutant is subjected to Al stress, whereas expression of these genes increases in the wild type (Sawaki et al. 2009; Liu et al. 2009). However, these Al-tolerance genes may not be involved in H^+ tolerance because the T-DNA inserted knockout mutant of *AtALMT1* is not sensitive to H^+ (Kobayashi et al. 2007). These results suggest that, although *STOP1* may be involved in the signal transduction pathway not only in H^+ tolerance but also in Al tolerance, H^+ tolerance-specific genes may exist under the control of *STOP1*.

Microarray analyses revealed that several genes responsible for H^+ tolerance in other organisms were also downregulated in the *stop1* mutant, including *CIPK23* and *AtTDT* (Sawaki et al. 2009). *CIPK23* encodes the regulatory kinase of the major K^+ transporter AKT1 (Lee et al. 2007). K^+ transport and homeostasis is a major pH-regulating mechanism in various organisms (Zhang and Kone 2002). Thus, *STOP1* may contribute to pH regulation through control of *CIPK23*. On the other hand, *AtTDT* encodes a tonoplast dicarboxylate transporter (a kind of malate transporter) and is also annotated as the gene for pH homeostasis in *Arabidopsis* (Hurth et al. 2005). Furthermore, Sawaki et al. (2009) revealed that dysfunction of *STOP1* downregulated biochemical pathways of the pH stat (Roberts et al. 1992; Sakano 1998) and the γ-aminobutyric acid (GABA) shunt (Bown and Shelp 1997; Bouché and Fromm 2004). Both metabolic pathways have been assumed to regulate cytosolic pH in the plant. These findings suggest that *STOP1* provides H^+ tolerance through activating these pathways together with the regulation of ion homeostasis and transport. It is known that ion homeostasis and pH-regulating metabolism also play important roles in low pH resistance in other organisms, such as *Escherichia coli* (Castanie-Cornet et al. 1999; Yohannes et al. 2004), fish (Hirata et al. 2003), and possibly in higher plants (Yan et al. 1992). Downregulation of genes involved in ion homeostasis and transport may be a cause of the H^+-hypersensitivity of the *stop1* mutant.

It has not yet been clarified whether expression of the *STOP1* gene and/or the *STOP1* signaling pathway is regulated by plant hormones. Recent research by Sawaki et al. (2009) suggests their plausible interaction. In the *stop1* mutant, some genes in the defense system are downregulated. It is well known that the local and systemic defense responses triggered by microorganisms are controlled through a signaling network in which the plant hormones ethylene, salicylic acid, and jasmonic acid play important roles (Glazebrook 2005; Kachroo and Kachroo 2007). Furthermore, Rudrappa et al. (2008) reported that expression of *AtALMT1*, which is under the control of *STOP1*, was induced in leaf pathogenic interactions. These observations suggest that *STOP1* is a pleiotropic gene that belongs to defense systems for both biotic and abiotic stresses and that its signaling pathway may be regulated by plant hormones.

Other than the *STOP1* gene itself, a homolog that closely resembles it was found in the *Arabidopsis* genome, and also in rice (*Oryza sativa*) and maize (*Zea mays*) (Iuchi et al. 2007). In the future, it will be interesting to examine whether these *STOP1* homologs regulate the expression of genes involved in H^+ and Al tolerance and whether their expressions are regulated by plant hormones.

6.3 Effects of Low pH on Hydroponically Cultured Lettuce Seedlings

Like other plant species, low-pH stress inhibits the root growth of lettuce seedlings. When lettuce seeds were sown and precultured on liquid medium at pH 6.0, transfer to fresh pH 4.0 medium led to formation of shorter roots compared to seedlings transferred to pH 6.0 medium (Inoue et al. 2000). Since the root is an important organ that absorbs water and nutrients (Clarkson 1985; Dolan et al. 1994; Hofer 1996; Peterson and Farquhar 1996; Ridge 1996), root growth inhibition should prevent shoot growth. However, interestingly, shoot growth was not affected by the acidity of the medium: Shoot fresh weight of the pH 4.0 seedlings was comparable to that of pH 6.0 seedlings. This discrepancy is explained by the presence of many root hairs formed on pH 4.0 seedlings.

Root hairs are long lateral tubular extensions of root epidermal cells. Their formation comprises at least two discrete phases: the initiation phase, which is characterized by the appearance of a small bulge in the outer periclinal cell wall, and the elongation phase, which is characterized by root hair development by tip growth from a selected site on the bulge. Many root hairs on pH 4.0 seedlings increase the surface area of the shorter root and hence compensate for the reduced absorption of water and nutrients from the root at pH 4.0 (Inoue et al. 2000). In fact, the pH 4.0 and pH 6.0 seedling roots showed similar water permeability per root despite the difference in their length (Azegami et al. 2000). When water permeability per 1 mm root was calculated, the pH 4.0 root had a twofold higher value than the pH 6.0 root. The pH of the medium is not a cause of this difference because a root-hairless mutant showed similar water permeability, regardless of the pH of the medium. Furthermore, when a fluorescent dye, which is water soluble but impermeable to the plasma membrane, was added to the medium, fluorescence localized at the surface of root hairs, especially at their tips. In contrast, at pH 6.0, fluorescence was observed at the boundary of epidermal cells (Azegami et al. 2000). These results indicate that the root hair has higher water permeability and surely compensates for absorption from a shorter root at pH 4.0. Therefore, we can recognize low-pH-induced root hair formation as one adaptation mechanism of lettuce seedlings to H^+ stress.

Plant response to low pH varies among species. Low pH also inhibits the root growth of *Arabidopsis* seedlings but does not induce formation of root hairs. Due to the reduced water absorption from the root, *Arabidopsis* shoot growth is suppressed at low pH. Reductions of density and length of root hairs under acidic conditions have also been observed across a range of perennial grasses and cereals (Haling et al. 2010a, b). On the other hand, the responses of *Corchorus olitorius* L. and *Malva verticillata* L. var. *crispa* (L.) Makino are similar to those of lettuce seedlings, in which the loss of absorption is compensated for by the development of root hairs and ultimately shoot growth occurs (Tamai et al. unpublished data). Therefore, although not universal, induction of root hair formation and its compensation for water absorption are surely a tolerance mechanism of plants against low-pH stress.

The root hair distribution pattern in the root epidermis varies among vascular plants (Clowes 2000; Dolan and Costa 2001; Kim et al. 2006). All lettuce root epidermal cells have the potential to form root hairs (Inoue and Hirota 2000). This feature contrasts with that of *Arabidopsis* roots, whose root epidermis has hair cells (trichoblasts) and non-hair cells (atrichoblasts) (Dolan et al. 1994; Galway et al. 1994). In the case of lettuce seedlings, the future hair-forming zone is located approximately 1.0 mm from the root tip at the time of transfer of seedlings to pH 4.0 medium. Hair formation initiates in the 1.5–2.0 mm area from the root tip in a highly synchronized manner 4 h after acidification, and then saturates 7 h after the transfer (Inoue and Hirota 2000). Because no root hairs form on seedlings at pH 6.0, we can regulate root hair formation in an "all or none" manner by controlling the pH of the medium. This morphological information and the high synchrony of root hair formation provide much help in investigating the kind of genetic, physiological, and morphological events that occur during root hair formation, even before root hair primordia are identified.

6.4 Role of Plant Hormones in Root Hair Formation

Genetic and physiological studies have demonstrated that the plant hormones ethylene and auxin promote root hair formation in *Arabidopsis*. Application of the ethylene precursor 1-aminocyclopropane-1-carboxylic acid (ACC) induces root hair formation at ectopic positions. In contrast, aminoethoxyvinylglycine (AVG), an ethylene biosynthesis inhibitor, or Ag^+, an ethylene action inhibitor, prevents root hair development (Masucci and Schiefelbein 1994; Tanimoto et al. 1995). Phenotypic analyses of the ethylene-signaling mutants *ein2*, *etr1* (Pitts et al. 1998), and *ctr1* (Dolan et al. 1994) support the positive effect of ethylene on root hair formation. On the other hand, the auxin response mutants *aux1*, *axr1*, *axr2*, and *axr3* (Leyser et al. 1996; Lincoln et al. 1990; Okada and Shimura 1994; Wilson et al. 1990) and the auxin homeostasis mutant *iar4* (Quint et al. 2009) affect root hair growth, suggesting that auxin also promotes the formation of root hairs.

Besides hormonal control studies, many genes have been identified that play a role in the formation of root hairs, including *CPC* (Wada et al. 1997), *GL2* (Masucci et al. 1996), *RHD2, 6* (Schiefelbein and Somerville 1990; Masucci and Schiefelbein 1994), *WER* (Lee and Schiefelbein 1999), and *TTG* (Galway et al. 1994). Masucci and Schiefelbein (1996) revealed that ethylene triggers root hair morphogenesis downstream from the patterning process regulated by the *TTG1/GL2* pathway.

As in the case of *Arabidopsis*, ethylene and auxin also participate in the low-pH-induced root hair formation in lettuce seedlings (Takahashi et al. 2003c). At pH 4.0, lettuce seedlings produce more ethylene than at pH 6.0. Addition of the ethylene inhibitors AVG or Ag^+, or the auxin competitive inhibitor 2-(*p*-chlorophenoxy)-2-methylpropionic acid (PCIB), to pH 4.0 medium suppressed root hair initiation. In contrast, addition of ACC or indole-3-acetic acid (IAA) to pH 6.0 medium induced root hair initiation. Furthermore, interaction between ethylene and auxin was studied, and it became clear that auxin is essential for root hair formation and that the increased amount of ethylene produced at pH 4.0 may promote the induction by auxin in hair-forming cells.

6.5 Role of Light in Root Hair Formation

Light perception is essential for plants to appropriately grow and develop throughout their life cycle. Studies with phytochrome-deficient mutants have revealed important roles for individual phytochromes in a range of light responses (Whitelam et al. 1993; van Tuinen et al. 1995; Neff et al. 2000; Takano et al. 2001; Weller et al. 1995, 2000, 2001). Although root hair formation occurs in roots in the soil in nature, interestingly, there is evidence that supports an involvement of light in this.

In contrast to seedlings grown in the light, no root hairs form on dark-grown lettuce seedlings, even when they are transferred to a pH 4.0 medium (De Simone et al. 2000c; Takahashi and Inoue 2008). Reduced numbers of root hairs in the dark have also been observed in *Arabidopsis* (Dolan 1997; De Simone et al. 2000a), suggesting that light is generally needed for root hair formation in plants. One cause of the suppression of root hair formation in the dark-grown lettuce seedlings may be reduced ethylene production in the dark. Compared to light-irradiated seedlings, dark-grown seedlings produce only about half the amount of ethylene (Takahashi et al. 2003b). Since ethylene is needed to promote auxin induction of root hair formation (Takahashi et al. 2003c), a low level of ethylene production in the dark may not be enough to facilitate it.

To determine the photoreceptors involved in low-pH-induced root hair formation in lettuce, De Simone et al. (2000c) used red/far-red light irradiation and found that root hair formation is regulated by phytochromes. Furthermore, microbeam irradiation experiments revealed that the photoperceptive site is the apical portion of the root (De Simone et al. 2000b). In nature, plant roots usually exist in the soil, and only the aerial portions of the plant are subjected to light. Therefore, it is reasonable that many of the signal transduction elements downstream of the phytochromes have been identified in the aerial portions of plants. The reason why lettuce root hair formation needs a light signal and why it is the root apical portion that perceives the light rather than the aerial portions is unknown. However, there is evidence to support the idea that functional photoreceptors exist in roots (Feldman and Briggs 1987; Johnson et al. 1994; Kiss et al. 2003; Hemm et al. 2004). In fact, significant levels of phytochrome gene expression have been observed in roots (Johnson et al. 1991; Somers and Quail 1995; Goosey et al. 1997).

Furthermore, there is direct evidence that shows the involvement of phytochromes in root hair formation in *Arabidopsis*. Expression of genes controlling root hair formation, such as *CPC* and *RHD3*, is upregulated by red light (Molas et al. 2006). De Simone et al. (2000a) found that light promotes root hair induction in *Arabidopsis* and that the effect is reduced in *phyA* and *phyB* mutants. *AtMYC2*, which interacts with light-responsive elements and thereby suppresses the expression of light-regulated genes (Yadav et al. 2005), has been shown to act as a negative regulator in the formation of root hairs (Sano et al. 2003). On the other hand, light also affects the length of root hairs. A *phyB* mutant has longer root hairs, and thus, *PHYB* is thought to affect the duration of tip growth (Reed et al. 1993; Schiefelbein 2000). A mutant of *HY5*, a bZIP transcription factor that acts downstream of phytochromes in light signaling pathways, has longer root hairs than the wild type (Oyama et al. 1997, 2002). Therefore, phytochrome may control not only the initiation of root hairs but also their elongation.

A number of mutants that show abnormal responses to auxin or ethylene are also affected in terms of the length of root hairs. This observation suggests that auxin and ethylene regulate the duration of tip growth (Pitts et al. 1998; Schiefelbein 2000). Since signaling of both light and plant hormones affects the length of root hairs, light may regulate root hair elongation through an auxin/ethylene response pathway. This idea is supported by the following experiment with lettuce seedlings. In the case of lettuce, dark-grown seedlings cannot form root hairs, but application of a microtubule-depolymerizing drug to dark-grown seedlings induces the formation of root hair primordia. However, tip growth from the primordia is still suppressed even though application of this drug to light-grown seedlings at pH 6.0 induces mature, elongated root hairs (Takahashi et al. 2003a). Application of IAA or ACC together with the microtubule-depolymerizing drug leads to recovery of the elongation of root hairs in dark-grown seedlings (Takahashi and Inoue 2008). These results indicate that elongation of root hairs needs auxin and ethylene and that light may regulate root hair elongation through the auxin/ethylene response pathways.

In fact, as mentioned in the following section, ethylene biosynthesis gene expression is upregulated in lettuce by light irradiation.

6.6 Ethylene Biosynthesis Gene Expression During Root Hair Formation in Lettuce

Ethylene is a gaseous plant hormone that plays multiple roles in plant growth and development throughout the plant life cycle. It is involved in seed germination, leaf abscission, organ senescence, and fruit ripening, as well as the synthesis of defense-related proteins in response to pathogen infection (Abeles et al. 1992). In higher plants, ethylene is synthesized via a well-defined pathway that starts with methionine. A direct precursor of ethylene is ACC, which is synthesized from *S*-adenosyl-methionine (SAM) in a reaction catalyzed by ACC synthase (ACS: EC 4.4.1.14). ACC is then converted to ethylene by ACC oxidase (ACO: EC 1.4.3) (Yang and Hoffman 1984). Genes encoding ACS and ACO have been isolated from numerous plant species, and each has been shown to form a multigene family. The members are differentially regulated at the transcription level in a tissue-specific manner during plant growth and development, as well as under different environmental conditions (ACS: van der Straeten et al. 1990; Rottmann et al. 1991; Zarembinski and Theologis 1993; Wang and Arteca 1995; Clark et al. 1997; Oetiker et al. 1997; Shiu et al. 1998; Arteca and Arteca 1999; Ge et al. 2000; Tsuchisaka and Theologis 2004; ACO: Liu et al. 1997; Kim et al. 1998; Petruzzelli et al. 2000; Yang et al. 2003; Rieu et al. 2005; Foo et al. 2006; Chen and McManus 2006; Binnie and McManus 2009).

Ethylene biosynthesis is regulated by both positive (Liu et al. 1985; Wang and Woodson 1989; Kim and Yang 1994; Hiwasa et al. 2003; Nakano et al. 2003) and negative (Lund et al. 1998; Nakajima et al. 1990; Yoon et al. 1997; Owino et al. 2006) feedback mechanisms (Yang and Hoffman 1984; Kende 1993). A step catalyzed by ACS is generally considered to be the rate-limiting step in ethylene biosynthesis (Yang and Hoffman 1984; Kende 1993). However, evidence is now available to indicate that ACO also plays a regulatory function in ethylene production (Blume and Grierson 1997; Leliévre et al. 1995; Dominguez and Vendrell 1994). Ethylene regulates the growth and development of roots (Lee et al. 1990; Clark et al. 1999; Buer et al. 2003; Stepanova et al. 2007; Gallie et al. 2009), but the manner in which ACS and ACO regulate ethylene biosynthesis has generally been studied using aboveground tissues or organs; thus, the data concerning roots are limited.

Two ACS genes (*Ls-ACS1, 2*; Takahashi et al. 2003b) and three ACO genes (*Ls-ACO1, 2, 3*; Takahashi et al. 2010b) have been isolated so far from lettuce. As in other plant species, these genes were differentially regulated during low-pH-induced root hair formation. The *Ls-ACS1* and *Ls-ACS2* mRNAs increased in roots after transfer to pH 4.0 medium, whereas their levels were constant in control seedlings at pH 6.0. *Ls-ACS2* mRNA reached a peak earlier than *Ls-ACS1* mRNA. IAA induced the accumulation of both mRNAs, whereas ACC induced only *Ls-ACS1* mRNA. These results suggest that acidification-induced auxin accumulation increases *Ls-ACS2* mRNA levels, which together with Ls-ACS2-induced ethylene raises the level of *Ls-ACS1* mRNA. Furthermore, light had an inductive effect on *Ls-ACS1*. This observation is consistent with that in which light-grown seedlings produce more ethylene than dark-grown ones (Takahashi et al. 2003b). On the other hand, the mRNA level of *Ls-ACO2* increased in roots after transfer to pH 4.0 medium and remained high until at least 4 h later. In contrast, *Ls-ACO1* and *Ls-ACO3* mRNA levels were constant and comparable to those at pH 6.0. Addition of ACC or IAA to the pH 6.0 medium induced the accumulation of *Ls-ACO2* mRNA (Takahashi et al. 2010b). Considering these observations, *Ls-ACS1* and *Ls-ACO2* play key roles among the ACS and ACO gene families, respectively, in ethylene production during low-pH-induced root hair formation.

One of the events that occur after exposure to pH 4.0 medium is the randomization of transverse cortical microtubule (CMT) arrays in future hair-forming cells, the rearrangement of which is a prerequisite for root hair initiation (Takahashi et al. 2003a, c). CMT randomization is induced by ethylene (Takahashi et al. 2003c). Considering that CMT randomization is complete within 30 min after acidification, ethylene used in CMT randomization may be produced by increased Ls-ACS1, Ls-ACS2, and Ls-ACO2. On the other hand, continuous expression of *Ls-ACS1* and *Ls-ACO2* may also contribute to the later stages after CMT randomization.

As mentioned in the previous section, a light signal is needed for lettuce root hair formation, which is under the control of phytochromes (De Simone et al. 2000c). Considering that ethylene promotes root hair formation, it is reasonable that *Ls-ACS1*, *Ls-ACO1*, and *Ls-ACO3* genes are photo-inducible (Takahashi et al. 2003b, 2010b) and that seedlings actually produce more ethylene in the light than in the dark (Takahashi et al. 2003b). The regulation of ethylene production by phytochrome has been observed in various plant species (Finlayson et al. 1998; Pierik et al. 2004, 2009; Foo et al. 2006). Ethylene production is usually suppressed by light through the suppression of *ACS* and/or *ACO* gene expression (Finlayson et al. 1999; Foo et al. 2006; Choudhury et al. 2008). In this respect, the relation between light and expression of *ACS* and *ACO* genes is somewhat different in lettuce root hair formation. However, the upregulation of *ACS* and *ACO* expression by light has also been observed in other species (Kathiresan et al. 1998; Rieu et al. 2005).

6.7 Ethylene Receptor Gene Expression During Root Hair Formation in Lettuce

To initiate ethylene signaling, synthesized ethylene has to be captured by its receptors. Although any ethylene control mechanisms should be understood from the aspect of ethylene signal transduction as well as ethylene biosynthesis, there are fewer studies of the former than of the latter. The reason is that cloning of the ethylene receptor genes is more recent than that of the *ACS* and *ACO* genes. It has only been about 20 years since the receptor gene was isolated for the first time (*Arabidopsis ETR1* gene; Chang et al. 1993). In particular, studies on gene expression and the role of ethylene receptors in the root are limited.

As in various other plant species, ethylene receptors form a gene family in lettuce. Four ethylene receptor genes (*Ls-ERS1, Ls-ETR1, 2, 3*) have been isolated so far (Takahashi et al. 2010a). *Ls-ERS1* and *Ls-ETR2* expression increased after transfer to pH 4.0 medium and was maintained at higher levels than at pH 6.0. Their expression is also induced by ACC or IAA treatment. In contrast, *Ls-ETR1* and *Ls-ETR3* expression was not affected either by a change of medium pH or ACC or IAA treatment. There are many ethylene receptor genes that are upregulated by ethylene, including *Arabidopsis ERS1*, *ERS2*, and *ETR2* (Hua et al. 1998); tomato *NR*, *LeETR4*, and *LeETR5* (Ciardi et al. 2000); strawberry *FaEtr2* (Trainotti et al. 2005); *Rumex RP-ERS1* (Vriezen et al. 1997); cucumber *CS-ETR2* and *CS-ERS* (Yamasaki et al. 2000); and mung bean *VR-ERS1* (Kim et al. 1999). Lettuce *Ls-ERS1* and *Ls-ETR2* expression increased in the ACC-treated or pH 4.0 seedlings, where ethylene production was enhanced (Takahashi et al. 2010a). Thus, the expression patterns of *Ls-ERS1* and *Ls-ETR2* are similar to those of many other ethylene receptor genes. In contrast to the effect of ethylene, the effect of auxin has not been well characterized in other plants, although a slight increase in ethylene receptor gene expression has been observed in auxin-treated soybean (Xie et al. 2007) and strawberry (Trainotti et al. 2005).

Ethylene receptors have been recognized as negative regulators of ethylene responses, since loss-of-function mutations in *ETR1*, *ETR2*, *ERS2*, and *EIN4* result in constitutive ethylene responses in *Arabidopsis* (Hua and Meyerowitz 1998). This is also supported by studies in tomato (Ciardi et al. 2000; Hackett et al. 2000; Tieman et al. 2000). Based on this interpretation, the expression of ethylene receptors was expected to decrease during root hair formation because it is promoted by ethylene. However, the mRNA levels of the ethylene receptor genes did not decrease after acidification; interestingly, expression of *Ls-ERS1* and especially *Ls-ETR2* increased. There are plausible explanations for this discrepancy. First, although increased ethylene receptor expression intensifies inhibition of the ethylene response, the larger amount of ethylene produced at pH 4.0 (Takahashi et al. 2003c) may turn off the negative regulation. A degradation of ethylene-bound receptors in a ligand-induced manner (Chen et al. 2007; Kevany et al. 2007) may also help to turn off negative regulation by the receptors. Indeed, if sufficient numbers of ethylene receptors were removed by loss-of-function mutations, the ethylene response was induced in the absence of ethylene (Hua and Meyerowitz 1998).

Alternatively, increased ethylene receptor expression may turn off the ethylene response temporarily once root hair initiation has progressed to a certain stage. Ethylene is required not only for the initiation of root hairs but also for their elongation (Takahashi and Inoue 2008), but the two processes are regulated by distinct mechanisms. Once the ethylene response is turned off by the increased expression of ethylene receptors, the ratio of the receptor members, especially those in subfamilies I and II, which have different enzymatic activities and different roles in signaling (Gamble et al. 1998, 2002; Wang et al. 2003; Moussatche and Klee 2004; Qu and Schaller 2004), may be modified at the protein level, allowing the later stages of root hair formation.

Furthermore, *Ls-ETR2*, which dramatically increases prior to root hair initiation, may play an important role in lettuce root hair formation by adjusting microtubule organization. Plett et al. (2009) suggest that *Arabidopsis* ETR2 plays a unique role in establishing cell shape by affecting microtubule stability. ETR2 shows the highest similarity to Ls-ETR2 among lettuce ethylene receptors (Takahashi et al. 2010a). The importance of microtubule organization dynamics in root hair formation has been confirmed in lettuce, as well as in *Arabidopsis* (Bao et al. 2001; Van Bruaene et al. 2004) and *Zea mays* (Baluška et al. 2000). In the next section, I will introduce the critical role of microtubules during lettuce root hair formation.

6.8 Reorganization of Cortical Microtubules in Root Epidermal Cells

The direction of the expansion of a plant cell is regulated by the oriented deposition of cellulose microfibrils around the cell, because the microfibrils serve as a hoop for the cell. Deposition of cellulose microfibrils is controlled through the CMTs. Cellulose synthases move along the plasma membrane on tracks provided by the underlying CMTs, and thus, cellulose microfibrils deposit parallel to the CMTs (for review: Paradez et al. 2006; Lloyd and Chan 2008; Crowell et al. 2010). The root hair is a tubular extension of the defined portion of the root epidermal cell. It is first recognized as a small bulge, which then elongates through a tip growth machinery to form the mature root hair. Therefore, it is reasonable that CMTs play important roles in the formation of root hairs.

In lettuce root epidermal cells, CMTs are perpendicular to the longitudinal axis of the cell before being subjected to low-pH stress, i.e., during culture at pH 6.0. The transverse CMT arrays begin disordering as early as 5 min after transfer to pH 4.0 medium and become random 30 min later. An important event for root hair initiation is not the "randomization" but the "disordering" of the transverse CMTs. When microtubule-depolymerizing drugs are applied to pH 6.0 medium, transverse CMTs disappear and root hairs are formed. In contrast, addition of a microtubule-stabilizing drug inhibits root hair formation even at pH 4.0 (Takahashi et al. 2003a). The transverse CMTs may not allow the cell to make room for the formation of root hair bulges because they restrict the orientation of cellulose microfibrils and create a tight hoop. When cellulose microfibrils become disorganized and relaxed through the disordering of CMTs, they may allow space for bulges to emerge. The shortening and swelling of epidermal cells at pH 4.0 may reflect a loss of directional elongation. Increasing evidence demonstrates the importance of changes in cell wall organization during bulge formation (Bernhardt and Tierney 2000; Baumberger et al. 2001; Favery et al. 2001; Wang et al. 2001; Baluška et al. 2000; Vissenberg et al. 2001).

As mentioned already, the plant hormones auxin and ethylene promote root hair formation in lettuce under low-pH stress. The auxin competitive inhibitor PCIB or ethylene biosynthesis or action inhibitors suppress CMT randomization and root hair initiation at pH 4.0. In contrast, addition of IAA or ACC to pH 6.0 medium induces CMT randomization and root hair formation. These observations indicate that auxin and ethylene regulate CMT randomization, an early event in root hair formation, as in root hair initiation and elongation (Takahashi et al. 2003c).

Not only auxin and ethylene decide the destination of CMT arrays; light also affects CMT organization because CMT arrays do not randomize in the dark, even at pH 4.0. However, addition of IAA or ACC to dark-grown seedlings induces CMT randomization and root hair formation. These observations indicate that light signals may be mediated through these two hormones (Takahashi and Inoue 2008). Interestingly, these hormones cannot fully compensate for the effect of light. During induction in the dark, CMT randomization occurs with only a slight delay compared to that in light. However, root hair initiation is delayed for several hours in the dark. These results suggest an additional influence of light through unknown factor(s) during root hair initiation (Takahashi and Inoue 2008).

6.9 Microtubule Organization and Biotic/Abiotic Stresses

CMTs have been shown to be involved in responses to biotic stress (Kobayashi et al. 1992, 1997; Cahill et al. 2002; Takemoto and Hardham 2004; Yuan et al. 2006; Hardham et al. 2008). They also play an important role in responses to abiotic stresses, including salt stress (Chinnusamy and Zhu 2003; Shoji et al. 2006; Wang et al. 2007), osmotic stress (Balancaflor and Hasenstein 1995), and cold acclimation (Wang and Nick 2001; Olinevich et al. 2002; Abdrakhamanova et al. 2003). In this respect, the need for CMT randomization in response to low pH, i.e., root hair formation under low-pH stress, is a new finding that illustrates the importance of CMT organization during abiotic stress.

The organization of microtubules is regulated by various factors. Phospholipase D (PLD) is one such factor. PLD is involved in plant response to abiotic stresses (Zhang et al. 2005) and contributes to the rearrangement of CMTs (Dhonukshe et al. 2003). The importance of PLD for organization of microtubules is supported by the observation that treatment with *n*-butanol, which decreases phosphatidic acid (PA) formation by all PLDs, disrupts the organization of interphase CMTs (Gardiner et al. 2003; Dhonukshe et al. 2003). In *Arabidopsis*, *PLDξ1* (*AtPLDξ1*) is likely involved in both initiation and maintenance of root hair morphogenesis. GL2 binds to the promoter of *AtPLDξ1*, prevents its expression, and makes a cell differentiate into a non-hair cell (Ohashi et al. 2003). PLD in animal cells and yeast regulates membrane trafficking events to and from the plasma membrane, perhaps by utilizing microtubules as rails for the trafficking (Liscovitch et al. 2000; Cockcroft 2001). Considering these findings, PLD may regulate vesicle trafficking by controlling microtubule organization during root hair formation. It may also regulate root hair morphogenesis through translocation of the membrane proteins involved in signal transduction for plant hormones, including auxin and ethylene.

Root hair elongation involves tip growth from a root hair bulge. The important role of PLD in polarized tip growth has also been implicated in pollen tube growth, another tip-growing system in plants (Zonia and Munnik 2004; Potocký et al. 2003). *n*-Butanol inhibits both pollen germination and tube growth. This inhibition can be overcome by application of PA (Potocký et al. 2003). Furthermore, the inhibition can be partially restored using a microtubule-stabilizing agent. These results suggest that PLD and PLD-derived PA affect microtubule dynamics generally during polar growth in plants (Gardiner et al. 2003; Dhonukshe et al. 2003; Zonia and Munnik 2004; Potocký et al. 2003).

Calcium and reactive oxygen species (ROS) are other cell constituents that interact with CMTs and participate in the plant response to salt stress (Xiong et al. 2002; Chinnusamy et al. 2005; Xu et al. 2006; Dixit and Cyr 2003). Ca^{2+} and ROS, as well as PA, auxin, and ethylene, can activate signal transduction pathways involving mitogen-activated protein kinases (MAPKs) in plant cells, which are triggered by biotic and abiotic stresses (Jonak et al. 2002; Munnik 2001). Interestingly, all of these components are involved in root hair formation. The role of PA, auxin, and ethylene has been mentioned already. ROS accumulation is observed in growing root hairs and is required to stimulate Ca^{2+} influx and maintain a tip-focused Ca^{2+} gradient during hair elongation (Foreman et al. 2003). Furthermore, MAPKs are thought to be involved in signal transduction in root hair formation (Foreman et al. 2003; Ohashi et al. 2003; Anthony et al. 2004; Rentel et al. 2004). Stress-induced MAPK (SIMK) plays an important role in predicting the place of root hair outgrowth and promoting tip growth (Šamaj et al. 2002). AtMPK6, the *Arabidopsis* ortholog of SIMK, is activated by ROS as well as by diverse biotic and abiotic factors (Jonak et al. 2002; Kovtun et al. 2000). These findings suggest that during the evolution of plants, they may have adopted some components for root hair development that are common to stress-related signaling pathways. In this respect, it is interesting that root hair formation occurs in lettuce subjected to the low-pH stress conditions.

6.10 Role of Manganese (Mn) in Root Hair Formation

The effects of various nutrients on root hair initiation and elongation have been investigated. For example, low-phosphate conditions cause an increase in root hair density in *Arabidopsis* seedlings (Foehse and Jungk 1983; Bates and Lynch 1996). Iron (Fe) deficiency also induces root hairs in seedlings of *Arabidopsis* (Moog et al. 1995) and sunflower (Landsberg 1996). Mutant analyses revealed that auxin and ethylene act in root hair formation in response to Fe deficiency, but the root hair formation triggered by phosphate deficiency is not mediated by these plant hormones (Schmidt et al. 2000; Schmidt and Schikora 2001). Hence, involvement of plant hormones in nutrient deficiency-induced root hair formation seems to vary, depending on the nutrient.

Manganese (Mn) is the second most prevalent transition metal in the Earth's crust. It is an essential trace element for metabolism in all living organisms: It works as a cofactor or as an activator for various enzymes in a range of biochemical pathways (Marschner 1995). In plants, Mn is also required as a tetra-Mn cluster in photosystem II for catalyzing the water-splitting reaction and oxygen evolution. In spite of the importance of Mn in plant life, there are few reports on the relationship between root hair formation and Mn.

In low-pH-induced root hair formation in lettuce seedlings, Konno et al. (2003) found that Mn deficiency induces root hair formation at pH 6.0. Both the number and length of root hairs increased under Mn deficiency conditions. In contrast, an increment in the Mn concentration suppressed root hair formation at pH 4.0. One plausible cause of this increased density of root hairs is inhibition of main root growth. But there was no difference between the length of the main root of seedlings grown in pH 6.0 medium without Mn, in pH 4.0 medium with excess Mn, and in pH 6.0 medium with a normal concentration of Mn (Konno et al. 2006). These observations indicate that lettuce root hair initiation can be controlled without affecting main root growth by changing the Mn concentration in the medium.

Because Mn concentrations in normal pH 6.0 and pH 4.0 media are the same, the above result poses a new question: Is Mn involved in root hair formation at pH 4.0? The answer is "Yes." The Mn content in roots of pH 4.0 seedlings was only 15% of that of pH 6.0 seedlings 24 h after lowering the pH of the medium (Konno et al. 2003). Although it is possible that the decreases in Mn uptake in pH 4.0 seedlings occurred after root hair formation was completed, this idea was ruled out through the following observations. A decreased (43%) Mn uptake became obvious in the pH 4.0 roots within 1 h of lowering the pH of the medium, which was earlier than the time of root hair initiation (4 h later; Konno et al. 2006). These results suggest that low-pH stress induces decreased Mn uptake in the root; in turn, the Mn deficiency in the root induced root hair initiation.

Induction of root hairs under Mn-deficient conditions seems to be an efficient plant strategy to increase Mn acquisition, since root hairs enhance absorption of water and nutrients. In fact, low-phosphate-induced root hair formation in *Arabidopsis* may contribute to increased phosphate acquisition (Bates and Lynch 2000). However, when the pH 4.0 seedlings, which formed root hairs, were transferred to pH 6.0 medium, Mn uptake was lower than in seedlings cultured continuously at pH 6.0 and was comparable to that of seedlings cultured continuously at pH 4.0 (Konno et al. 2006). These results suggest that root hairs formed at low pH do not compensate for the decreased Mn uptake.

Besides being required in the tetra-Mn cluster in photosystem II, Mn^{2+} is also required for autophosphorylation of ETR1, a member of the ethylene receptor family in *Arabidopsis* (Gamble et al. 1998). Recently, Bisson et al. (2009) revealed that ETR1 interacts at the ER network with EIN2, a central and critical element in ethylene signaling. Bisson and Groth (2010) speculated that without ethylene, receptor complexes are kept in their phosphorylated state due to their intrinsic autokinase activity (Voet-van-Vormizeele and Groth 2008), and they may combine with CTR1, a negative regulator of ethylene signaling (Kieber et al. 1993; Huang et al. 2003). Upon ethylene binding, the autokinase activity of the receptors is inhibited, converting them to nonphosphorylated status (Voet-van-Vormizeele and Groth 2008). This change results in inactivation of CTR1 and a tight interaction of the ethylene receptor complexes with EIN2, and induces ethylene responses. Based on this machinery, a reduction of Mn uptake and content in lettuce seedlings at pH 4.0 or under Mn-deficient conditions leads to inactivation of the autokinase activity of ethylene receptors, which in turn induces ethylene responses, i.e., root hair formation. At pH 4.0, increased ethylene production (Takahashi et al. 2003c, 2010a) may also facilitate inhibition of autokinase activity of the receptors.

Induction of root hair formation by Mn deficiency is also observed in *Arabidopsis*. Mn deficiency increased root hair density 2.2-fold (Ma et al. 2001). Besides normal root hairs from trichoblasts, Mn deficiency also induced the abnormal differentiation of atrichoblasts into root hairs (Yang et al. 2008). This result suggests that Mn deficiency changes the fate of epidermal cells. However, none of the genes that function in cell fate specification were markedly affected by Mn deficiency (Yang et al. 2008), suggesting that the normal patterning mechanism is bypassed by Mn deficiency. Ethylene can also alter cell fate without changing the expression of key determinants of epidermal cell fate (Masucci and Schiefelbein 1996). This common point may support the above discussion concerning activation of the ethylene signaling pathway through Mn deficiency in the root. Future studies are needed to investigate the involvement of the ethylene signaling pathway in lettuce root hair formation induced by Mn deficiency.

6.11 Role of Chlorogenic Acid and Sugar in Root Hair Formation

Although the root and shoot are separated, their communication is indispensable for living plants. Concerning low-pH-induced root hair formation in lettuce, the existence of unknown factors, present in the shoot but essential for roots to form hairs, has been suspected. When lettuce seedlings were decapitated and transferred to pH 4.0 medium, root hair formation was suppressed. However, addition of shoot extract to the medium restored root hair formation in the decapitated seedlings (Narukawa et al. 2009). Since leaves consume large amounts of water for photosynthesis and for cooling through transpiration, it may be reasonable that the shoot has some influence on root hair formation. However, few reports focus on shoot–root communication during root hair formation. Although Yang et al. (2007) found that direct damage to the shoot apical meristem diminished the density and length of root hairs in *Arabidopsis*, the nature of the signal transmitted from shoot to root has not yet been identified.

Recently, Narukawa et al. (2009) found that chlorogenic acid (CGA), a common polyphenol in higher plants, acts as a signal transmitted from shoot to root in low-pH-induced lettuce root hair formation. Application of CGA to decapitated lettuce seedlings restored root hair formation at pH 4.0. On the other hand, when CGA was added to pH 6.0 medium, intact seedlings formed root hairs. Decapitation reduced the CGA content in roots. Furthermore, application of CGA biosynthesis inhibitor to intact seedlings reduced the amount of root hairs at pH 4.0. These observations indicate that either CGA itself, or a substrate or signal for CGA synthesis, moves from shoot to root to induce root hairs.

Moreover, the relationship between CGA and light has been clarified (Narukawa et al. 2010). In contrast to the above experiments performed under continuous light, CGA content was reduced in dark-grown seedling roots, but significantly increased with light irradiation. Application of CGA to dark-grown seedlings at pH 4.0 restored root hair formation. These results indicate that light induces root hair formation at pH 4.0 through the synthesis of CGA.

Sugar plays an important role during CGA synthesis (Narukawa et al. 2010). In the light, sucrose treatment induced CGA synthesis and root hair formation in decapitated seedlings at pH 4.0, but when the CGA synthesis inhibitor was present, sucrose could not induce root hair formation. In the dark, the level of CGA synthesis was low, and no root hairs formed even when sucrose was applied. These results suggest that light induces CGA synthesis from sugar. In fact, there are many reports that show the induction of CGA synthesis through light exposure in many species (Zucker 1963; Aerts and Baumann 1994; Percival and Baird 2000; Abbasi et al. 2007).

In contrast to intact seedlings, decapitated, dark-grown pH 4.0 seedlings could not form root hairs even when CGA was supplied. Similarly, application of sugar could not induce root hairs on these seedlings. These results suggest that factor(s) other than CGA are needed for root hair formation. Interestingly, simultaneous application of CGA and sucrose induced root hair formation on decapitated, dark-grown pH 4.0 seedlings, indicating that sugar plays a role not only as a starting material for CGA synthesis but also as another essential factor for root hair formation. One possible explanation for the additional role of sugar is that it may be needed for cell wall synthesis during root hair formation. The importance of cell wall synthesis in root hair formation is supported by experiments with *Arabidopsis* mutants that lack enzymes for cell wall synthesis: In such mutants, elongation of root hairs was suppressed, although root hair primordia were formed (Favery et al. 2001; Wang et al. 2001; Kim et al. 2007). Because application of sucrose to decapitated dark-grown seedlings could not lead to formation of root hair primordia, sugar may be needed in root hair initiation with CGA other than as a starting material for cell wall synthesis.

Since CGA plays an important role in root hair formation at pH 4.0 in the light, it was thought that CGA content may be lower in pH 6.0 seedlings than in pH 4.0 seedlings and that this may explain why the former cannot form root hairs. However, light irradiation induced CGA synthesis in pH 6.0 seedlings as well as in pH 4.0 seedlings. Furthermore, application of CGA to intact dark-grown pH 6.0

seedlings failed to induce root hair formation (Narukawa et al. 2010). Therefore, the amount of CGA was not a sufficient condition for root hair formation, although it is needed. The pH 4.0 medium may induce other factor(s) that promote root hair formation in conjunction with CGA and sugar. This agrees with the observation that no root hairs form on decapitated seedlings at pH 6.0 in the dark, even when CGA and sugar coexist (Narukawa et al. 2010).

Acidification of the cell wall is known to occur at the root hair initiation site, and preventing acidification arrests the initiation process of root hairs in *Arabidopsis* (Bibikova et al. 1998). Acidification of the cell may modify expression of genes for the production and signal transduction of ethylene, as mentioned above (Takahashi et al. 2003b, 2010a, b). Furthermore, acidification may affect the activity of expansin protein (McQueen-Mason et al. 1992) and cell wall plasticity (Wu et al. 1996). Cell wall-modifying enzymes, including expansins and xyloglucan endotransglycosylase, are also known to show a root hair-specific localization pattern in various plant species (Baluška et al. 2000; Vissenberg et al. 2001; Kim et al. 2007; Won et al. 2010; Zhiming et al. 2011). The importance of changes in cell wall organization during bulge formation is supported by experiments of Bernhardt and Tierney (2000), who isolated *At-PRP3*, a gene encoding a proline-rich structural cell wall protein. This gene is expressed during the later stages of root epidermal cell differentiation and is regulated by developmental pathways leading to root hair outgrowth. Baumberger et al. (2001) reported that LRX1, a chimeric leucine-rich repeat/extensin cell wall protein, is also required for root hair morphogenesis. In addition, root hair elongation defects were observed in cellulose synthase-like mutants (Favery et al. 2001; Wang et al. 2001; Kim et al. 2007; Bernal et al. 2008; Singh et al. 2008).

6.12 Conclusions

Although acid rain and acid soil constitute a serious environmental problem, previous studies have concentrated on the toxicity of and tolerance to heavy metals solubilized by the acidity. In comparison, studies concerning H^+ stress itself are few. We have found that root hair formation under low-pH stress is a tolerance mechanism against acid conditions that is common to certain plant species. Low-pH-induced root hair formation in lettuce seedlings is regulated by various factors: ethylene, auxin, light, CMT randomization, Mn, CGA, and sugar. Among these factors, the relationships between the former four members have been well investigated; they interact with each other in a complicated manner. However, the relationships between these four factors and the rest of the factors have not yet been adequately elucidated. Future studies are needed to answer such questions as whether Mn deficiency or CGA application induces CMT randomization, and whether the plant hormones ethylene or auxin are involved in signal transduction in Mn deficiency or CGA application. At the same time, more in-depth studies are needed concerning CMT randomization. An exciting question is whether CMT randomization is a result of the rearrangement of preexisting CMT arrays or a result of the random construction of new CMT arrays. It should also be clarified how CMT randomization is regulated by auxin and ethylene at the molecular level. Furthermore, another interesting question is whether a homolog of the *STOP1* gene, which was recently determined to be involved in H^+ tolerance in *Arabidopsis*, exists in the lettuce genome and whether it is involved in low-pH-induced root hair formation. Answering these questions will provide insights into the detailed regulatory mechanisms of low-pH-induced root hair formation in lettuce seedlings, i.e., how they adapt to low-pH stress conditions. Since plant hormones and microtubule organization also play important roles in responses to other biotic and abiotic stresses, the above studies would also reveal some part of the common defense mechanisms that plants have against various stresses.

References

Abbasi BH, Tian CL, Murch SJ, Saxena PK, Liu CZ (2007) Light-enhanced caffeic acid derivatives biosynthesis in hairy root cultures of *Echinacea purpurea*. Plant Cell Rep 26:1367–1372

Abdrakhamanova A, Wang QY, Khokhlova L, Nick P (2003) Is microtubule disassembly a trigger for cold acclimation? Plant Cell Physiol 44:676–686

Abeles FB, Morgan PW, Saltveit ME Jr (1992) Ethylene in plant biology. Academic, San Diego

Aerts RJ, Baumann TW (1994) Distribution and utilization of chlorogenic acid in *Coffea* seedlings. J Exp Bot 45:497–503

Anthony RG, Henriques R, Helfer A, Mészáros T, Rios G, Testerink C, Munnik T, Deák M, Koncz C, Bögre L (2004) A protein kinase target of a PDK1 signalling pathway is involved in root hair growth in *Arabidopsis*. EMBO J 11:572–581

Arteca JM, Arteca RN (1999) A multi-responsive gene encoding 1-aminocyclopropane-1-carboxylate synthase (*ACS6*) in mature *Arabidopsis* leaves. Plant Mol Biol 39:209–219

Azegami H, Imai K, Yanagihara Y, Inoue Y (2000) Root hairs, which were induced by low pH in lettuce seedling, contributed on the increment of water permeability of root. J Plant Res 114 (Suppl):87

Balancaflor EB, Hasenstein KH (1995) Growth and microtubule orientation of *Zea mays* roots subjected to osmotic stress. Int J Plant Sci 156:774–783

Baluška F, Salaj J, Mathur J, Braun M, Jasper F, Šamaj J, Chua NH, Peter W, Barlow PW, Volkmann D (2000) Root hair formation: F-actin-dependent tip growth is initiated by local assembly of profilin-supported F-actin meshworks accumulated within expansin-enriched bulges. Dev Biol 227:618–632

Bao Y, Kost B, Chua NH (2001) Reduced expression of α-tubulin genes in *Arabidopsis thaliana* specifically affects root growth and morphology, root hair development and root gravitropism. Plant J 28:145–157

Bates TR, Lynch JP (1996) Stimulation of root hair elongation in *Arabidopsis thaliana* by low phosphorus availability. Plant Cell Environ 19:529–538

Bates TR, Lynch JP (2000) Plant growth and phosphorus accumulation of wild type and two root hair mutants of *Arabidopsis thaliana* (Brassicaceae). Am J Bot 87:958–963

Baumberger N, Ringli C, Keller B (2001) The chimeric leucine-rich repeat/extensin cell wall protein LRX1 is required for root hair morphogenesis in *Arabidopsis thaliana*. Genes Dev 15:1128–1139

Bernal AJ, Yoo CM, Mutwil M, Jensen JK, Hou G, Blaukopf C, Sørensen I, Blancaflor EB, Scheller HV, Willats WG (2008) Functional analysis of the cellulose synthase-like genes *CSLD1*, *CSLD2*, and *CSLD4* in tip-growing Arabidopsis cells. Plant Physiol 148:1238–1253

Bernhardt C, Tierney ML (2000) Expression of AtPRP3, a proline-rich structural cell wall protein from Arabidopsis, is regulated by cell-type-specific developmental pathways involved in root hair formation. Plant Physiol 122:705–714

Bibikova TN, Jacob T, Dahse I, Gilroy S (1998) Localized changes in apoplastic and cytoplasmic pH are associated with root hair development in *Arabidopsis thaliana*. Development 125:2925–2934

Binnie JE, McManus MT (2009) Characterization of the 1-aminocyclopropane-1-carboxylic acid (ACC) oxidase multigene family of *Malus domestica* Borkh. Phytochemistry 70:348–360

Bisson MMA, Groth G (2010) New insight in ethylene signaling: autokinase activity of ETR1 modulates the interaction of receptors and EIN2. Mol Plant 3:882–889

Bisson MMA, Bleckmann A, Allekotte S, Groth G (2009) EIN2, the central regulator of ethylene signalling, is localized at the ER membrane where it interacts with the ethylene receptor ETR1. Biochem J 424:1–6

Blume B, Grierson D (1997) Expression of ACC oxidase promoter–GUS fusions in tomato and *Nicotiana plumbaginifolia* regulated by developmental and environmental stimuli. Plant J 12:731–746

Bouché N, Fromm H (2004) GABA in plants: just a metabolite? Trends Plant Sci 9:110–115

Bown AW, Shelp BJ (1997) The metabolism and functions of γ-aminobutyric acid. Plant Physiol 115:1–5

Buer CS, Wasteneys GO, Masle J (2003) Ethylene modulates root-wave responses in Arabidopsis. Plant Physiol 132:1085–1096

Cahill D, Rookes J, Michalczyk A, McDonald K, Drake A (2002) Microtubule dynamics in compatible and incompatible interactions of soybean hypocotyl cells with *Phytophthora sojae*. Plant Pathol 51:629–640

Castanie-Cornet MP, Penfound TA, Smith D, Elliott JF, Foster JW (1999) Control of acid resistance in *Escherichia coli*. J Bacteriol 181:3525–3535

Chang C, Kwok SF, Bleecker AB, Meyerowitz EM (1993) *Arabidopsis* ethylene-response gene *ETR1*: similarity of product to two-component regulators. Science 262:539–544

Chen BCM, McManus MT (2006) Expression of 1-aminocyclopropane-1-carboxylate (ACC) oxidase genes during the development of vegetative tissues in white clover (*Trifolium repens* L.) is regulated by ontological cues. Plant Mol Biol 60:451–467

Chen YF, Shakeel SN, Bowers J, Zhao XC, Etheridge N, Shaller GE (2007) Ligand-induced degradation of the ethylene receptor ETR2 through a proteasome-dependent pathway in *Arabidopsis*. J Biol Chem 282:24752–24758

Chinnusamy V, Zhu JK (2003) Plant salt tolerance. Top Curr Genet 4:241–270

Chinnusamy V, Jagendorf A, Zhu JK (2005) Understanding and improving salt tolerance in plants. Crop Sci 45:437–448

Choudhury SR, Roy S, Sengupta DN (2008) Characterization of transcriptional profiles of *MA-ACS1* and *MA-ACO1* genes in response to ethylene, auxin, wounding, cold and different photoperiods during ripening in banana fruit. J Plant Physiol 165:1865–1878

Ciardi JA, Tieman DM, Lund ST, Jones JB, Stall RE, Klee HJ (2000) Response to *Xanthomonas campestris* pv. *vesicatoria* in tomato involves regulation of ethylene receptor gene expression. Plant Physiol 123:81–92

Clark DG, Richards C, Hilioti Z, Lind-Iversen S, Brown K (1997) Effect of pollination on accumulation of ACC synthase and ACC oxidase transcripts, ethylene production and flower petal abscission in geranium (*Pelargonium x hortorum* L.H. Bailey). Plant Mol Biol 34:855–865

Clark DG, Gubrium EK, Barrett JE, Nell TA, Klee HJ (1999) Root formation in ethylene insensitive plants. Plant Physiol 121:53–60

Clarkson DT (1985) Factors affecting mineral nutrient acquisition by plants. Annu Rev Plant Physiol 36:77–115

Clowes FAL (2000) Pattern in root meristem development in angiosperms. New Phytol 146:83–94

Cockcroft S (2001) Signalling roles of mammalian phospholipase D1 and D2. Cell Mol Life Sci 58:1674–1687

Crowell EF, Gonneau M, Vernhettes S, Höfte H (2010) Regulation of anisotropic cell expansion in higher plants. C R Biol 333:320–324

De Simone S, Oka Y, Inoue Y (2000a) Effect of light on root hair formation in *Arabidopsis thaliana* phytochrome-deficient mutants. J Plant Res 113:63–69

De Simone S, Oka Y, Inoue Y (2000b) Photoperceptive site of the photoinduction of root hairs in lettuce (*Lactuca sativa* L. cv. Grand Rapids) seedlings under low pH conditions. J Plant Res 113:55–62

De Simone S, Oka Y, Nishioka N, Tadano S, Inoue Y (2000c) Evidence of phytochrome mediation in the low-pH-induced root hair formation process in lettuce (*Lactuca sativa* L. cv. Grand Rapids) seedlings. J Plant Res 113:45–53

Dhonukshe P, Laxalt AM, Goedhart J, Gadella TWJ, Munnik T (2003) Phospholipase D activation correlates with microtubule reorganization in living plant cells. Plant Cell 15:2666–2679

Dixit R, Cyr R (2003) Cell damage and reactive oxygen species production induced by fluorescence microscopy: effect on mitosis and guidelines for non-invasive fluorescence microscopy. Plant J 36:280–290

Dolan L (1997) The role of ethylene in the development of plant form. J Exp Bot 48:201–210

Dolan L, Costa S (2001) Evolution and genetics of root hair stripes in the root epidermis. J Exp Bot 52:413–417

Dolan L, Duckett CM, Grierson C, Linstead P, Schneider K, Lawson E, Dean C, Poethig S, Roberts K (1994) Clonal relationships and cell patterning in the root epidermis of *Arabidopsis*. Development 120:2465–2474

Dominguez M, Vendrell M (1994) Effect of ethylene treatment on ethylene production, EFE activity and ACC levels in peel and pulp of banana fruit. Postharv Biol Technol 4:167–177

Favery B, Ryan E, Foreman J, Linstead P, Boudonck K, Steer M, Shaw P, Dolan L (2001) *KOJAK* encodes a cellulose synthase-like protein required for root hair cell morphogenesis in *Arabidopsis*. Genes Dev 15:79–89

Feldman LJ, Briggs WR (1987) Light-regulated gravitropism in seedling roots of maize. Plant Physiol 83:241–243

Finlayson SA, Lee I-J, Morgan PW (1998) Phytochrome B and the regulation of circadian ethylene production in Sorghum. Plant Physiol 116:17–25

Finlayson SA, Lee I-J, Mullet JE, Morgan PW (1999) The mechanism of rhythmic ethylene production in Sorghum. The role of phytochrome B and simulated shading. Plant Physiol 119:1083–1089

Foehse D, Jungk A (1983) Influence of phosphate and nitrate supply on root hair formation of rape, spinach and tomato plants. Plant Soil 74:359–368

Foo E, Ross JJ, Davies NW, Reid JB, Weller JL (2006) A role for ethylene in the phytochrome-mediated control of vegetative development. Plant J 46:911–921

Foreman J, Demidchik V, Bothwell JHF, Mylona P, Miedema H, Torres MA, Linstead P, Costa S, Brownlee C, Jones JDG, Davies JM, Dolan L (2003) Reactive oxygen species produced by NADPH oxidase regulate plant cell growth. Nature 422:442–446

Gallie DR, Geisler-Lee J, Chen J, Jolley B (2009) Tissue-specific expression of the ethylene biosynthetic machinery regulates root growth in maize. Plant Mol Biol 69:195–211

Galway ME, Masucci JD, Lloyd AM, Walbot V, Davis RW, Schiefelbein JW (1994) The *TTG* gene is required to specify epidermal cell fate and cell patterning in the *Arabidopsis* root. Dev Biol 166:740–754

Gamble RL, Coonfield ML, Schaller GE (1998) Histidine kinase activity of the ETR1 ethylene receptor from *Arabidopsis*. Proc Natl Acad Sci USA 95:7825–7829

Gamble RL, Qu X, Schaller GE (2002) Mutational analysis of the ethylene receptor ETR1. Role of the histidine kinase domain in dominant ethylene insensitivity. Plant Physiol 128:1428–1438

Gardiner J, Collings DA, Harper JDI, Marc J (2003) The effects of the phospholipase D-antagonist 1-butanol on seedling development and microtubule organisation in *Arabidopsis*. Plant Cell Physiol 44:687–696

Ge L, Liu JZ, Wong WS, Hsiao WLW, Chong K, Xu ZK, Yang SF, Kung SD, Li N (2000) Identification of a novel multiple environmental factor-responsive 1-aminocyclopropane-1-carboxylate synthase gene, NT-ACS2, from tobacco. Plant Cell Environ 23:1169–1182

Glazebrook J (2005) Contrasting mechanisms of defense against biotrophic and necrotrophic pathogens. Annu Rev Phytopathol 43:205–227

Goosey L, Palecanda L, Sharrock RA (1997) Differential patterns of expression of the Arabidopsis *PHYB*, *PHYD* and *PHYE* phytochrome genes. Plant Physiol 115:959–969

Hackett RM, Ho C-W, Lin Z, Foote HCC, Fray RG, Grierson D (2000) Antisense inhibition of the *Nr* gene restores normal ripening to the tomato *Never-ripe* mutant, consistent with the ethylene receptor-inhibition model. Plant Physiol 124:1079–1085

Haling RE, Richardson AE, Culvenor RA, Lambers H, Simpson RJ (2010a) Root morphology, root-hair development and rhizosheath formation on perennial grass seedlings is influenced by soil acidity. Plant Soil 335:457–468

Haling RE, Simpson RJ, Delhaize E, Hocking PJ, Richardson AE (2010b) Effect of lime on root growth, morphology and the rhizosheath of cereal seedlings growing in an acid soil. Plant Soil 327:199–212

Hardham AR, Takemoto D, White RG (2008) Rapid and dynamic subcellular reorganization following mechanical stimulation of *Arabidopsis* epidermal cells mimics responses to fungal and oomycete attack. BMC Plant Biol 8:63

Hemm MR, Rider SD, Ogas J, Murry DJ, Chapple C (2004) Light induces phenylpropanoid metabolism in *Arabidopsis* roots. Plant J 38:765–778

Hirata T, Kaneko T, Ono T, Nakazato T, Furukawa N, Hasegawa S, Wakabayashi S, Shigekawa M, Chang MH, Romero MF, Hirose S (2003) Mechanism of acid adaptation of a fish living in a pH 3.5 lake. Am J Physiol Regul Integr Comp Physiol 284:R1199–R1212

Hiwasa K, Kinugasa Y, Amano S, Hashimoto A, Nakano R, Inaba A, Kubo Y (2003) Ethylene is required for both the initiation and progression of softening in pear (*Pyrus communis* L.) fruit. J Exp Bot 54:771–779

Hoekenga OA, Maron LG, Piñeros MA, Cancado GMA, Shaff J, Kobayashi Y, Ryan PR, Dong B, Delhaize E, Sasaki T, Matsumoto H, Yamamoto Y, Koyama H, Kochian LV (2006) *AtALMT1*, which encodes a malate transporter, is identified as one of several genes critical for aluminum tolerance in *Arabidopsis*. Proc Natl Acad Sci USA 103:9738–9743

Hofer R-M (1996) Root hairs: cell biology and development. In: Waisei Y, Eshel A, Kafkafi U (eds) Plant roots: the hidden half, 2nd edn. Marcel Dekker, New York, pp 111–126

Horst WJ, Wang Y, Eticha D (2010) The role of the root apoplast in aluminium-induced inhibition of root elongation and in aluminium resistance of plants: a review. Ann Bot 106:185–197

Hua J, Meyerowitz EM (1998) Ethylene responses are negatively regulated by a receptor gene family in *Arabidopsis thaliana*. Cell 94:261–271

Hua J, Sakai H, Nourizadeh S, Chen QG, Bleecker AB, Ecker JR, Meyerowitz EM (1998) *EIN4* and *ERS2* are members of the putative ethylene receptor gene family in Arabidopsis. Plant Cell 10:1321–1332

Huang Y, Li H, Hutchison CE, Laskey J, Kieber JJ (2003) Biochemical and functional analysis of CTR1, a protein kinase that negatively regulates ethylene signaling in *Arabidopsis*. Plant J 33:221–233

Hurth MA, Suh SJ, Kretzschmar T, Geis T, Bregante M, Gambale F, Martinoia E, Neuhaus HE (2005) Impaired pH homeostasis in Arabidopsis lacking the vacuolar dicarboxylate transporter and analysis of carboxylic acid transport across the tonoplast. Plant Physiol 137:901–910

Inoue Y, Hirota K (2000) Low pH-induced root hair formation in lettuce (*Lactuca sativa* L. cv. Grand Rapids) seedlings: determination of root hair-forming site. J Plant Res 113: 245–251

Inoue Y, Yamaoka K, Kimura K, Sawai K, Arai T (2000) Effects of low pH on the induction of root hair formation in young lettuce (*Lactuca sativa* L. cv. Grand Rapids). J Plant Res 113:39–44

Iuchi S, Koyama H, Iuchi A, Kobayashi Y, Kitabayashi S, Ikka T, Hirayama T, Shinozaki K, Kobayashi M (2007) Zinc finger protein STOP1 is critical for proton tolerance in *Arabidopsis* and coregulates a key gene in aluminum tolerance. Proc Natl Acad Sci USA 104:9900–9905

Jardim SN (2007) Comparative genomics of grasses tolerant to aluminum. Genet Mol Res 6:1178–1189

Johnson EM, Pao LI, Feldman LJ (1991) Regulation of phytochrome message abundance in root caps of maize. Plant Physiol 95:544–550

Johnson E, Bradley M, Harberd NP, Whitelam GC (1994) Photoresponses of light-grown *phyA* mutants of *Arabidopsis* (Phytochrome A is required for the perception of daylength extensions). Plant Physiol 105:141–149

Jonak C, Ökrész L, Bögre L, Hirt H (2002) Complexity, cross talk and integration of plant MAP kinase signalling. Curr Opin Plant Biol 5:415–424

Kachroo A, Kachroo P (2007) Salicylic acid-, jasmonic acid- and ethylene-mediated regulation of plant defense signaling. Genet Eng (NY) 28:55–83

Kathiresan A, Nagarathna KC, Moloney MM, Reid DM, Chinnappa CC (1998) Differential regulation of 1-aminocyclopropane-1-carboxylate synthase gene family and its role in phenotypic plasticity in *Stellaria longipes*. Plant Mol Biol 36:265–274

Kende H (1993) Ethylene biosynthesis. Annu Rev Plant Physiol Plant Mol Biol 44:283–307

Kevany BM, Tieman DM, Taylor MG, Cin VD, Klee HJ (2007) Ethylene receptor degradation controls the timing of ripening in tomato fruit. Plant J 51:458–467

Kieber JJ, Rothenberg M, Roman G, Feldmann KA, Ecker JR (1993) *CTR1*, a negative regulator of the ethylene response pathway in Arabidopsis, encodes a member of the Raf family of protein kinases. Cell 72:427–441

Kim WT, Yang SF (1994) Structure and expression of cDNAs encoding 1-aminocyclopropane-1-carboxylate oxidase homologs isolated from excised mung bean hypocotyls. Planta 194:223–229

Kim YS, Choi D, Lee MM, Lee SH, Kim WT (1998) Biotic and abiotic stress-related expression of 1-aminocyclopropane-1-carboxylate oxidase gene family in *Nicotiana glutinosa* L. Plant Cell Physiol 39:565–573

Kim JH, Lee JH, Joo S, Kim WT (1999) Ethylene regulation of an ERS1 homolog in mung bean seedlings. Physiol Plant 106:90–97

Kim DW, Lee SH, Choi S-B, Won S-K, Heo Y-K, Cho M, Park Y-I, Cho H-T (2006) Functional conservation of a root hair cell-specific *cis*-element in angiosperms with different root hair distribution patterns. Plant Cell 18:2958–2970

Kim CM, Park SH, Je BI, Park SH, Park SJ, Piao HL, Eun MY, Dolan L, Han CD (2007) *OsCSLD1*, a cellulose synthase-like D1 gene, is required for root hair morphogenesis in rice. Plant Physiol 143:1220–1230

Kinraide TB (1998) Three mechanisms for the calcium alleviation of mineral toxicities. Plant Physiol 118:513–520

Kinraide TB (2003) Toxicity factors in acidic forest soils: attempts to evaluate separately the toxic effects of excessive Al^{3+} and H^{+} and insufficient Ca^{2+} and Mg^{2+} upon root elongation. Eur J Soil Sci 54:323–333

Kiss JZ, Mullen JL, Correll MJ, Hangarter RP (2003) Phytochromes A and B mediate red-light-induced positive phototropism in roots. Plant Physiol 131:1411–1417

Kobayashi I, Kobayashi Y, Yamaoka N, Kunoh H (1992) Recognition of a pathogen and a nonpathogen by barley coleoptile cells. III. Responses of microtubules and actin filaments in barley coleoptile cells to penetration attempts. Can J Bot 70:1815–1823

Kobayashi Y, Kobayashi I, Funaki Y, Fujimoto S, Takemoto T, Kunoh H (1997) Dynamic reorganization of microfilaments and microtubules is necessary for the expression of non-host resistance in barley coleoptile cells. Plant J 11:525–537

Kobayashi Y, Hoekenga OA, Itoh H, Nakashima M, Saito S, Shaff JE, Maron LG, Piñeros MA, Kochian LV, Koyama H (2007) Characterization of *AtALMT1* expression in aluminum-inducible malate release and its role for rhizotoxic stress tolerance in Arabidopsis. Plant Physiol 145:843–852

Kochian LV (1995) Cellular mechanisms of aluminum toxicity and resistance in plants. Annu Rev Plant Physiol Plant Mol Biol 46:237–260

Kochian LV, Hoekenga OA, Piñeros MA (2004) How do crop plants tolerate acid soils? Mechanisms of aluminum tolerance and phosphorous efficiency. Annu Rev Plant Biol 55:459–493

Konno M, Ooishi M, Inoue Y (2003) Role of manganese in low-pH-induced root hair formation in *Lactuca sativa* cv. Grand Rapids seedlings. J Plant Res 116:301–307

Konno M, Ooishi M, Inoue Y (2006) Temporal and positional relationships between Mn uptake and low-pH-induced root hair formation in *Lactuca sativa* cv. Grand Rapids seedlings. J Plant Res 119:439–447

Kovtun Y, Chiu WL, Tena G, Sheen J (2000) Functional analysis of oxidative stress-activated mitogen-activated protein kinase cascade in plants. Proc Natl Acad Sci USA 97:2940–2945

Koyama H, Toda T, Yokota S, Zuraida D, Hara T (1995) Effects of aluminum and pH on root growth and cell viability in *Arabidopsis thaliana* strain Landsberg in hydroponic culture. Plant Cell Physiol 36:201–205

Koyama H, Toda T, Hara T (2001) Brief exposure to low-pH stress causes irreversible damage to the growing root in *Arabidopsis thaliana*: pectin-Ca interaction may play an important role in proton rhizotoxicity. J Exp Bot 52:361–368

Landsberg E-C (1996) Hormonal regulation of iron-stress response in sunflower roots: a morphological and cytological investigation. Protoplasma 194:69–80

Larsen PB, Geisler MJB, Jones CA, Williams KM, Cancel JD (2005) *ALS3* encodes a phloem-localized ABC transporter-like protein that is required for aluminum tolerance in Arabidopsis. Plant J 41:353–363

Lee MM, Schiefelbein J (1999) WEREWOLF, a MYB-related protein in *Arabidopsis*, is a position-dependent regulator of epidermal cell patterning. Cell 99:473–483

Lee JS, Chang W-K, Evans ML (1990) Effects of ethylene on the kinetics of curvature and auxin redistribution in gravistimulated roots of *Zea mays*. Plant Physiol 94:1770–1775

Lee SC, Lan WZ, Kim BG, Li L, Cheong YH, Pandey GK, Lu G, Buchanan BB, Luan S (2007) A protein phosphorylation/dephosphorylation network regulates a plant potassium channel. Proc Natl Acad Sci USA 104:15959–15964

Leliévre JM, Tichit L, Larrigaudiére C, Vendrell M, Pech JC (1995) Cold-induced accumulation of 1-aminocyclopropane 1-carboxylate oxidase protein in Granny-Smith apples. Postharv Biol Technol 5:11–17

Leyser HMO, Pickett FB, Dharmasiri S, Estelle M (1996) Mutations in *AXR3* gene of *Arabidopsis* result in altered auxin responses including ectopic expression of the *SAUR-ACI* promoter. Plant J 10:403–414

Lincoln C, Britton JH, Estelle M (1990) Growth and development of the *axr1* mutants of *Arabidopsis*. Plant Cell 2:1071–1080

Liscovitch M, Czarny M, Fiucci G, Tang X (2000) Phospholipase D: molecular and cell biology of a novel gene family. Biochem J 345:401–415

Liu Y, Hoffman NE, Yang SF (1985) Promotion by ethylene of the capability to convert 1-aminocyclopropane-1-carboxylic acid to ethylene in preclimacteric tomato and cantaloupe fruits. Plant Physiol 77:407–411

Liu J-H, Lee-Tamon SH, Reid DM (1997) Differential and wound-inducible expression of 1-aminocylopropane-1-carboxylate oxidase genes in sunflower seedlings. Plant Mol Biol 34:923–933

Liu JP, Magalhaes JV, Shaff J, Kochian LV (2009) Aluminum-activated citrate and malate transporters from the MATE and ALMT families function independently to confer Arabidopsis aluminum tolerance. Plant J 57:389–399
Lloyd C, Chan J (2008) The parallel lives of microtubules and cellulose microfibrils. Curr Opin Plant Biol 11:641–646
Lund ST, Stall RE, Klee HJ (1998) Ethylene regulates the susceptible response to pathogen infection in tomato. Plant Cell 10:371–382
Ma JF (2000) Role of organic acids in detoxification of aluminum in higher plants. Plant Cell Physiol 41:383–390
Ma JF, Hiradate S, Matsumoto H (1998) High aluminum resistance in buckwheat. II. Oxalic acid detoxifies aluminum internally. Plant Physiol 117:753–759
Ma Z, Bielenberg DG, Brown KM, Lynch JP (2001) Regulation of root hair density by phosphorus availability in *Arabidopsis thaliana*. Plant Cell Environ 24:459–467
Magalhaes JV, Liu J, Guimaraes CT, Lana UG, Alves VM, Wang YH, Schaffert RE, Hoekenga OA, Piñeros MA, Shaff JE, Klein PE, Carneiro NP, Coelho CM, Trick HN, Kochian LV (2007) A gene in the multidrug and toxic compound extrusion (MATE) family confers aluminum tolerance in sorghum. Nat Genet 39:1156–1161
Marschner H (1995) Mineral nutrition of higher plants, 2nd edn. Academic, Cambridge, UK
Masucci JD, Schiefelbein JW (1994) The *rhd6* mutation of *Arabidopsis thaliana* alters root-hair initiation through an auxin- and ethylene-associated process. Plant Physiol 106:1335–1346
Masucci JD, Schiefelbein JW (1996) Hormones act downstream of *TTG* and *GL2* to promote root hair outgrowth during epidermis development in the Arabidopsis root. Plant Cell 8:1505–1517
Masucci JD, Rerie WG, Foreman DR, Zhang M, Galway ME, Marks MD, Schiefelbein JW (1996) The homeobox gene *GLABRA 2* is required for position-dependent cell differentiation in the root epidermis of *Arabidopsis thaliana*. Development 122:1253–1260
McQueen-Mason S, Durachko DM, Cosgrove DJ (1992) Two endogenous proteins that induce cell wall extension in plants. Plant Cell 4:1425–1433
Miyake K (1916) The toxic action of aluminum salts upon the growth of the rice plant. J Biol Chem 25:23–28
Molas ML, Kiss JZ, Correll MJ (2006) Gene profiling of the red light signaling pathways in roots. J Exp Bot 12:3217–3229
Moog PR, van der Kooij TAW, Brüggemann W, Schiefelbein JW, Kuiper PJ (1995) Responses to iron deficiency in *Arabidopsis thaliana*: the Turbo iron reductase does not depend on the formation of root hairs and transfer cells. Planta 195:505–513
Moussatche P, Klee HJ (2004) Autophosphorylation activity of the Arabidopsis ethylene receptor multigene family. J Biol Chem 279:48734–48741
Munnik T (2001) Phosphatidic acid: an emerging plant lipid second messenger. Trends Plant Sci 6:227–233
Nakajima N, Mori H, Yamazaki K, Imaseki H (1990) Molecular cloning and sequence of a complementary DNA encoding 1-aminocyclopropane-1-carboxylate synthase induced by tissue wounding. Plant Cell Physiol 31:1021–1029
Nakano R, Ogura E, Kubo Y, Inaba A (2003) Ethylene biosynthesis in detached young persimmon fruit is initiated in calyx and modulated by water loss from the fruit. Plant Physiol 131:276–286
Narukawa M, Kanbara K, Tominaga Y, Aitani Y, Fukuda K, Kodama T, Murayama N, Nara Y, Arai T, Konno M, Kamisuki S, Sugawara F, Iwai M, Inoue Y (2009) Chlorogenic acid facilitates root hair formation in lettuce seedlings. Plant Cell Physiol 50:504–514
Narukawa M, Watanabe K, Inoue Y (2010) Light-induced root hair formation in lettuce (*Lactuca sativa* L. cv. Grand Rapids) roots at low pH is brought by chlorogenic acid synthesis and sugar. J Plant Res 123:789–799
Neff MN, Fankhauser C, Chory J (2000) Light: an indicator of time and place. Genes Dev 14:257–271

Oetiker JH, Olson DC, Shiu OY, Yang SF (1997) Differential induction of seven 1-aminocyclopropane-1-carboxylate synthase genes by elicitor in suspension cultures of the tomato (*Lycopersicon esculentum*). Plant Mol Biol 34:275–286

Ohashi Y, Oka A, Rodrigues-Pousada R, Possenti M, Ruberti I, Morelli G, Aoyama T (2003) Modulation of phospholipid signaling by GLABRA2 in root-hair pattern formation. Science 300:1427–1430

Okada K, Shimura Y (1994) Modulation of root growth by physical stimuli. In: Arabidopsis. Cold Spring Harbor Laboratory Press, Cold Spring Harbor, NY, pp 665–684

Olinevich OV, Khokhlova LP, Raudaskoski M (2002) The microtubule stability increases in abscisic acid-treated and cold-acclimated differentiating vascular root tissues of wheat. J Plant Physiol 159:465–472

Owino WO, Manabe Y, Mathooko FM, Kubo Y, Inaba A (2006) Regulatory mechanisms of ethylene biosynthesis in response to various stimuli during maturation and ripening in fig fruit (*Ficus carica* L.). Plant Physiol Biochem 44:335–342

Oyama T, Shimura Y, Okada K (1997) The *Arabidopsis HY5* gene encodes a bZIP protein that regulates stimulus-induced development of root and hypocotyl. Genes Dev 11:2983–2995

Oyama T, Shimura Y, Okada K (2002) The *IRE* gene encodes a protein kinase homologue and modulates root hair growth in *Arabidopsis*. Plant J 30:289–299

Panda SK, Baluska F, Matsumoto H (2009) Aluminum stress signaling in plants. Plant Signal Behav 4:592–597

Paradez A, Wright A, Ehrhardt DW (2006) Microtubule cortical array organization and plant cell morphogenesis. Curr Opin Plant Biol 9:571–578

Percival GC, Baird L (2000) Influence of storage upon light-induced chlorogenic acid accumulation in potato tubers (*Solanum tuberosum* L.). J Agric Food Chem 48:2476–2482

Peterson LR, Farquhar ML (1996) Root hairs: specialized tubular cells extending root surface. Bot Rev 62:1–35

Petruzzelli L, Coraggio I, Leubner-Metzger G (2000) Ethylene promotes ethylene biosynthesis during pea seed germination by positive feedback regulation of 1-aminocyclo-propane-1-carboxylic acid oxidase. Planta 211:144–149

Pierik R, Cuppens MLC, Voesenek LACJ, Visser EJW (2004) Interactions between ethylene and gibberellins in phytochrome-mediated shade avoidance responses in tobacco. Plant Physiol 136:2928–2936

Pierik R, Djakovic-Petrovic T, Keuskamp DH, de Wit M, Voesenek LACJ (2009) Auxin and ethylene regulate elongation responses to neighbor proximity signals independent of gibberellin and DELLA proteins in Arabidopsis. Plant Physiol 149:1701–1712

Pitts RJ, Cernac A, Esteile M (1998) Auxin and ethylene promote root hair elongation in *Arabidopsis*. Plant J 16:553–560

Plett JM, Mathur J, Regan S (2009) Ethylene receptor ETR2 controls trichome branching by regulating microtubule assembly in *Arabidopsis thaliana*. J Exp Bot 60:3923–3933

Poschenrieder C, Gunsé B, Corrales I, Barceló J (2008) A glance into aluminum toxicity and resistance in plants. Sci Total Environ 400:356–368

Potocký M, Eliáš M, Profotová B, Novotná Z, Valentová O, Žárský V (2003) Phosphatidic acid produced by phospholipase D is required for tobacco pollen tube growth. Planta 217: 122–130

Qu X, Schaller GE (2004) Requirement of the histidine kinase domain for signal transduction by the ethylene receptor ETR1. Plant Physiol 136:2961–2970

Quint M, Barkawi LS, Fan K-T, Cohen JD, Gray WM (2009) Arabidopsis *IAR4* modulates auxin response by regulating auxin homeostasis. Plant Physiol 150:748–758

Rangel AF, Mobin M, Rao IM, Horst WJ (2005) Proton toxicity interferes with the screening of common bean (*Phaseolus vulgaris* L.) genotypes for aluminium resistance in nutrient solution. J Plant Nutr Soil Sci 168:607–616

Reed JW, Nagpal P, Poole DS, Furuya M, Chory J (1993) Mutations in the gene for the red/far-red light receptor phytochrome B alter cell elongation and physiological responses throughout Arabidopsis development. Plant Cell 5:147–157

Rentel MC, Lecourieux D, Ouaked F, Usher SL, Petersen L, Okamoto H, Knight H, Peck SC, Grierson CS, Hirt H, Knight MR (2004) OXI1 kinase is necessary for oxidative burst-mediated signalling in *Arabidopsis*. Nature 427:858–861

Ridge RW (1996) Root hairs: cell biology and development. In: Waisei Y, Eshel A, Kafkafi U (eds) Plant roots: the hidden half, 2nd edn. Marcel Dekker, New York, pp 127–147

Rieu I, Cristescu SM, Harren FJM, Huibers W, Voesenek LACJ, Mariani C, Vriezen WH (2005) *RP-ACS1*, a flooding-induced 1-aminocyclopropane-1-carboxylate synthase gene of *Rumex palustris*, is involved in rhythmic ethylene production. J Exp Bot 56:841–849

Roberts JK, Hooks MA, Miaullis AP, Edwards S, Webster C (1992) Contribution of malate and amino acid metabolism to cytoplasmic pH regulation in hypoxic maize root tips studied using nuclear magnetic resonance spectroscopy. Plant Physiol 98:480–487

Rottmann WE, Petre GF, Oeller PW, Keller JA, Shen NF, Nagy BP, Tayler LP, Campbell AD, Theologis A (1991) 1-aminocyclopropane-1-carboxylate acid synthase in tomato is encoded by a multigene family whose transcription is induced during fruit and floral senescence. J Mol Biol 222:937–961

Rudrappa T, Czymmek KJ, Paré PW, Bais HP (2008) Root-secreted malic acid recruits beneficial soil bacteria. Plant Physiol 148:1547–1556

Ryan P, Delhaize E, Jones D (2001) Function and mechanism of organic anion exudation from plant roots. Annu Rev Plant Physiol Plant Mol Biol 52:527–560

Sakano K (1998) Revision of biochemical pH-stat: involvement of alternative pathway metabolisms. Plant Cell Physiol 39:467–473

Šamaj J, Ovecka M, Hlavacka A, Lecourieux F, Meskiene I, Lichtscheidl I, Lenart P, Salaj J, Volkmann D, Bögre L, Baluška F, Hirt H (2002) Involvement of the mitogen-activated protein kinase SIMK in regulation of root hair tip growth. EMBO J 21:3296–3306

Sano R, Nagasaka R, Inoue K, Shirano Y, Hayashi H, Shibata D, Sato S, Kato T, Tabata S, Okada K, Wada T (2003) Analysis of bHLH (MYC) genes involved in root hair and trichome differentiation. In: 14th International conference on Arabidopsis research

Sasaki T, Yamamoto Y, Ezaki B, Katsuhara M, Ahn SJ, Ryan PR, Delhaize E, Matsumoto H (2004) A wheat gene encoding an aluminum-activated malate transporter. Plant J 37:645–653

Sawaki Y, Iuchi S, Kobayashi Y, Kobayashi Y, Ikka T, Sakurai N, Fujita M, Shinozaki K, Shibata D, Kobayashi M, Koyama H (2009) STOP1 regulates multiple genes that protect Arabidopsis from proton and aluminum toxicities. Plant Physiol 150:281–294

Schiefelbein JW, Somerville C (1990) Genetic control of root hair development in *Arabidopsis thaliana*. Plant Cell 2:235–243.

Schiefelbein JW (2000) Constructing a plant cell. The genetic control of root hair development. Plant Physiol 124:1525–1531

Schmidt W, Schikora A (2001) Different pathways are involved in phosphate and iron stress-induced alterations of root epidermal cell development. Plant Physiol 125:2078–2084

Schmidt W, Tittel J, Schikora A (2000) Role of hormones in the induction of iron deficiency responses in Arabidopsis roots. Plant Physiol 122:1109–1118

Shiu OY, Oetiker JH, Yip WK, Yang SF (1998) The promoter of *LE-ACS7*, an early flooding-induced 1-aminocyclopropane-1-carboxylate synthase gene of the tomato, is tagged by a *Sol3* transposon. Proc Natl Acad Sci USA 95:10334–10339

Shoji T, Suzuki K, Abe T, Kaneko Y, Shi H, Zhu J-K, Rus A, Hasegawa PM, Hashimoto T (2006) Salt stress affects cortical microtubule organization and helical growth in Arabidopsis. Plant Cell Physiol 47:1158–1168

Singh SK, Fischer U, Singh M, Grebe M, Marchant A (2008) Insight into the early steps of root hair formation revealed by the *procuste1* cellulose synthase mutant of *Arabidopsis thaliana*. BMC Plant Biol 8:57

Somers DE, Quail PH (1995) Phytochrome-mediated light regulation of *PHYA*- and *PHYB-GUS* transgenes in *Arabidopsis thaliana* seedlings. Plant Physiol 107:523–534

Stepanova AN, Yun J, Likhacheva AV, Alonso JM (2007) Multilevel interactions between ethylene and auxin in *Arabidopsis* roots. Plant Cell 19:2169–2185

Takahashi H, Inoue Y (2008) Stage-specific crosstalk between light, auxin, and ethylene during low-pH-induced root hair formation in lettuce (*Lactuca sativa* L.) seedlings. Plant Growth Regul 56:31–41

Takahashi H, Hirota K, Kawahara A, Hayakawa E, Inoue Y (2003a) Randomization of cortical microtubules in root epidermal cells induces root hair initiation in lettuce (*Lactuca sativa* L.) seedlings. Plant Cell Physiol 44:350–359

Takahashi H, Iwasa T, Shinkawa T, Kawahara A, Kurusu T, Inoue Y (2003b) Isolation and characterization of the ACC synthase genes from lettuce (*Lactuca sativa* L.), and the involvement in low pH-induced root hair initiation. Plant Cell Physiol 44:62–69

Takahashi H, Kawahara A, Inoue Y (2003c) Ethylene promotes the induction by auxin of the cortical microtubule randomization required for low-pH-induced root hair initiation in lettuce (*Lactuca sativa* L.) seedlings. Plant Cell Physiol 44:932–940

Takahashi H, Nakamura A, Harigaya W, Fujigasaki R, Iwasa T, Inoue Y (2010a) Increased expression of ethylene receptor genes during low pH-induced root hair formation in lettuce (*Lactuca sativa* L.) seedlings: direct and indirect induction by ethylene and auxin, respectively. Plant Root 4:53–64

Takahashi H, Shinkawa T, Nakai S, Inoue Y (2010b) Differential expression of ACC oxidase genes during low-pH-induced root hair formation in lettuce (*Lactuca sativa* L.) seedlings. Plant Growth Regul 62:137–149

Takano M, Kanegae H, Shinomura T, Miyao A, Hirochika H, Furuya M (2001) Isolation and characterization of rice phytochrome A mutants. Plant Cell 13:521–534

Takemoto D, Hardham AR (2004) The cytoskeleton as a regulator and target of biotic interactions in plants. Plant Physiol 136:3864–3876

Tanimoto M, Roberts K, Dolan L (1995) Ethylene is a positive regulator of root hair development in *Arabidopsis thaliana*. Plant J 8:943–948

Tieman DM, Taylor MG, Ciardi JA, Klee HJ (2000) The tomato ethylene receptors NR and LeETR4 are negative regulators of ethylene response and exhibit functional compensation within a multigene family. Proc Natl Acad Sci USA 97:5663–5668

Trainotti L, Pavanello A, Casadoro G (2005) Different ethylene receptors show an increased expression during the ripening of strawberries: does such an increment imply a role for ethylene in the ripening of these non-climacteric fruits? J Exp Bot 56:2037–2046

Tsuchisaka A, Theologis A (2004) Unique and overlapping expression patterns among the Arabidopsis 1-amino-cyclopropane-1-carboxylate synthase gene family members. Plant Physiol 2:2982–3000

Van Bruaene N, Joss G, Van Oostveldt P (2004) Reorganization and in vivo dynamics of microtubules during Arabidopsis root hair development. Plant Physiol 136:3905–3919

Van Der Straeten D, Van Wiemeersch L, Goodman HM, Van Montagu M (1990) Cloning and sequence of two different cDNAs encoding 1-aminocyclopropane-1-carboxylate synthase in tomato. Proc Natl Acad Sci USA 87:4859–4863

van Tuinen A, Kerckchoffs LHJ, Nagatani A, Kendrick RE, Koornneef M (1995) Far-red light-insensitive, phytochrome A-deficient mutants of tomato. Mol Gen Genet 246:133–141

Vissenberg K, Fry SC, Verbelen J-P (2001) Root hair initiation is coupled to a highly localized increase of xyloglucan endotransglycosylase action in Arabidopsis roots. Plant Physiol 127:1125–1135

Voet-van-Vormizeele J, Groth G (2008) Ethylene controls autophosphorylation of the histidine kinase domain in ethylene receptor ETR1. Mol Plant 1:380–387

von Uexküll HR, Mutert E (1995) Global extent, development and economic impact of acid soils. Plant Soil 171:1–15

Vriezen WH, van Rijn CPE, Voesenek LACJ, Mariani C (1997) A homolog of the *Arabidopsis thaliana ERS* gene is actively regulated in *Rumex palustris* upon flooding. Plant J 11: 1265–1271

Wada T, Tachibana T, Shimura Y, Okada K (1997) Epidermal cell differentiation in *Arabidopsis* determined by a *Myb* homolog, *CPC*. Science 277:1113–1116

Wang T-W, Arteca RN (1995) Identification and characterization of cDNAs encoding ethylene biosynthetic enzymes from *Pelargonium* x *hortorum* cv Snow Mass leaves. Plant Physiol 109:627–636

Wang QY, Nick P (2001) Cold acclimation can induce microtubular cold stability in a manner distinct from abscisic acid. Plant Cell Physiol 42:999–1005

Wang H, Woodson WR (1989) Reversible inhibition of ethylene action and interruption of petal senescence in carnation flowers by norbornadiene. Plant Physiol 89:434–438

Wang X, Cnops G, Vanderhaeghen R, De Block S, Van Montagu M, Van Lijsebettens M (2001) *AtCSLD3*, a cellulose synthase-like gene important for root hair growth in Arabidopsis. Plant Physiol 126:575–586

Wang W, Hall AE, O'Malley R, Bleecker AB (2003) Canonical histidine kinase activity of the transmitter domain of the ETR1 ethylene receptor from *Arabidopsis* is not required for signal transmission. Proc Natl Acad Sci USA 100:352–357

Wang C, Li J, Yuan M (2007) Salt tolerance requires cortical microtubule reorganization in Arabidopsis. Plant Cell Physiol 48:1534–1547

Watanabe T, Okada K (2005) Interactive effects of Al, Ca and other cations on root elongation of rice cultivars under low pH. Ann Bot 95:379–385

Weller JL, Nagatani A, Kendrick RE, Murfet IC, Reid JB (1995) New *lv* mutants of pea are deficient in phytochrome B. Plant Physiol 108:525–532

Weller JL, Schreuder MEL, Smith H, Koornneef M, Kendrick RE (2000) Physiological interactions of the phytochromes A, B1 and B2 in the control of development in tomato. Plant J 24:345–356

Weller JL, Beauchamp N, Kerckhoffs LHJ, Platten JD, Reid JB (2001) Interaction of phytochrome A and B in the control of de-etiolation and flowering in pea. Plant J 26:283–294

Whitelam GC, Johnson E, Peng J, Carol P, Anderson ML, Cowl JS, Harberd NP (1993) Phytochrome A null mutants of Arabidopsis display a wild-type phenotype in white light. Plant Cell 5:757–768

Wilson AK, Pickett FB, Turner JC, Estelle M (1990) A dominant mutation in *Arabidopsis* confers resistance to auxin, ethylene, and abscisic acid. Mol Gen Genet 222:377–383

Won S-K, Choi S-B, Kumari S, Cho M, Lee SH, Cho H-T (2010) Root hair-specific EXPANSIN B genes have been selected for Graminaceae root hairs. Mol Cells 30:369–376

Wu Y, Sharp RE, Durachko DM, Cosgrove DJ (1996) Growth maintenance of the maize primary root at low water potentials involves increases in cell-wall extension properties, expansin activity, and wall susceptibility to expansins. Plant Physiol 111:765–772

Xie ZM, Lei G, Hada W, Tian AG, Zhang JS, Chen SY (2007) Cloning and expression of putative ethylene receptor genes in soybean plant. Prog Nat Sci 17:1152–1160

Xiong L, Schumaker KS, Zhu J-K (2002) Cell signaling during cold, drought, and salt stress. Plant Cell 14:S165–S183

Xu C, Liu C, Guo H, Li Z, Jiang Y, Zhang D, Yuan M (2006) Photosensitive breakage of fluorescence-labeled microtubules and its mechanism. Acta Phys Sin 55:206–210 (in Chinese)

Yadav V, Mallappa C, Gangappa SN, Bhatia S, Chattopadhyay S (2005) A basic helix-loop-helix transcription factor in Arabidopsis, MYC2, acts as a repressor of blue light-mediated photomorphogenic growth. Plant Cell 17:1953–1966

Yamasaki S, Fujii N, Takahashi H (2000) The ethylene-regulated expression of *CS-ETR2* and *CS-ERS* genes in cucumber plants and their possible involvement with sex expression in flowers. Plant Cell Physiol 41:608–616

Yan F, Schubert S, Mengel K (1992) Effect of low root medium pH on net proton release, root respiration, and root growth of corn (*Zea mays* L.) and broad bean (*Vicia faba* L.). Plant Physiol 99:415–421

Yang SF, Hoffman NE (1984) Ethylene biosynthesis and its regulation in higher-plants. Annu Rev Plant Physiol 35:155–189

Yang C-Y, Chu FH, Wang YT, Chen Y-T, Yang SF, Shaw J-F (2003) Novel broccoli 1-aminocyclopropane-1-carboxylate oxidase gene (*Bo-ACO3*) associated with the late stage of postharvest floret senescence. J Agric Food Chem 51:2569–2575

Yang JL, Zheng SJ, He YF, Matsumoto H (2005) Aluminium resistance requires resistance to acid stress: a case study with spinach that exudes oxalate rapidly when exposed to Al stress. J Exp Bot 56:1197–1203

Yang G, Wu L, Chen L, Pei B, Wang Y, Zhan F, Wu Y, Yu Z (2007) Targeted irradiation of shoot apical meristem of *Arabidopsis* embryos induces long-distance bystander/abscopal effects. Radiat Res 167:298–305

Yang TJW, Perry PJ, Ciani S, Pandian S, Schmidt W (2008) Manganese deficiency alters the patterning and development of root hairs in *Arabidopsis*. J Exp Bot 59:3453–3464

Yohannes E, Barnhart DM, Slonczewski JL (2004) pH-dependent catabolic protein expression during anaerobic growth of *Escherichia coli* K-12. J Bacteriol 186:192–199

Yokota S, Ojima K (1995) Physiological response of root tip of alfalfa to low pH and aluminium stress in water culture. Plant Soil 171:163–165

Yoon IS, Mori H, Kim JH, Kang BG, Imaseki H (1997) *VR-ACS6* is an auxin-inducible 1-aminocyclopropane-1-carboxylate synthase gene in mungbean (*Vigna radiata*). Plant Cell Physiol 38:217–224

Yuan H-Y, Yao L-L, Jia Z-Q, Li Y, Li Y-Z (2006) *Verticillium dahliae* toxin induced alterations of cytoskeletons and nucleoli in *Arabidopsis thaliana* suspension cells. Protoplasma 229:75–82

Zarembinski TI, Theologis A (1993) Anaerobiosis and plant growth hormones induce two genes encoding 1-aminocyclopropane-1-carboxylate synthase in rice (*Oryza sativa* L.). Mol Biol Cell 4:363–373

Zhang W, Kone BC (2002) NF-κB inhibits transcription of the H^+-K^+-ATPase α_2-subunit gene: role of histone deacetylases. Am J Physiol Renal Physiol 283:F904–F911

Zhang W, Yu L, Zhang Y, Wang X (2005) Phospholipase D in the signaling networks of plant response to abscisic acid and reactive oxygen species. Biochim Biophys Acta 1736:1–9

Zheng SJ, Yang JL, He YF, Yu XH, Zhang L, You JF, Shen RF, Matsumoto H (2005) Immobilization of aluminum with phosphorus in roots is associated with high aluminum resistance in buckwheat. Plant Physiol 138:297–303

ZhiMing Y, Bo K, XiaoWei H, ShaoLei L, YouHuang B, WoNa D, Ming C, Hyung-Taeg C, Ping W (2011) Root hair-specific expansins modulate root hair elongation in rice. Plant J 66:725–734

Zonia L, Munnik T (2004) Osmotically induced cell swelling versus cell shrinking elicits specific changes in phospholipid signals in tobacco pollen tubes. Plant Physiol 134:813–823

Zucker M (1963) The influence of light on synthesis of protein and of chlorogenic acid in potato tuber tissue. Plant Physiol 38:575–580

Chapter 7
Cytokinin Metabolism

Somya Dwivedi-Burks

Abstract Cytokinins play an important role in various physiological functions in the plant. This chapter is an introduction to the mechanisms involved in cytokinin metabolism and highlights the need for transforming our approach from an organismal level to a holistic environmental level.

Keywords Cytokinins • Cytokinin metabolism • Cytokinin *N*-glucosylation

List of Abbreviations

ABA	Abscisic acid
BA	N^6-benzyladenine
BA7G	N^6-benzyladenine-7-glucoside
BA9G	N^6-benzyladenine-9-glucoside
*cis*Z	*Cis*-zeatin
*cis*Z7G	*Cis*-zeatin-7-glucoside
*cis*Z9G	*Cis*-zeatin-9-glucoside
*cis*ZOG	*Cis*-zeatin-*O*-glucoside
*cis*ZR	*Cis*-zeatin riboside
CK	Cytokinin
CKX	Cytokinin oxidase/dehydrogenase
DHZ	Dihydrozeatin
DHZ7G	Dihydrozeatin-7-glucoside
DHZ9G	Dihydrozeatin-9-glucoside
DHZOG	Dihydrozeatin-*O*-glucoside
DHZR	Dihydrozeatin riboside

S. Dwivedi-Burks (✉)
Pennsylvania State University York Campus, 1031 Edgecomb Avenue, York PA 17403, USA
e-mail: sud16@psu.edu

N.A. Khan et al. (eds.), *Phytohormones and Abiotic Stress Tolerance in Plants*,
DOI 10.1007/978-3-642-25829-9_7,

GA_3	Gibberellic acid
N-GT	*N*-glucosyltransferase
IP	N^6-(Δ^2-isopentenyl)adenine
iP7G	N^6-(Δ^2-isopentenyl)adenine 7-glucoside
Z	*Trans*-zeatin
Z7G	*Trans*-zeatin-7-glucoside
Z9G	*Trans*-zeatin-9-glucoside
ZR	*Trans*-zeatin riboside
ZOG	*Trans*-zeatin-*O*-glucoside

7.1 Introduction

The form and coordinated functioning of plants result from efficient communication between cells, tissues, and organs. Hormones are the signaling molecules that help in regulation and coordination of metabolism, growth, and morphogenesis in higher plants. The plant hormone cytokinin is known as a regulator of cell division and various other important processes such as leaf senescence, nutrient mobilization, apical dominance, branching patterns, chlorophyll degradation, root initiation, breaking of bud and seed dormancy, formation and activity of shoot apical meristems, and reproductive development. Although cytokinins play a critical role in plant growth and development, the mechanism of their action is poorly understood (Brzobohaty et al. 1994; Mok and Mok 2001; Davies 2004; Gan and Amasino 1995; Taiz and Zeiger 2004; Lexa et al. 2003).

Since the discovery of cytokinins in the 1950s, new insights about cytokinin function have been obtained from the recent discovery of some of the genes involved in cytokinin biosynthesis, metabolism, and signal transduction (Kieber 2000). For example, the discovery of the genes for cytokinin oxidase, which degrades cytokinins, has allowed scientists to produce plants that have lower concentrations of endogenous cytokinin (Houba-Herin et al. 1999; Morris et al. 1999). Transgenic tobacco plants that expressed various cytokinin oxidase (AtCKX) genes derived from *Arabidopsis* showed many phenotypic changes, such as reduced size of shoot apical meristem, reduced vasculature in leaves, enhanced leaf thickness, more adventitious roots, more root branching, and enlarged root apical meristems. AtCKX overexpressing plants also challenged our perception of the relationship between leaf senescence and cytokinins. It was thought that senescence occurred at a faster rate in plants with reduced cytokinin content. However, older leaves in the AtCKX overexpressing plants stayed green, suggesting that cytokinins are not the only signal governing leaf senescence (Schmülling et al. 2003).

Plants have another pathway for decreasing endogenous cytokinins through the conjugation of the cytokinin molecule to a glucose sugar. Binding of a glucose moiety to the purine ring at the 7*N* or 9*N* position produces stable conjugates that exhibit no significant biological activity. The enzyme(s) responsible for this conjugation pathway has been named *N*-glucosyltransferase (N-GT). This pathway has been demonstrated through feeding studies with radiolabeled cytokinins in many plant species, but the enzymes and genes have not been elucidated (Auer 1997). We do not know if there are one or more enzymes at work in this metabolism pathway or how they are regulated across time and space.

7.1.1 Structural Variation and Biological Activity

Cytokinin biochemistry is very complex, as over 30 different structural variations have been found in plants. Natural cytokinins are adenine derivatives with at least one substituent at the 6*N* position. On the basis of this 6*N* substituent, they can be classified into two types: (1) aromatic cytokinins where the purine base has an aromatic side chain and (2) isoprenoid cytokinins where an isoprenoid group is attached to the side chain of the purine base. Common natural isoprenoid cytokinins are 6*N*-(Δ^2-isopentenyl)adenine (iP), *t*Z, *cis*-zeatin (*c*Z), and dihydrozeatin (DHZ). The cytokinins *t*Z, iP, and their respective sugar conjugates occur more commonly, but this varies depending on plant species, tissue, and developmental stage. The aromatic cytokinins that are found in some plants are *ortho*-topolin, *meta*-topolin, and their methoxy derivatives and BA (See Fig. 7.1). There is also a wide range of synthetic derivatives of cytokinins that have not been discovered in nature so far (Strnad et al. 1997; Sakakibara 2006).

The classical bioassays using tobacco pith and moss suggested that the active cytokinins were the free-base cytokinins such as iP and *t*Z, whereas *ct*ZR showed lower level of activity. Discrepancy in bioassay results led to the possibility of interconversion of free bases into other metabolites during experiments. Discovery of cytokinin receptor genes provided a better understanding of relative activity of cytokinins. An in vitro assay using ^{3}H-labeled cytokinin and the cytokinin receptor AHK4 (CRE1) showed that *t*Z and iP could bind to the receptor but not the nucleosides. Therefore, free-base cytokinins are regarded as active forms (Yamada et al. 2004; Sakakibara 2006).

7.1.2 Biosynthesis of Cytokinins

Biosynthesis of isoprenoid cytokinins can take place by two pathways: one is derived from tRNA degradation and another is derived from isopentylation of adenine nucleotides. The biosynthetic reaction substrates of two cytokinins, iP and iPR, are dimethylallyl diphosphate (DMAPP) and adenosine-5′-monophosphate (AMP). The resulting product was iPRMP. The reaction was catalyzed by isopentyltransferase (iPT). The first characterization of an IPT gene was carried out in *Agrobacterium tumefaciens* (Taiz and Zeiger 2004).

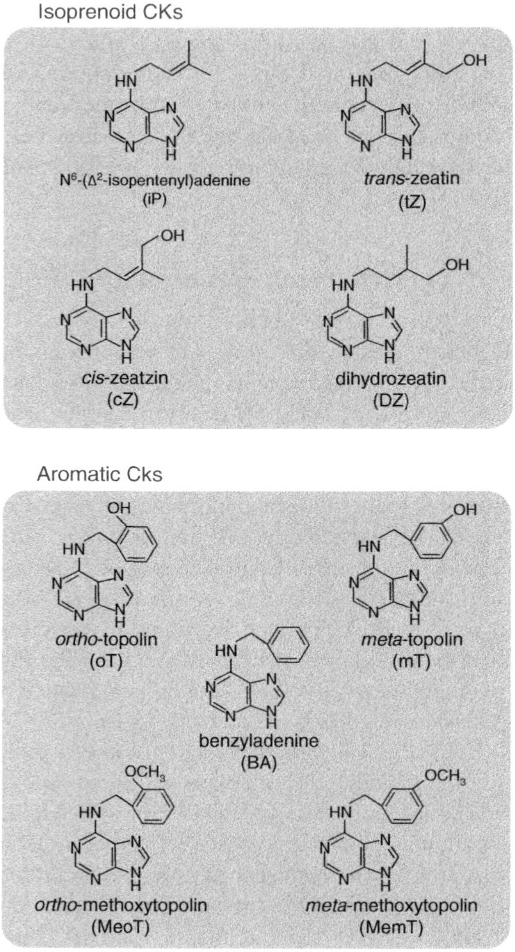

Fig. 7.1 Structures of representative naturally occurring active cytokinins, iP 6*N*-(Δ-isopentenyl) adenine, *tZ trans*-zeatin, *cZ cis*-zeatin, *DZ* dihydrozeatin, *oT ortho*-topolin, *mT meta*-topolin, *BA* benzyladenine, *MeoT ortho*-methoxytopolin, *MemT meta*-methoxytopolin (with permission from Dr. H. Sakakibara)

The current model of isoprenoid cytokinin biosynthesis as shown in Fig. 7.2 shows that the side chains of iP and *tZ* originate from the methylerythritol phosphate (MEP) pathway. A large fraction of *cZ* side chain is derived from the mevalonate (MVA) pathway. Plant IPTs use ATP or ADP as isoprenoid acceptors and form iPRTP and iPRDP, respectively. The metabolic pool of iPRMP and iPRDP is created by dephosphorylation of iPRTP and iPRDP with the help of phosphatase enzyme, phosphorylation of iPR by adenosine kinase, and conjugation of phosphoribosyl moieties to iP by adenine ribosyltransferase (APRT) enzyme. In addition to iP, APRT also utilizes other cytokinin nucleobases. Cytokinin nucleotides are converted into the corresponding *tZ* nucleotides by enzymes CYP735A1 and CYP735A2, which were recently identified in *Arabidopsis* (Sakakibara 2006).

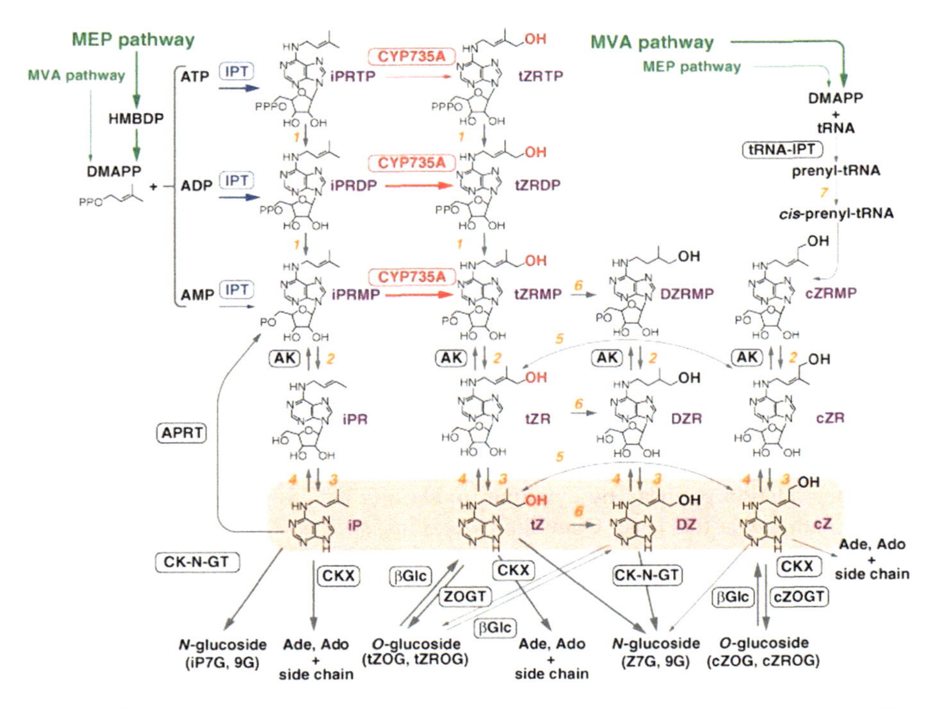

Fig. 7.2 Current model of isoprenoid cytokinin biosynthesis and metabolism in *Arabidopsis*. The width of the *arrowheads* and *lines* in the *green*, *blue*, and *red arrows* indicates the strength of metabolic flow. Flows indicated by *black arrows* are not well characterized to date (with permission from H. Sakakibara). (*1*) *t*ZRDP *t*ZR 5-diphosphate, (*2*) *t*ZRTP *t*ZR 5-triphosphate, (*3*) 5-ribonucleotide phosphohydrolase, (*4*) adenosine nucleosidase, (*5*) purine nucleoside phosphorylase, (*6*) zeatin reductase, (*7*) CK *cis*-hydroxylase

A. tumefaciens has two IPT genes, DMAPP encoding gene (*tmr*) and *trans-zeatin synthesizing* (*tzs*). *tmr* gets attached to the host cell, whereas *tzs* acts in the bacterial cells. The IPT activity of the recombinant *tmr* was identified and showed that *tmr* exclusively used AMP as the acceptor. The reaction pathway of iPRMP found in higher plants is similar to the pathway in *Agrobacterium*. The IPT genes were first identified in *Arabidopsis*. The *Arabidopsis* genome encodes nine genes (AtIPT1–AtIPT9) which were similar to the bacterial IPT. The nine IPT genes can be categorized into two groups on the basis of their phylogeny. The first group consists of genes AtIPT2 and AtIPT9, which share similarity to the gene encoding enzyme tRNA-isopentyltransferase. The second group of genes consisted of AtIPT2 and genes AtIPT3–AtIPT8. This group of gene was similar to the gene encoding the IPT (iPRMP-forming enzyme).

A recent breakthrough in this field was the discovery of *LOG* (*LONELY GUY*) gene in rice provided as evidence against the long-held idea of multistep cytokinin biosynthesis reactions. The gene *LOG* encodes for an enzyme that converts inactive cytokinin nucleotides to the free-base forms by phosphoribohydrolase activity.

The enzyme utilizes cytokinin nucleosides 5′-monophosphates, but not the di- or triphosphates or any cytokinin ribosides or bases (Kurakawa et al. 2007).

7.1.3 Cytokinin Metabolism

Plant cells must carefully regulate hormone levels in order to control their development appropriately. Cytokinin metabolism pathways can be broadly classified into two types: (1) modification of the adenine ring and (2) modification of the side chain. The cytokinin nucleobases and their corresponding nucleosides and nucleotides share the steps in metabolism with the purine metabolic pathway (salvage pathway) (Mok and Mok 2001). Exogenously supplied cytokinins get converted into corresponding nucleotides and nucleosides in the plant tissue (Taiz and Zeiger 2004). There are two major pathways for downregulation of cytokinins. The degradation pathway by cytokinin oxidase (CKX) is the most clearly understood pathway at this time. Common cytokinins, such as Z and iP, are catabolized by CKX through side chain cleavage to their adenine or adenosine derivatives, which results in an irreversible loss of biological activity (Motyka et al. 2003; Schmülling et al. 2003). The cytokinin nucleotides, *O*-glucosides, aromatic cytokinins, and cytokinins with saturated side chains are not CKX substrates (Armstrong 1994). The *c*Z and *t*Z can be enzymatically interconverted by zeatin *cis–trans* isomerase. Recently, a comprehensive study using models representing the monocot and dicot plants showed that the levels of *c*Z did not correlate with the evolutionary relationship between plants, but the levels of ontogenesis the plant undergoes when subjected to a stress that may cause cessation of normal growth and development (Gajdošová et al. 2011). The conversion of *t*Z to *O*-glucosides is reversible and utilizes enzymes zeatin-*O*-glucosyltransferase and β-glucosidase. Cytokinin nucleobases can also be converted to *N*-glucosides by cytokinin *N*-glucosyltransferase (as shown in Fig. 7.3).

7.1.3.1 *N*-Glucosylation

A second pathway for cytokinin downregulation in plants is *N*-glucosylation (Bowles et al. 2006; Brzobohaty et al. 1994; Mok and Mok 2001). Glycosylation may occur on the purine ring (*N*-glucosylation) or on the side chain (*O*-glucosylation and xylosylation). The glucosylation of the purine ring occurring at 3, 7, and 9 positions yields *N*-glucosides. The 3*N*-glucosylation is a reversible process, which allows release of free CK bases or ribosides from their glycoconjugates. The 7- and 9-glucosides show limited or no activity in most bioassays. Lim and Bowles (2004) they are irreversible and are considered to be deactivation products (Vankova 1999). *N*-glucosylation is the dominant deactivation pathway, and it has been studied extensively in radish. Research work on this species dates back 30 years when Summons (1979) discovered that zeatin-7-

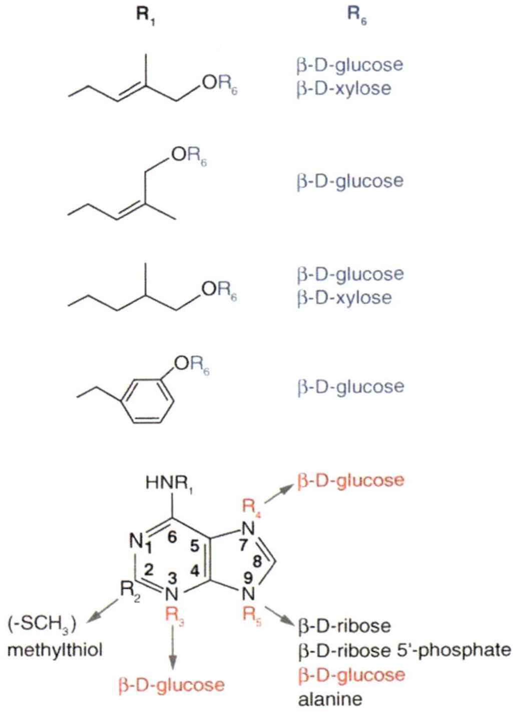

Fig. 7.3 Cytokinin conjugates with sugars and sugar phosphates. *O*-glycosylation of side chain (colored in *blue*) is catalyzed by zeatin *O*-glucosyltranferase or *O*-xylosyltransferase. *N*-glucosylation of adenine moiety (colored in *red*) is catalyzed by cytokinin *N*-glucosyltransferase (with permission from H. Sakakibara)

glucoside (Z7G) was the most abundant endogenous compound (Mok and Mok 1994). Soon after this discovery, various glucosides were detected in different cytokinins, such as iP and DHZ (McGaw et al. 1984). These enzymes were partially purified by Parker and Letham (1973). *N*-glucosyltransferase catalyzed both 7- and 9-glucosides but showed an affinity toward the 7 position (Entsch et al. 1983). N-GT enzyme(s) use isoprenoid as well as aromatic cytokinins as their substrates (Martin et al. 1999; Mok and Mok 2001). Singh et al. (1988) studied the metabolic fate of individual cytokinins with respect to their chemical structure. *N*-glucosides and *O*-glucosides are found in many plant species (Mok and Mok 1994; Motyka et al. 1996; Redig et al. 1996a, b). Various plant species including *Arabidopsis* showed the formation of 7- and 9-glucosides when exogenously supplied with BA for long periods of time (Auer 1997; Vlasakova et al. 1998). Strnad et al. (1994) reported formation of *N*-glucosides in aromatic cytokinins.

Hou et al. (2004) identified *Arabidopsis* genes encoding enzymes capable of cytokinin *N*-glucosylation. The glucosyltransferase activity was measured using glycosyltransferases (UGTs) expression plasmids that were transformed into *Escherichia coli* XL-1 Blue individually for recombinant protein expression. From a screening of >100 UGTs, two recombinants were found that could glucosylate cytokinins at 7*N* and 9*N* positions. The experiments failed to elucidate the functioning of these recombinant UGTs in planta. Moreover, hormone metabolism was also not clearly explained by this study.

7.1.4 Cytokinin Genes and Their Functions

At present, only a small number of genes have been characterized that help us to understand cytokinin biology. In Table 7.1, the genes involved in cytokinin biosynthesis, signal transduction, and metabolism are described. Gene discovery is providing new avenues for manipulating and studying cytokinins. Over the years, mutants have been selected that were thought to be specific for cytokinins, but later it was discovered that they were also responsible for other hormone pathways. Rashotte et al. (2003) for example, *ckr1/ein 1* (*cytokinin resistant*) was later found to be taking part in signal transduction of ethylene. In another instance, the *ror-1* (*roscovitine-resistant-1*) mutant, which was initially perceived to be putatively involved in cytokinin *N*-glucosylation, was later found to be a potential regulator of cytokinin metabolism, specifically the aromatic cytokinins (Dwivedi et al. 2010).

Table 7.1 Important genes involved in cytokinin regulation and metabolism isolated in *Arabidopsis*

Gene name	Function	References
WOL1/CRE1 Cytokinin response	Cytokinin signaling	Mähönen et al. (2000)
AHK 2, 3, and 4 Arabidopsis histidine kinase	Cytokinin signaling	Inoue et al. (2001) Suzuki et al. (2001)
CKX (7 genes) Cytokinin Oxidase genes	Degradation	Bilyeu et al. (2001) Houba-Herin et al. (1999)
AtIPT (9 genes) Arabidopsis isopentyltransferase	Biosynthesis	Kakimoto (2001) Takei et al. (2001)
cis-ZOG1 (*cis*-zeatin-*O*-glucosyltransferase) ZOX (zeatin-O-xylosyltransferase)	Metabolism	Martin et al. (2001)
APT1 Adenine phosphoribosyltransferase	Metabolism/adenine salvage pathway	Moffatt and Hiroshi (2002)
ARR (10 genes) Arabidopsis response regulator gene	Cytokinin primary response genes	Imamura et al. (1998) Brandstatter and Kieber (1998) D'Agostino et al. (2000) Salomé (2005)
UGT76C1 and C2	*N*-glucosylation of cytokinin	Hou et al. (2004)
ROR-1 Roscovitine-resistant-1	Putative aromatic cytokinin regulation gene	Dwivedi et al. (2010)

7.1.5 Future Direction

Our current knowledge about cytokinin metabolism is derived chiefly from methods used for cytokinin extraction, identification, and characterization of involved enzymes. Gas chromatography and mass spectroscopy are promising techniques for metabolic studies. Advancement in these techniques to enhance detection sensitivity and specificity can lead to more precise results. Isolation of genetic mutants can be helpful in understanding cytokinin regulation and metabolism. However, the use of mutants has been limited because of the complex nature of cytokinins in regulating processes of plant development. Molecular biology tools such as activation tagging have facilitated discovery of genes. Completion of *Arabidopsis* genome has helped in the discovery of genes for cytokinin degradation (CKX) and biosynthesis (AtIPT1, AtIPT3–AtIPT8). Additionally, most of the findings in the field of cytokinin research in the past two decades have predominantly been related to identification of genes using molecular techniques. These studies are valuable as they have provided genetic basis to numerous physiological developments observed in the plant in response to cytokinins. But studies of this nature did not take into the account the pool of cytokinin which is already present in the environment. Walters and McRoberts (2006) a review by Stirk and van Staden (2010) highlighted on different significant sources of cytokinins. The sources of cytokinins in a terrestrial environment were senescing plant debris, bacteria, fungi, microalgae (cyanophyta), algae (cholorophyta), nematodes, insects, and agrochemicals containing natural and synthetic cytokinins. Similarly, the aquatic and the marine environments have algae and bacteria which synthesize endogenous cytokinins. Because in a natural environment plants are constantly regulating the levels of cytokinins to maintain homeostasis, the fore mentioned sources of exogenous cytokinins are indeed playing a role, perhaps even a beneficial one. Therefore, our experimental approach to study cytokinin metabolism should be an association of analytical, biochemical, genetic, and ecological studies. This joint effort would certainly be helpful in providing answers to questions associated with the movement of this important hormone not only at a plant level but also on an environmental scale.

References

Argueso CT, Ferreira FJ, Kieber JJ (2009) Environmental perception avenues: the interaction of cytokinin and environmental response pathways. Plant Cell Environ 32(9):1147–1160

Armstrong D (1994) Cytokinin oxidase and the regulation of cytokinin degradation. Cytokinins: chemistry, activity and function. CRC, Boca Raton, FL, p 356

Auer CA (1997) Cytokinin conjugation: recent advances and patterns in plant evolution. Plant Growth Regul 23:17–32

Bilyeu KD, Cole JL, Laskey JG, Riekhof WR, Esparza TJ (2001) Molecular and biochemical characterization of a cytokinin oxidase from maize. Plant Physiol 125:378–386

Brzobohaty B, Moore I, Palme K (1994) Cytokinin metabolism: implications for regulation of plant growth and development. Plant Mol Biol 26(5):1483–1497

Bouchez O, Huard C, Lorraine S, Roby D, Balague C (2007) Ethylene is one of the key element of cell death and defense response control in the *Arabidopsis* lesion mimic mutant *vad1*. Plant Physiol 145:465–477

Bowles D, Lim E-K, Poppenberger B, Vaistij F (2006) Glycosyltransferases of lipophylic small molecules. Annu Rev Plant Biol 57:567–597

Brandstatter I, Kieber JJ (1998) Two genes with similarity to bacterial response regulators are rapidly and specifically induced by cytokinin in *Arabidopsis*. Plant Cell 10:1009–1020

Davies PJ (2004) The plant hormones: their nature, occurrence and function. In: Davies PJ (ed) Plant hormones: biosynthesis, signal transduction action! 3rd edn. Kluwer, Dordrecht, Netherlands, p 750

D'Agostino IB, Deruère J, Kieber JJ (2000) Characterization of the response of the Arabidopsis ARR gene family to cytokinin. Plant Physiol 124:1706–1717

Dwivedi S, Vanková R, Motyka V, Herrera C, Zizkova E, Auer CA (2010) Characterization of *Arabidopsis thaliana* mutant ror-1(roscovitine-resistant) and its utilization in understanding of the role of cytokinin N-glucosylation pathway in plants. Plant Growth Regul 61(3):231–242

Enstch B, Parker CW, Letham DS, Summons RE (1983) An enzyme from lupin seeds form alanine derivatives of cytokinins. Phytochemistry 22:375–381

Gajdošová S1, Spíchal L, Kamínek M, Hoyerová K, Novák O, Dobrev P, Galuszka P, Klíma P, Gaudinová A, Žižková E, Hanuš J, Dančá M, Trávníček B, Pešek B, Krupička M, Vanková R, Strnad M, Motyka V (2011) Distribution, biological activities, metabolism, and the conceivable function of cis-zeatin-type cytokinins in plants. J Exp Bot 62(8):2827–2840

Galuska P, Frebort I, Sebela M, Strnad M, Peč P (1999) Cytokinin oxidase: the key enzyme in the biodegradation of cytokinins. In: Strnad M, Pec P, Beck E (eds) Advances in regulation of plant growth and development. Peres, Prague, pp 245–248

Gan S, Amasino RM (1995) Inhibition of leaf senescence by autoregulated production of cytokinins. Science 270:1986–1988

Hou B, Lim E-K, Higgins GS, Bowles DJ (2004) *N*-Glucosylation of cytokinins by glycosyltransferases of *Arabidopsis thaliana*. J Biol Chem 279:47822–47832

Houba-Herin N, Pethe C, d'Alayer J, Laloue M (1999) Cytokinin oxidase from *Zea mays*: purification, cDNA cloning and expression in moss protoplasts. Plant J 17:615–626

Imamura A, Hanaki N, Umeda H, Nakamura A, Suzuki T, Ueguchi C, Mizuno T (1998) Response regulators implicated in His-to-Asp phosphotransfer signaling in *Arabidopsis*. Proc Natl Acad Sci 95:2691–2696

Inoue T, Higuchi M, Hashimoto Y, Seki M, Kobayashi M (2001) Identification of CRE1 as a cytokinin receptor from *Arabidopsis*. Nature 409:1060–1063

Kakimoto T (2001) Identification of plant cytokinin biosynthetic enzymes as dimethylallyldiphosphate: ATP/ADP isopentenyltransferases. Plant Cell Physiol 42:677–685

Kieber JJ (2000) Cytokinins. In: Somerville CR, Meyerowitz EM (eds) The Arabidopsis book. American Society of Plant Biologists, Rockville, MD, pp 1–25, Doi:10.1199/tab.0009, http://www.aspb.org/publications/arabidopsis/

Kurakawa T, Ueda N, Maekawa M, Kobayashi K, Kojima M, Nagato Y, Sakakibara H, Kyozuka J (2007) Direct control of shoot meristem activity by a cytokinin-activating enzyme. Nature 445:652–655

Letham DS, Parker CW, Duke CC, Summons RE, MacLeod JK (1982) *O*-Glucosylzeatin and related compounds—a new group of cytokinin metabolites. Ann Bot 41:261–263

Lexa M, Genkov T, Malbeck J, Machackova I, Brzobohaty B (2003) Dynamics of endogenous cytokinin pools in tobacco seedlings: a modeling approach. Ann Bot 5:585–597

Lim EK, Bowles DJ (2004) A class of plant glycosyltransferases involved in cellular homeostasis. EMBO J 23(15):2915–2922

Ljung K, Őstin A, Lioussane L, Sandberg G (2004) Developmental regulation of IAA turnover in Scots pine seedlings. Plant Physiol 125:464–475

Mähönen AP, Bonke M, Kauppinen L, Riikonen M, Benfey PN, Helariutta Y (2000) A novel two-component hybrid molecule regulates vascular morphogenesis of the Arabidopsis root. Genes Dev 14(23):2938–2943

Martin RC, Mok MC, Mok DW (1999) A gene encoding the cytokinin enzyme zeatin*O*-xylosyltransferase of *Phaseolus vulgaris*. Plant Physiol 120:553–558

Martin RC, Mok MC, Habben JE, Mok DW (2001) A maize cytokinin gene encoding an *O*-glucosyltransferase specific to *cis*-zeatin. Proc Natl Acad Sci 98:5922–5926

McGaw BA, Heald JK, Horgan R (1984) Dihydrozeatin metabolism in radish seedlings. Phytochemistry 23:1373–1377

Moffatt B, Hiroshi A (2002) Purine and pyrimidine nucleotide synthesis and metabolism. The Arabidopsis book. American Society of Plant Biologists, Rockville, MD (doi:10.1199/tab.0009, http://www.aspb.org/publications/arabidopsis/)

Mok DW, Mok MC (2001) Cytokinin metabolism and action. Ann Rev Plant Physiol 52:89–118

Mok DW, Mok MC (1994) Cytokinins: chemistry, activity and function. CRC, Boca Raton, FL, p 356

Morris RO, Bilyeu KD, Laskey JG, Cheikh N (1999) Isolation of a gene encoding a glycosylated cytokinin oxidase from maize. Biochem Biophys Res Commun 255:328–333

Motyka V, Faiss M, Strnad M, Kaminek M, Scmulling T (1996) Changes in cytokinin content and cytokinin oxidase activity in response to derepression of *ipt* gene transcription in transgenic tobacco calli and plants. Plant Physiol 112:1035–1043

Motyka V, Vankova R, Capkova V, Petrasek J, Kaminek M, Schmulling T (2003) Cytokinin-induced upregulation of cytokinin oxidase activity in tobacco includes changes in enzyme glycosylation and secretion. Physiol Plant 117:11–21

Parker CW, Letham DS (1973) Regulators of cell division in plant tissues. XVI. Metabolism of zeatin by radish cotyledons and hypocotyls. Planta 114:199–218

Rashotte AM, Susan DB, Jennffer PC, Kieber JJ (2003) Expression profiling of cytokinin action in *Arabidopsis*. Plant Physiol 132:1–14

Redig P, Motyka V, Van Onckelen HA, Kaminek M (1996a) Regulation of cytokinin oxidase activity in tobacco callus expressing the T-DNA *ipt* gene. Physiol Plant 99:89–96

Redig P, Schmulling T, Van Onckelen HA (1996b) Analysis of cytokinin metabolism in *ipt* transgenic tobacco by liquid chromatography-tandem mass spectrometry. Anal Biochem 112:141–148

Sakakibara H (2006) Cytokinins: activity, biosynthesis, and translocation. Annu Rev Plant Biol 57:431–549

Salomé PA (2005) Arabidopsis response regulators *ARR3* and *ARR4* play cytokinin-independent roles in control of circadian period. Plant Cell 18:55–69

Schmülling T, Werner T, Riefler M, Krupková E, Bartrina Y, Manns I (2003) Structure and function of cytokinin oxidase/dehydrogenase genes of maize, rice, *Arabidopsis* and other species. J Plant Res 116:241–252

Singh S, Letham DS, Jameson PE, Zhang R, Parker CW, Badenoch-Jones J, Noodé LD (1988) Cytokinin biochemistry in relation to leaf senescence. Cytokinin metabolism in soybean explants. Plant Physiol 88:788–794

Strik WA, van Staden J (2010) Flow of cytokinins through the environment. Plant Growth Regul 62:101–116

Strnad M, Peters W, Hanus J, Beck E (1994) *Ortho-topolin-9-glucoside,* an aromatic cytokinin from *Populus* X *canadensis* Moench cv. Robusta leaves. Phytochemistry 37:1059–1062

Strnad M (1997) The aromatic cytokinins. Physiol Plant 101:674–688

Summons RE (1979) Mass spectrometric analysis of cytokinins in plant tissues. Quantitation of cytokinins in Zea mays kernels using deuterium labeled standards. Biol Mass Spectrom 6(9):23–29

Suzuki T, Miwa K, Ishikawa K, Yamada H, Aiba H, Mizuno T (2001) The *Arabidopsis* sensor His-kinase, AHK4 can respond to cytokinins. Plant Cell Physiol 42:107–113

Taiz L, Zeiger E (2004) Cytokinins. Plant physiology, 2nd edn. Sinauer, Sunderland, MA, p 700

Takei K, Sakakibara H, Sugiyama T (2001) Identification of genes encoding adenylate isopentenyltransferase, a cytokinin biosynthesis enzyme, in *Arabidopsis thaliana*. J Biol Chem 276:26405–26410

To JPC, Kieber JJ (2008) Cytokinin signaling: two-components and more. Trends Plant Sci 13(2):86–92

Vanková R (1999) Cytokinin glycoconjugates: distribution, metabolism and function. In M Strnad, P Pec, E Beck, eds, Advances in Regulation of Plant Growth and Development. Peres Company, Prague, pp 67–78

Vlasakova V, Brezinova A, Holik J (1998) Study of cytokinin metabolism using HPLC with radioisotope detection. J Pharm Biomed Anal 17:39–44

Wagner S, Bader ML, Drew D, de Gier J-W (2006) Rationalizing membrane protein over expression. Prot Purif 24:365–371

Walters DR, McRoberts N (2006) Plants and biotrophs: a pivotal role for cytokinins? Trends Plant Sci 11:581–586

Yamada H, Koizumi N, Nakamichi N, Kiba T, Yamashino T, Mizuno T (2004) Rapid response of *Arabidopsis* T87 cultured cells to cytokinin through His-to-Asp phosphorelay signal transduction. Biosci Biotech Biochem 68(9):1966–1976

Chapter 8
Origin of Brassinosteroids and Their Role in Oxidative Stress in Plants

Andrzej Bajguz

Abstract Brassinosteroids (BRs) are a class of plant polyhydroxysteroids that have been recognized as a kind of phytohormones and play essential roles in plant development. BRs occur at low concentrations in lower and higher plants. Natural 70 BRs identified so far have a common 5α-cholestan skeleton, and their structural variations come from the kind and orientation of oxygenated functions in rings A and B. As regards the B-ring oxidation, BRs are divided into the following types: 7-oxalactone, 6-oxo, 6-deoxo and 6-hydroxy. These steroids can be also classified as C_{27}, C_{28} or C_{29} BRs depending on the alkyl substitution on the C-24 in the side chain. In addition to free BRs, sugar and fatty acid conjugates have been also identified in plants. Plant growth and developmental processes as well as environmental responses require the action and cross talk of BRs and reactive oxygen species (ROS). ROS can partake in signalling, although these events will be modulated by the complement of antioxidants in, or even around, the cell. ROS can interact with other signal molecules, including BRs in regulation of these physiological responses. BRs can modify the synthesis of antioxidants and the activity of basic antioxidant enzymes, and some of these enzymes are also implicated in catabolism of plant hormone. However, it is still unclear whether endogenous BRs directly or indirectly modulate the responses of plants to oxidative stress. The recent progress made in understanding the response of BRs in plants under oxidative stress.

8.1 Introduction

Brassinosteroids (BRs) are a group of plant hormones, which is represented by brassinolide (BL) and castasterone (CS), and their derivatives. BRs are essential for normal plant growth, reproduction and development. They play critical roles in a

A. Bajguz (✉)
Institute of Biology, University of Bialystok, Swierkowa 20 B, 15-950 Bialystok, Poland
e-mail: abajguz@uwb.edu.pl

N.A. Khan et al. (eds.), *Phytohormones and Abiotic Stress Tolerance in Plants*,
DOI 10.1007/978-3-642-25829-9_8, © Springer-Verlag Berlin Heidelberg 2012

variety of physiological responses in plants, including stem elongation, pollen tube growth, leaf bending and epinasty, root growth inhibition, ethylene biosynthesis, proton pump activation, vascular differentiation, nucleic acid and protein synthesis and photosynthesis (Hayat et al. 2010). BRs play a significant role in amelioration of various abiotic and biotic stresses, such as cold stress, water deficit, salt injury, oxidative damage, thermal stress, heavy metal stress and pathogen infection. Despite the correlation between oxidative stress and BR level in plants, the physiological rationale for such alteration in BR level is little known (Bajguz and Hayat 2009).

8.2 Origin and Structures of Brassinosteroids

8.2.1 Occurrence of Brassinosteroids

Brassinosteroids (BRs) occur at low concentrations throughout the plant kingdom. They are widely distributed in lower and higher plants. BRs have been identified in lower plants, such as green algae *Hydrodictyon reticulatum* and *Chlorella vulgaris*, a pteridophyte (*Equisetum arvense*) and a bryophyte (*Marchantia polymorpha*). BRs have been detected in all organs of higher plants, such as pollen, anthers, seeds, leaves, stems, roots, flowers and grain. Pollen and immature seeds are the richest sources of BRs than shoots and leaves. Furthermore, young growing tissues contain higher levels of BRs than mature tissues. They also occur in the insect and crown galls of *Castanea crenata*, *Distylium racemosum* or *Catharanthus roseus*. These plants have higher levels of BRs than the normal tissues (Bajguz and Tretyn 2003; Bajguz 2009).

8.2.2 Chemical Structures of Brassinosteroids

Natural BRs have 5α-cholestane skeleton, and their structural variations come from the kind and orientation of oxygenated functions in A- and B-ring (Fig. 8.1). These modifications are produced by oxidation and reduction reactions during biosynthesis. Generally, BRs are divided into free (64) and conjugated (5) compounds. These steroids can be classified as C_{27}, C_{28} or C_{29} BRs (10, 38 and 16 compounds, respectively) depending on the alkyl substitutions in the side chain (Figs. 8.2–8.4). These side chain structures are all common in plants sterols. According to the cholestane side chain (Fig. 8.1), BRs are divided into 11 types with different substituents at C-23, C-24 and C-25: 23-oxo, 24 *S*-methyl, 24*R*-methyl, 24-methylene, 24 *S*-ethyl, 24-ethylidene, 24-methylene-25-methyl, 24-methyl-25-methyl, without substituent at C-23, without substituent at C-24 and

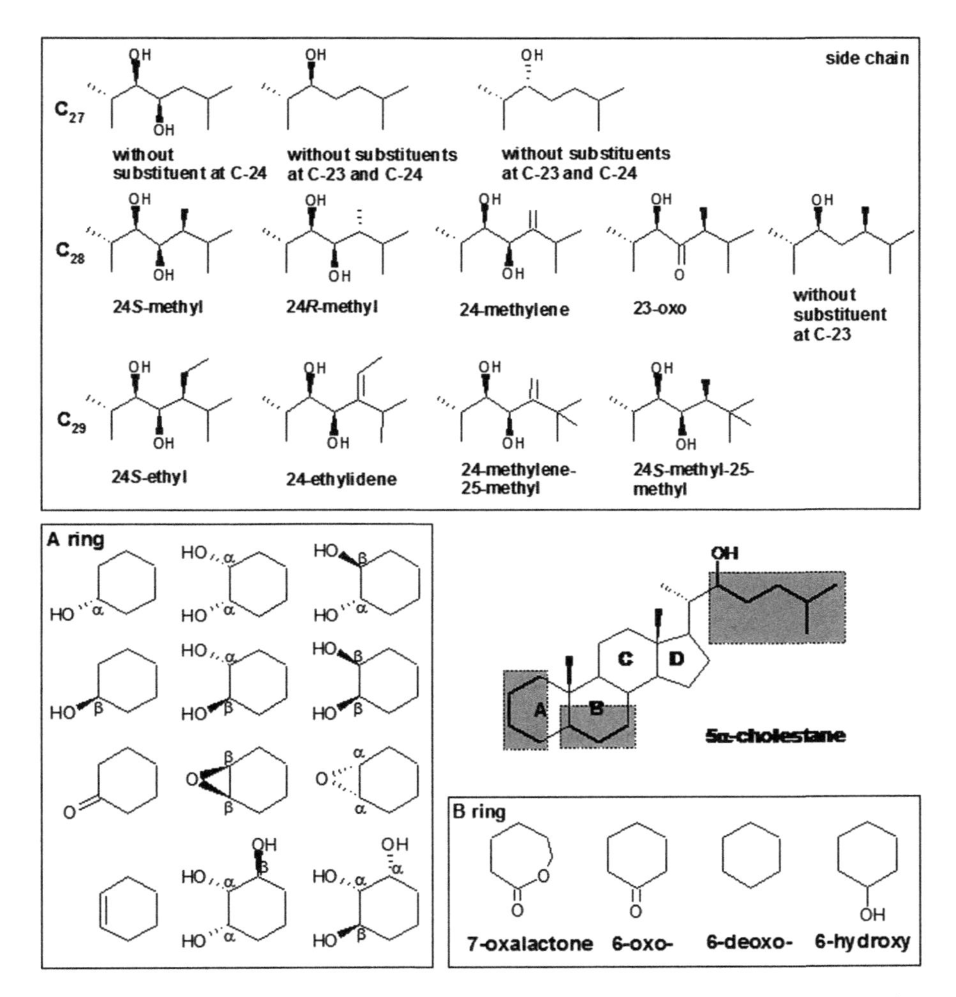

Fig. 8.1 Different substituents in the A- and B-ring and side chain of naturally occurring brassinosteroids

without substituents at C-23, C-24 (Yokota 1999; Abe et al. 2001; Bajguz and Tretyn 2003).

With respect to the A-ring, BRs with an α-hydroxyl, β-hydroxyl or ketone at position C-3 are precursors of BRs having 2α,3α-vicinal hydroxyls (Fig. 8.1). As regards the B-ring oxidation, BRs are divided into 7-oxalactone, 6-oxo, 6-deoxo and 6-hydroxy. 7-Oxalactone BRs (such as brassinolide, BL) have stronger biological activity than 6-oxo types (such as castasterone, CS), and non-oxidized BRs reveal no activity in biological tests. BRs with either a 3α-hydroxyl (e.g. typhasterol, TY), 3β-hydroxyl or 3-oxo in the A-ring are precursors of BRs carrying 2α,3α-vicinal diol; those with 2α,3β-, 2β,3α- or 2β,3β-vicinal diol can be metabolites of active BRs with a 2α,3α-vicinal diol. Decreasing order of activity

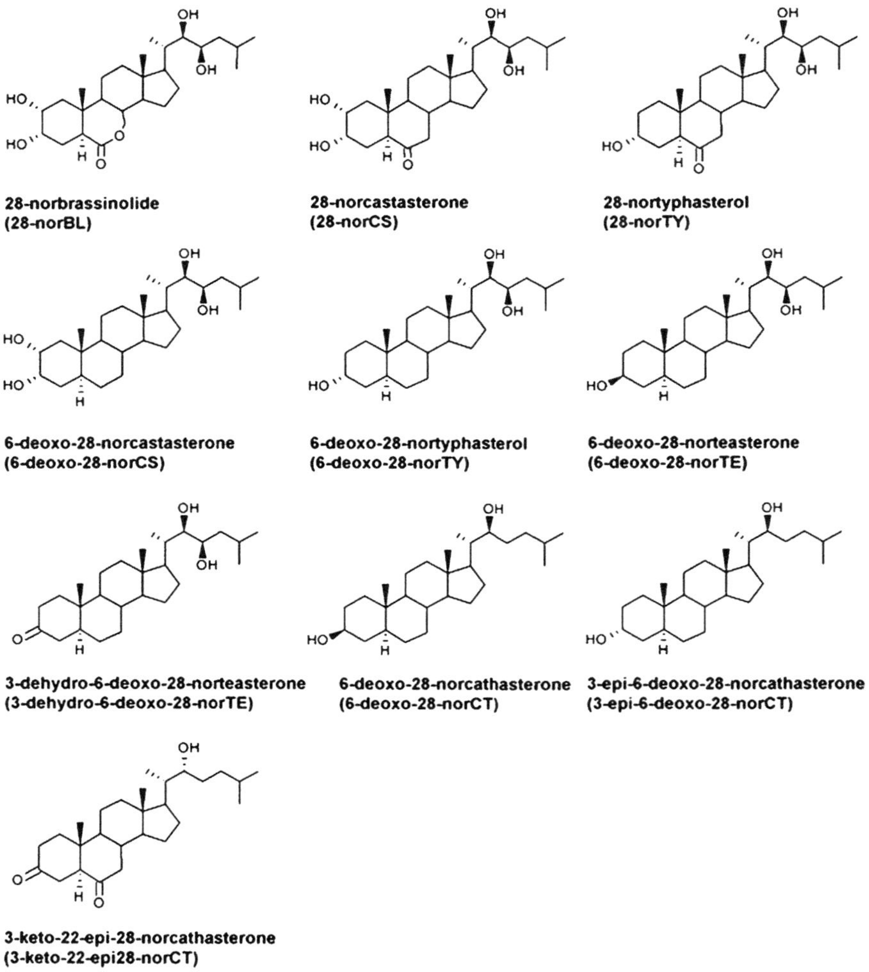

28-norbrassinolide
(28-norBL)

28-norcastasterone
(28-norCS)

28-nortyphasterol
(28-norTY)

6-deoxo-28-norcastasterone
(6-deoxo-28-norCS)

6-deoxo-28-nortyphasterol
(6-deoxo-28-norTY)

6-deoxo-28-norteasterone
(6-deoxo-28-norTE)

3-dehydro-6-deoxo-28-norteasterone
(3-dehydro-6-deoxo-28-norTE)

6-deoxo-28-norcathasterone
(6-deoxo-28-norCT)

3-epi-6-deoxo-28-norcathasterone
(3-epi-6-deoxo-28-norCT)

3-keto-22-epi-28-norcathasterone
(3-keto-22-epi28-norCT)

Fig. 8.2 Chemical structures of C_{27} brassinosteroids

(2α,3α > 2α,3β > 2β,3α > 2β,3β) shown by structure–activity relationship suggests that α-oriented hydroxyl group at C-2 is essential for a greater biological activity of BRs in plants. BRs, which have 3α,4α-diols, are more active than 2α,3α. This fact is in strong contrast with the structure requirements mentioned above. The higher activity of unnatural 3α,4α-diols could be explained by twisting and distortion of the molecule due to the seven- or eight-membered B-ring and also by the position of a carbonyl group relative to the A-ring diol. With the exception of some not fully characterized BRs with an oxo group at C-23, all bioactive BRs possess a

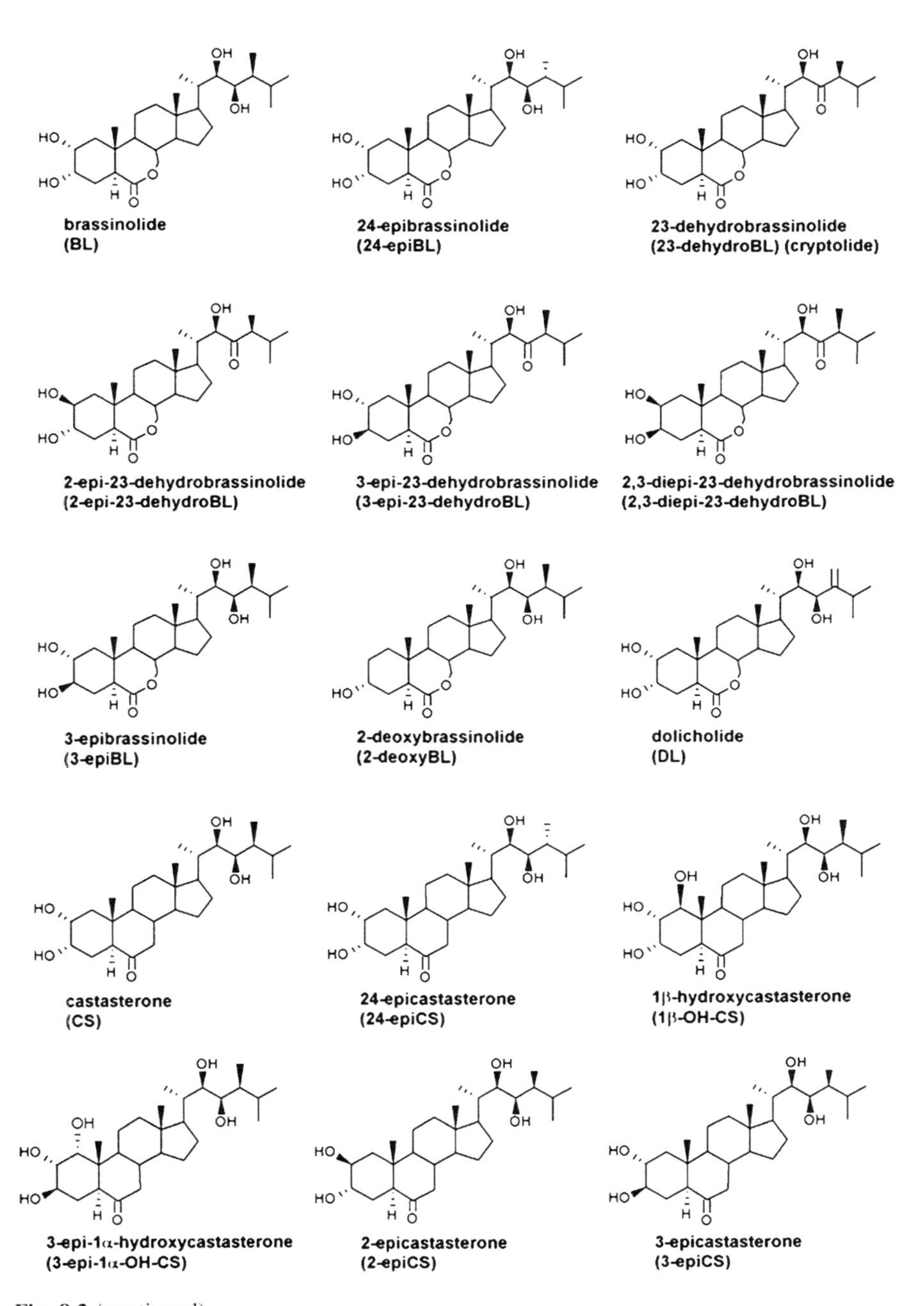

Fig. 8.3 (continued)

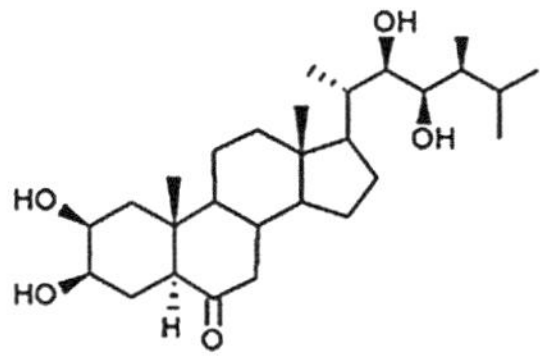

2,3-diepicastasterone
(2,3-diepiCS)

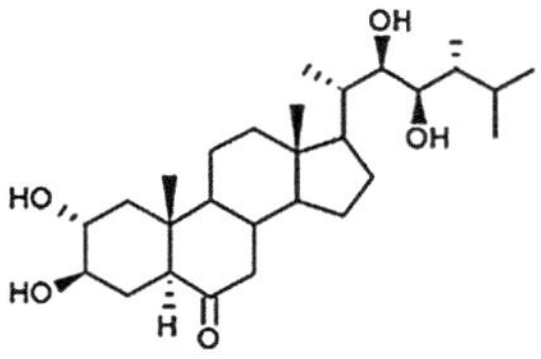

3,24-diepicastasterone
(3,24-diepiCS)

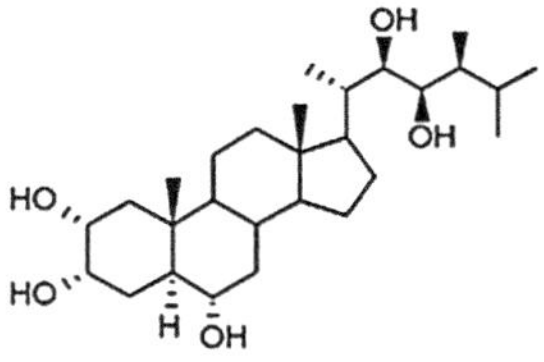

6-α-hydroxycastasterone
(6-α-OH-CS)

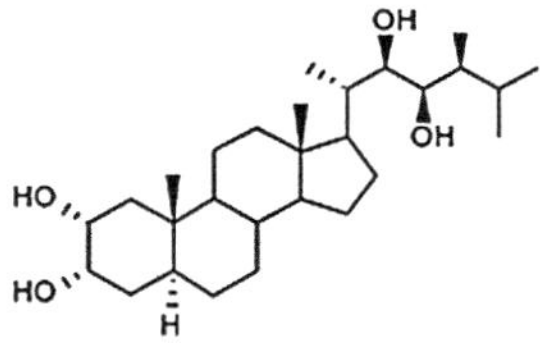

6-deoxocastasterone
(6-deoxoCS)

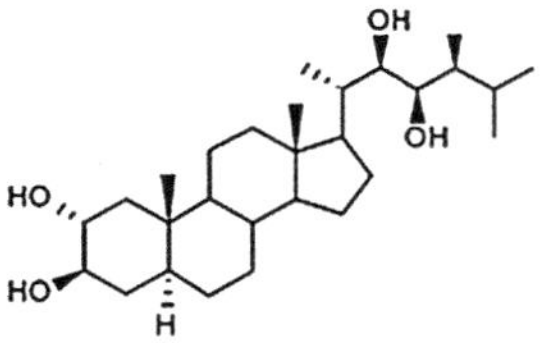

3-epi-6-deoxocastasterone
(3-epi-6-deoxoCS)

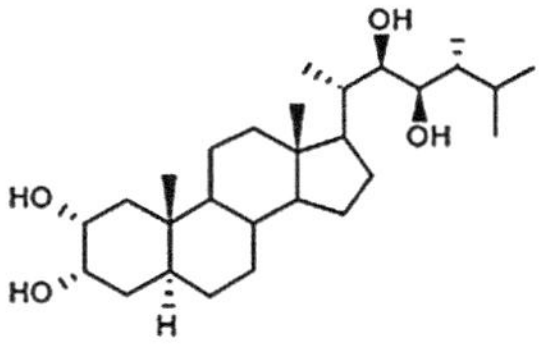

6-deoxo-24-epicastasterone
(6-deoxo-24-epiCS)

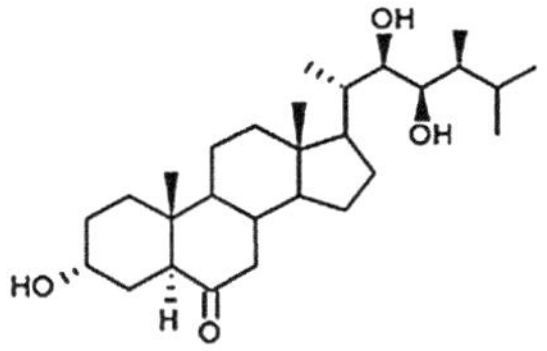

typhasterol
(TY)

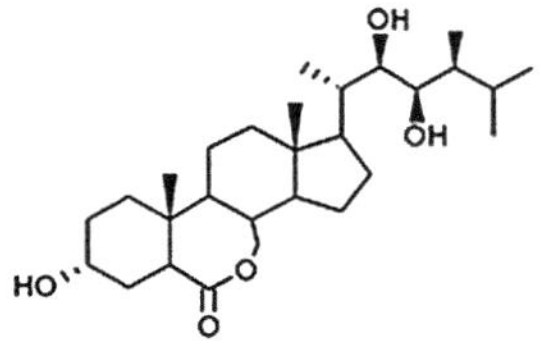

7-oxotyphasterol
(7-oxTY)

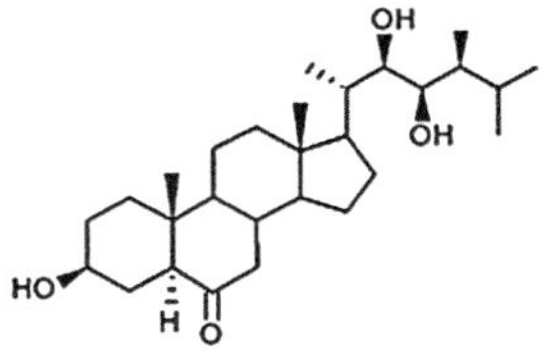

teasterone
(TE)

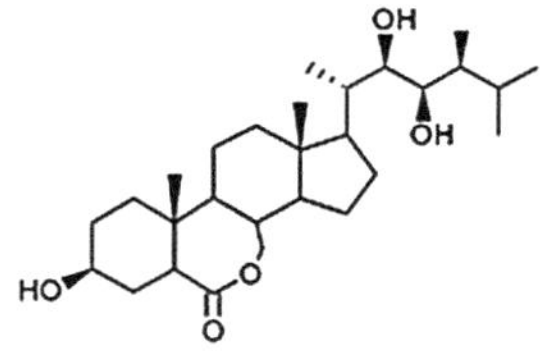

7-oxoteasterone
(7-oxTE)

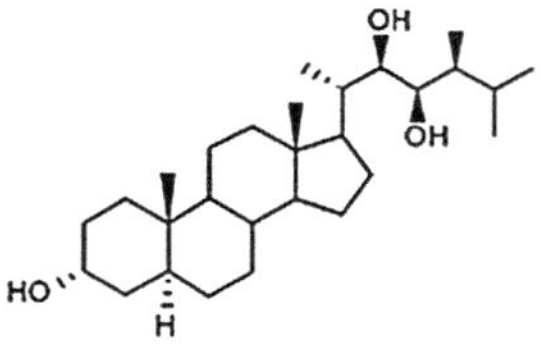

6-deoxotyphasterol
(6-deoxoTY)

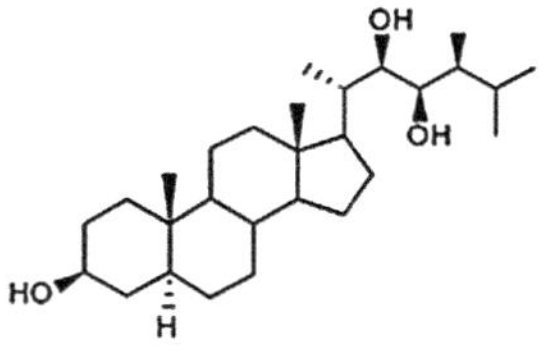

6-deoxoteasterone
(6-deoxoTE)

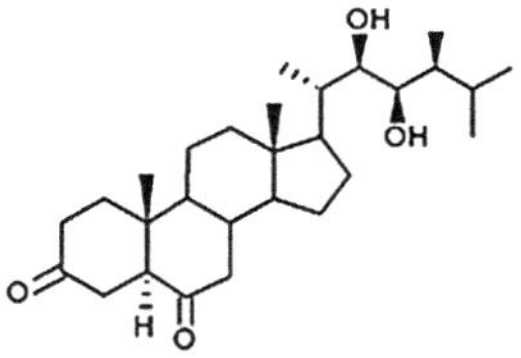

3-dehydroteasterone
(3-DT)

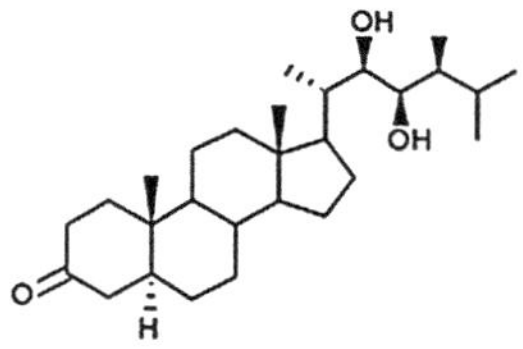

6-deoxo-3-dehydroteasterone
(6-deoxo-3-DT)

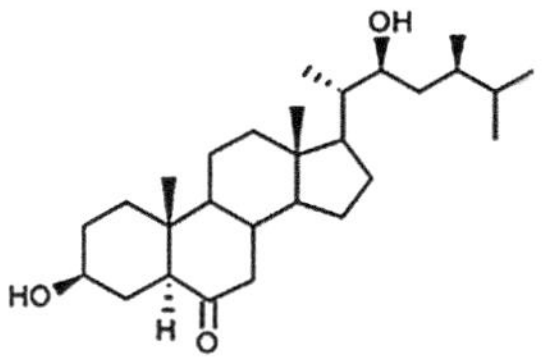

cathasterone
(CT)

Fig. 8.3 (continued)

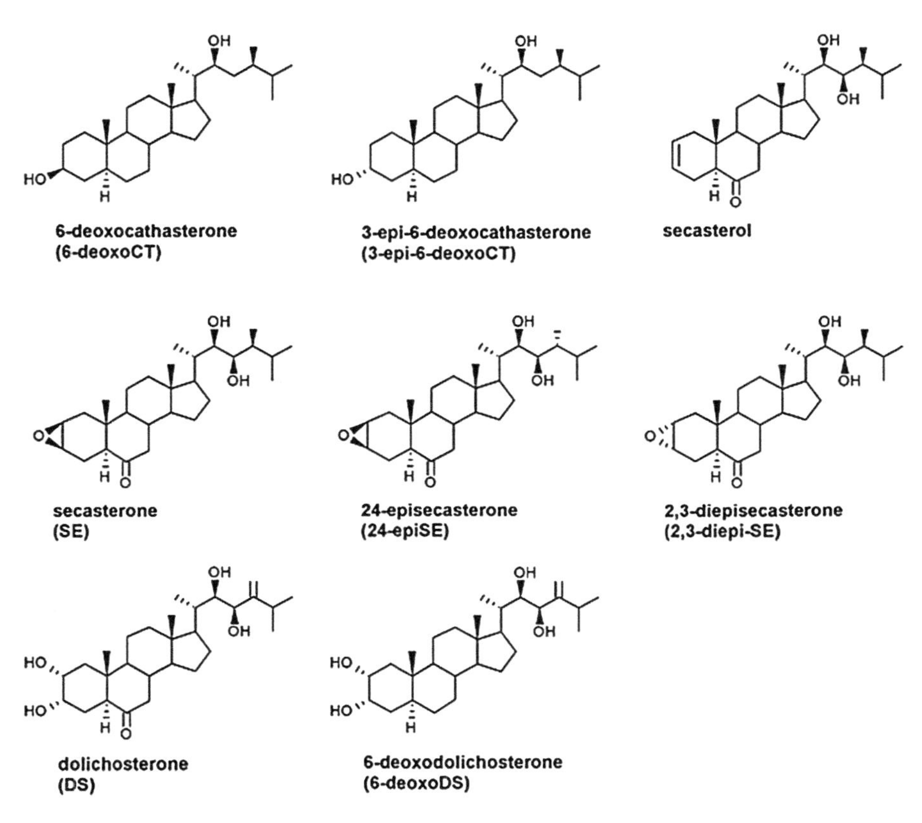

Fig. 8.3 Chemical structures of C_{28} brassinosteroids

vicinal 22*R*,23*R* diol structural functionality, which appears essential for a high biological activity (Yokota 1999; Bajguz and Tretyn 2003).

In addition to free BRs, also five sugar and fatty acid conjugates were identified in plants so far (Fig. 8.5). 25-Methyldolichosterone-23-β-D-glucoside (25-MeDS-Glu) and its 2β isomer from *Phaseolus vulgaris* seeds and teasterone-3β-D-glucoside (TE-3-Glu), teasterone-3-laurate (TE-3-La) and teasterone-3-myristate (TE-3-My) from *Lilium longiflorum* pollen have been isolated as endogenous BRs (Abe et al. 2001).

8.2.3 Biosynthetic Pathways of Brassinosteroids

Brassinolide (BL), the most important BR in plants, has been shown to be synthesized via two pathways from campesterol (early and late C-6 oxidation pathways). Sterols are synthesized via the non-mevalonate pathway in lower plants or the mevalonate pathway of isoprenoid metabolism in higher plants. Campesterol, one of the major plant sterols, is the precursor of BRs, which is primarily derived from isopentenyl diphosphate. Although metabolic experiments with labelled C_{27} BRs have not yet been performed, the natural occurrence of C_{27} BRs in tomato (6-deoxo-28-norCT, 6-deoxo-28-norTE, 6-deoxo-28-norTY, 6-deoxo-28-norCS and 28-norCS) suggests an in vivo biosynthetic pathway to 6-deoxo-28-norCT. Based on these findings, a biosynthetic pathway of C_{27} BRs has been suggested: cholestanol→6-deoxo-28-norCT→6-deoxo-28-norTE→3-dehydro-6-deoxo-28-norTE→6-deoxo-28-norTY→6-deoxo-28-norCS→28-norCS in tomato seedlings. The reactions, named the late C-6 oxidation pathway for C_{27} BRs, have been demonstrated in Fig. 8.6. Furthermore, the reactions of biosynthesis of C_{28} BRs, named the early and late C-6 oxidation pathways, have been also showed in Fig. 8.6. In addition to the early and late C-6 oxidation pathways, cross-links between both branches exist. Conversion of CS to BL is the final biosynthetic step of BRs. Unfortunately, the biosynthesis of C_{29} BRs is still unclear (Choe 2006).

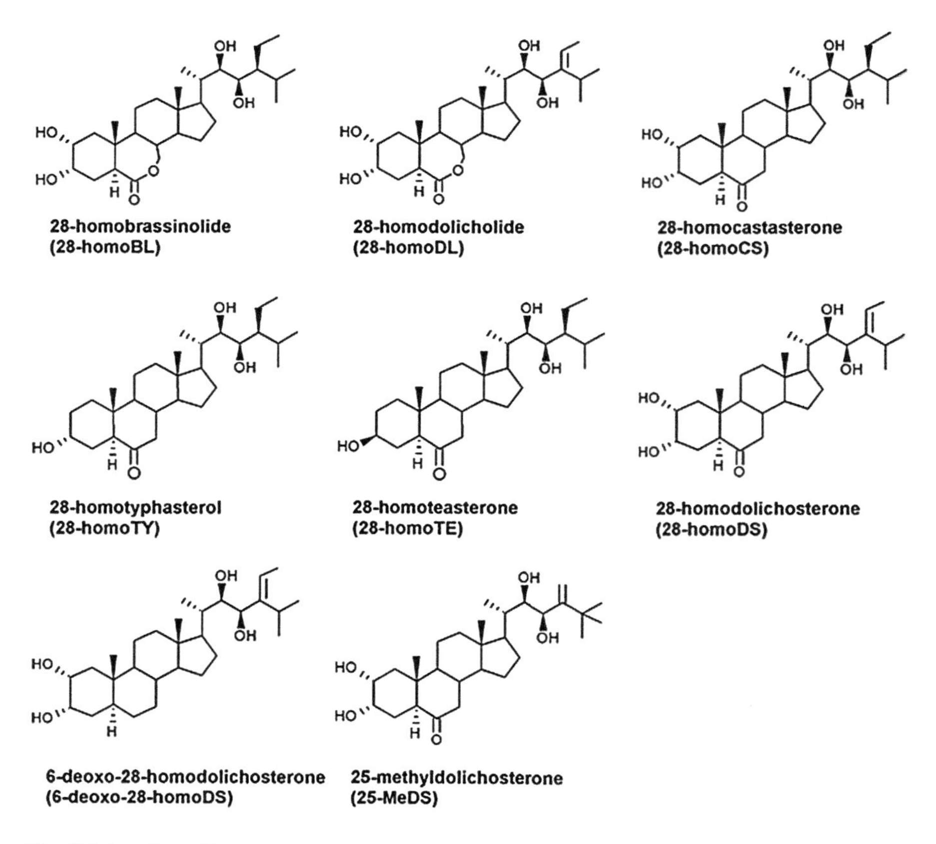

Fig. 8.4 (continued)

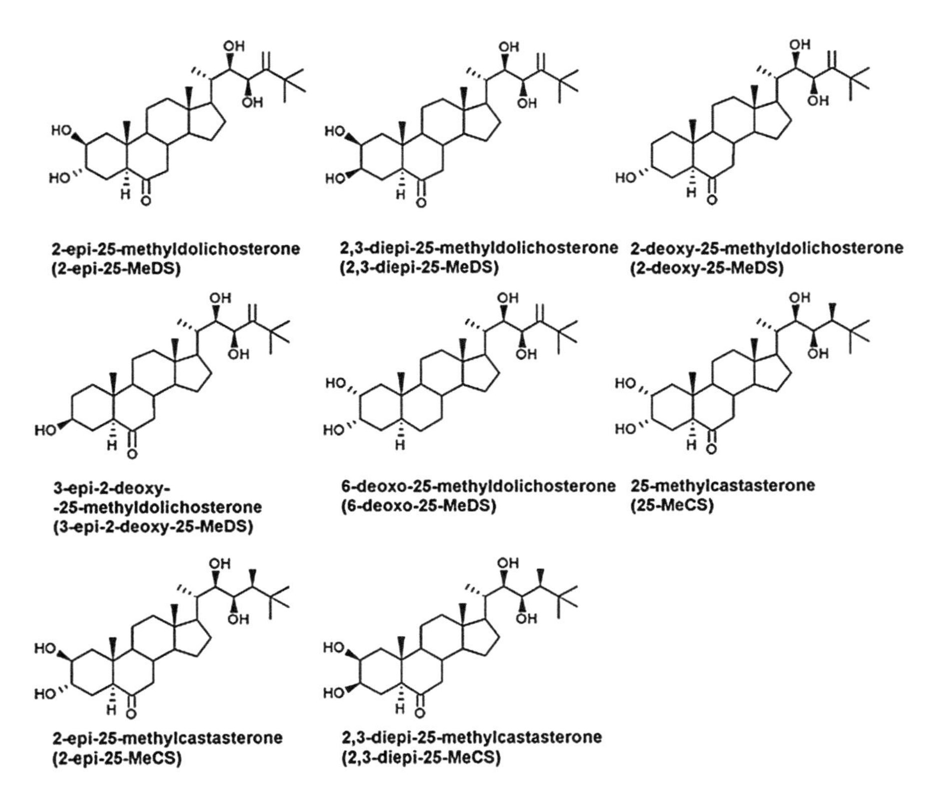

Fig. 8.4 Chemical structures of C_{29} brassinosteroids

8.3 Brassinosteroids and Oxidative Stress

Adaptations to abiotic stresses and pathogens in plants involve various changes in the biochemistry of the cell. These changes include the evolution of new metabolic pathways, the accumulation of low molecular weight metabolites, the synthesis of special proteins, detoxification mechanisms and changes in phytohormone levels. One of the biochemical changes that occurs when plants are subjected to environmental stresses is the production of ROS such as the superoxide radical (O_2^-), hydrogen peroxide (H_2O_2) and the hydroxyl radical (OH^-). Although, under normal growth conditions, the production of ROS in cells is very low, there are many stresses which can enhance the production of ROS. These stresses include drought stress and desiccation, salt stress, chilling, heat shock, heavy metals, ultraviolet radiation, ozone, mechanical stress, nutrient deprivation, pathogen attack and high light stress. ROS can have a detrimental effect on normal metabolism through oxidative damage to lipids, proteins and nucleic acids. However, many

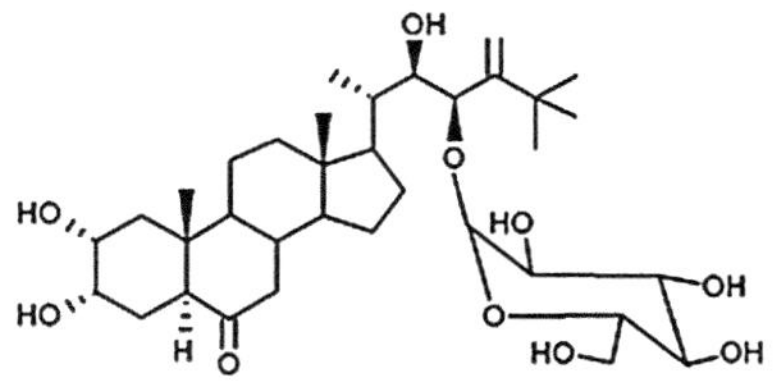

23-*O*-β-glucopyranosyl-25-methyldolichosterone (25-MeDS-Glu)

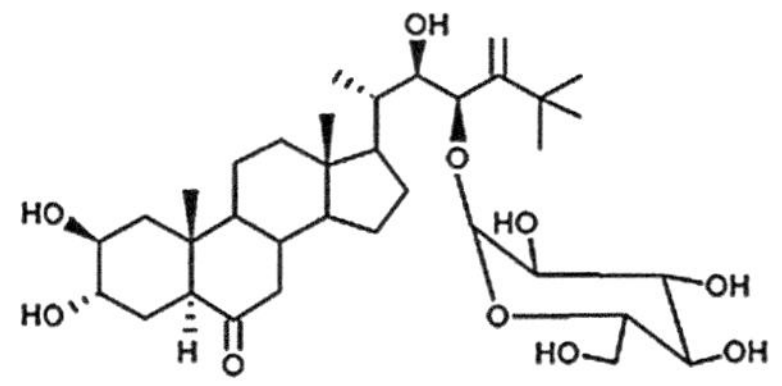

23-*O*-β-glucopyranosyl-2-epi-25-methyldolichosterone (2-epi-25-MeDS-Glu)

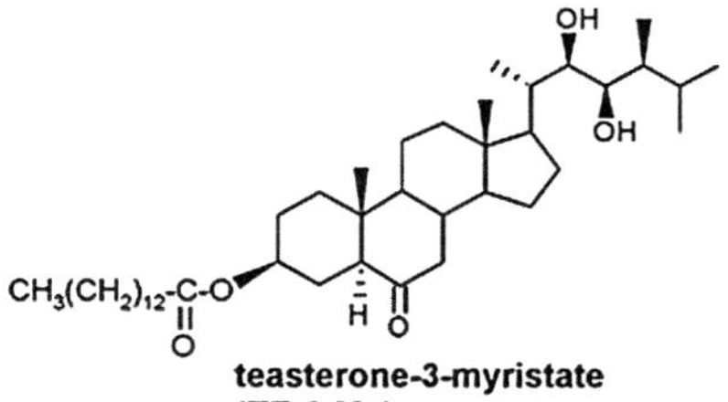

teasterone-3-myristate (TE-3-My)

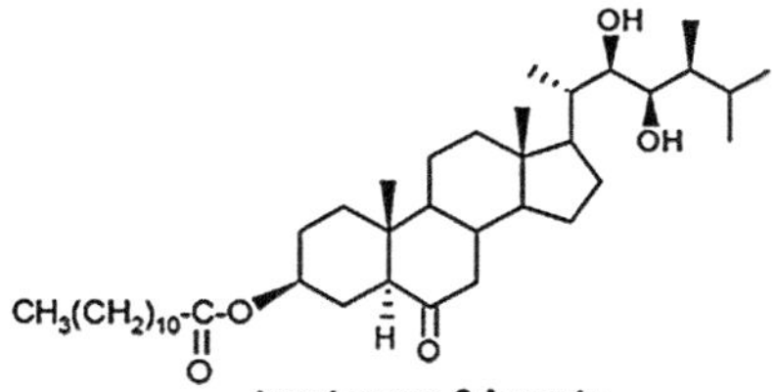

teasterone-3-laurate (TE-3-La)

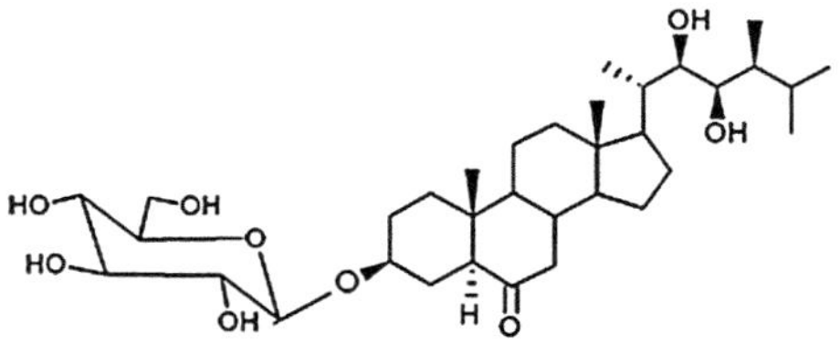

teasterone-3-*O*-β-D-glucoside (TE-3-Glu)

Fig. 8.5 Chemical structures of brassinosteroid conjugates

plants that live in heavy metal pollutions have evolved several adaptations. Recent studies have demonstrated that to control the level of ROS and to protect the cells, plants possess low molecular weight antioxidants (ascorbic acid, reduced glutathione, carotenoids, tocopherols) and antioxidant enzymes such as superoxide dismutase (SOD; E.C. 1.15.1.1), ascorbate peroxidase (APX; E.C. 1.11.1.11) and catalase (CAT; E.C. 1.11.1.6) (Mittler 2002; Sharma and Dietz 2009; Triantaphylidès and Havaux 2009). Some reports indicate that ROS can interact with phytohormones in regulation of these physiological responses. It is suggested that plant growth regulators can modify the synthesis of antioxidants and the activity of basic antioxidant enzymes and some of these enzymes are also implicated in phytohormone catabolism (Synková et al. 2004).

However, it is little known about the content of BRs in plants in response to different stresses. *Chlorella vulgaris* cultures treated with BL in the absence or presence of heavy metals showed no differences in the endogenous level of BL. On

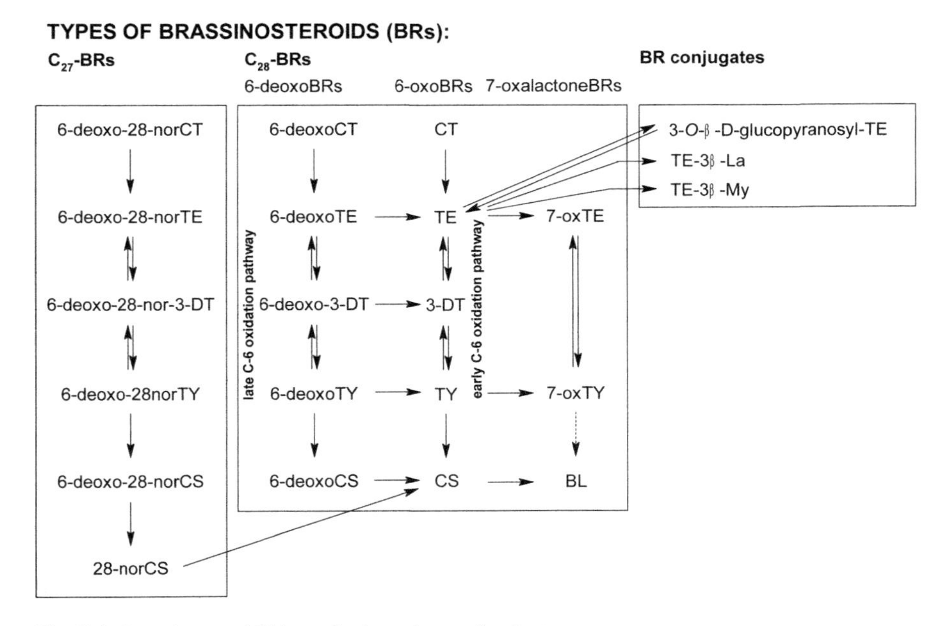

Fig. 8.6 Brassinosteroid biosynthetic pathways in plants

the other hand, treatment with heavy metals results in BL level very similar to that of control cell cultures. It suggests that the activation of BR biosynthesis, via an increase of endogenous BL, is not essential for the growth and development of *Chlorella vulgaris* cells in response to heavy metal stress. Simultaneously, BL enhanced the content of indole-3-acetic acid, zeatin and abscisic acid in cultures treated with heavy metals (Bajguz 2011).

It has been showed that co-application of both 24-epiBL and copper to *Raphanus sativus* seedlings resulted in reduction of putrescine and spermidine contents (Choudhary et al. 2009). However, copper treatment increased levels of polyamines when compared to control seedlings. 24-epiBL treatment alone decreased levels of spermidine. Supplementation of metal treatment with 24-epiBL further recorded decrease in free and bound IAA. Free and bound IAA showed maximum increase in 24-epiBL treatment alone, while maximum naphthaleneacetic acid (NAA) levels (free and bound) were recorded in metal-treated seedlings only. In controls and 24-epiBL-treated seedlings, free and bound NAA was not detected. Moreover, a decrease in both forms of NAA was recorded in seedlings given treatments of copper metal and 24-epiBL combinations. Enhanced contents of free and bound forms of NAA under metal treatment may suggest its effectiveness in oxidative stress management than IAA. Results of that study suggest significant effects of 24-epiBL on endogenous contents of polyamines and auxins under heavy metal stress, thereby indicating the modulation of stress management by BRs via regulating the contents of polyamines and auxins (Choudhary et al. 2009).

It was also reported that lead increased ABA and cytokinin level while decreased GA_3 level in germinating chickpea seeds. On the other hand, high concentrations of zinc decreased content of cytokinins and GA_3, while low concentration increased the content of these hormones (Sharma and Kumar 2002; Hsu and Kao 2003; Atici et al. 2005).

Exogenous application of BRs modified antioxidant enzymes, such as superoxide dismutase, catalase, glutathione peroxidase and ascorbate peroxidase, and non-enzymatic antioxidants, such as ascorbic acid, tocopherols, carotenoids, glutathione, etc., in plants under different stress conditions (Li et al. 1998; Núñez et al. 2003; Vardhini and Rao 2003; Őzdemir et al. 2004).

When maize (*Zea mays*) seedlings treated with brassinolide (BL) were subjected to water stress, the activities of SOD, CAT and APX, as well as ascorbic acid and carotenoid contents increased (Li et al. 1998). On the other hand, BRs enhanced the activity of CAT and reduced the activities of peroxidase and ascorbic acid oxidase under osmotic stress conditions in sorghum (*Sorghum vulgare*) (Vardhini and Rao 2003).

Rice seedlings exposed to saline stress and treated with BR showed a significant increase in the activities of CAT, SOD and glutathione reductase (GR) and a slight increase in APX (Núñez et al. 2003). A comparison between salt-sensitive and salt-resistant varieties of rice treated with 24-epiBL was also reported (Őzdemir et al. 2004). 24-epiBL treatment, at least in part, improved the tolerance of salt-sensitive seedlings to short-term salt stress. The difference in alteration of antioxidant enzyme activities in salt-sensitive rice cultivar may suggest that the higher salt stress resistance in sensitive seedlings induced by 24-epiBL may be due in part to a maintained higher activity of APX under salt stress conditions.

Algae *Chlorella vulgaris* respond to heavy metals (cadmium, copper and lead) by induction of several antioxidants, including several enzymatic systems, and the synthesis of low molecular weight compounds. Treatment with BL was effective in increasing the activity of antioxidant enzymes (CAT, GR and APX) and the content of ascorbic acid, carotenoids and glutathione (Bajguz 2010). BRs also enhanced activity of the antioxidant enzymes (catalase, peroxidase and superoxidase) and proline content in chickpea (*Cicer arietinum*) treated with cadmium (Hasan et al. 2008). It has been found a significant correlation with the degree of improvements, which was measured in terms of nodulation, nitrogen fixation, pigment composition, carbonic anhydrase and nitrate reductase activities. A similar pattern of response together with an elevation in the photosynthesis was observed in the plants of mustard exposed to cadmium through nutrient medium (Hayat et al. 2007). The foliar spray of either with 24-epiBL or 28-homoBL significantly enhanced the growth, photosynthesis, antioxidant enzymes and proline content in aluminium stressed mung bean plants. 24-epiBL enhanced the level of antioxidant system (SOD, CAT, peroxidase and glutathione reductase and proline), both under stress and stress-free conditions. The influence of 24-epiBL on the antioxidant system was more pronounced under stress situation, suggesting that the elevated level of antioxidant system, at least in part, increased the tolerance of mustard plants to saline and/or nickel stress, thus protected the photosynthetic machinery and the plant growth (Ali et al. 2008).

The activities of the enzymes catalase, peroxidase, carbonic anhydrase and nitrate reductase also exhibited a significant enhancement in mustard plants grown under nickel stress (Alam et al. 2007). These plants also exhibited an elevation in the relative water content and photosynthetic performance.

The influence of 24-epiBL on some enzymatic antioxidants in tomato leaf disc under high (40°C) temperature was reported (Mazorra et al. 2002). BR increased the activity of CAT, peroxidase and SOD in respond to high temperature in tomato leaves. It has been shown that the *det2 Arabidopsis* mutant, which is blocked in the biosynthetic pathway of BRs (Choe 2006), had significantly thicker leaves with thickened cuticle layers and cell walls in the epidermal and mesophyll layers, increased stomatal density and compacted leaf structure with less intercellular space than the wild type when grown under normal oxygen conditions. Addition of BL to the growth medium resulted in leaves that were more similar in morphology to those of Columbia.

Interestingly, the *det2 Arabidopsis* mutant was also found to be insensitive to the dwarfing effects of very low O_2 (2.5 kPa) and furthermore did not show the purpling stress response of wild type. BRs could be as possible mediators of particular aspects of the developmental response to hypoxia, a conclusion supported by the fact that BRs require molecular oxygen at several steps in their biosynthesis (Ramonell et al. 2001). It has been also demonstrated that *ATPA2* and *ATP24a* genes encoding peroxidases were constitutively upregulated in the *det2 Arabidopsis* mutant (Goda et al. 2002). Furthermore, the oxidative stress-related genes encoding monodehydroascorbate reductase and thioredoxin, the cold and drought stress response genes *COR47* and *COR78*, and the heat stress-related genes *HSP83*, *HSP70*, *HSF3*, *Hsc70-3* and *Hsc70-G7* have been identified by microarray analysis of either BR-deficient or BR-treated plants (Müssig et al. 2002). The enhanced oxidative stress resistance in *det2* plants was correlated with a constitutive increase in SOD activity and increased transcript levels of the defence gene *CAT*. Therefore, a possible explanation for the fact that the *det2* mutant exhibited an enhanced oxidative stress resistance is that the long-term BR deficiency in the *det2* mutant results in a constant in vivo physiological stress that, in turn, activates the constitutive expression of some defence genes and, consequently, the activities of related enzymes. If this is the case, it may suggest that endogenous BRs in wild-type plants somehow act to repress the transcription or post-transcription activities of the defence genes to ensure the normal growth and development of plants. However, it is still unclear whether BRs directly or indirectly modulate the responses of *Arabidopsis* to oxidative stress (Cao et al. 2005).

Brassinosteroids are implicated in plant responses to abiotic environmental stresses and to undergo profound changes in plants interacting with pathogens. Brassinosteroid-regulated stress response as a result of complex sequence of biochemical reactions such as activation or suppression of key enzymatic reactions, induction of protein synthesis and the production of various chemical defence compounds. Unfortunately, it is little known about the correlation between oxidative stress and endogenous level of BRs in plants.

References

Abe H, Soeno K, Koseki N-N, Natsume M (2001) Conjugated and unconjugated brassinosteroids. In: Baker DR, Umetsu NK (eds) Agrochemical discovery. Insect, weed, and fungal control, ACS Symposium series 774. ACS, Washington, DC

Alam MM, Hayat S, Ali B, Ahmad A (2007) Effect of 28-homobrassinolide treatment on nickel toxicity in *Brassica juncea*. Photosynthetica 45:139–142

Ali B, Hayat S, Fariduddin Q, Ahmad A (2008) 24-Epibrassinolide protects against the stress generated by salinity and nickel in *Brassica juncea*. Chemosphere 72:1387–1392

Atici Ö, Ağar G, Battal P (2005) Changes in phytohormone contents in chickpea seeds germinating under lead or zinc stress. Biol Plant 49:215–222

Bajguz A (2010) An enhancing effect of exogenous brassinolide on the growth and antioxidant activity in *Chlorella vulgaris* cultures under heavy metals stress. Environ Exp Bot 68:175–179

Bajguz A (2009) Isolation and characterization of brassinosteroids from algal cultures of *Chlorella vulgaris* Beijerinck (Trebouxiophyceae). J Plant Physiol 166:1946–1949

Bajguz A (2011) Suppression of *Chlorella vulgaris* growth by cadmium, lead and copper stress and its restoration by endogenous brassinolide. Arch Environ Contam Toxicol 60:406–416

Bajguz A, Hayat S (2009) Effects of brassinosteroids on the plant responses to environmental stresses. Plant Physiol Biochem 47:1–8

Bajguz A, Tretyn A (2003) The chemical characteristic and distribution of brassinosteroids in plants. Phytochemistry 62:1027–1046

Cao S, Xu Q, Cao Y, Qian K, An K, Zhu Y, Binzeng H, Zhao H, Kua B (2005) Loss-of-function mutations in *DET2* gene lead to an enhanced resistance to oxidative stress in *Arabidopsis*. Physiol Plant 123:57–66

Choe S (2006) Brassinosteroid biosynthesis and inactivation. Physiol Plant 126:539–548

Choudhary SP, Bhardwaj R, Gupta BD, Dutt P, Kanwar M, Arora M (2009) Epibrassinolide regulated synthesis of polyamines and auxins in *Raphanus sativus* L. seedlings under Cu metal stress. Braz J Plant Physiol 21:25–32

Goda H, Shimada Y, Asami T, Fujioka S, Yoshida S (2002) Microarray analysis of brassinosteroid-regulated genes in *Arabidopsis*. Plant Physiol 130:1319–1334

Hasan SA, Hayat S, Ali B, Ahmad A (2008) 28-homobrassinolide protects chickpea (*Cicer arietinum*) from cadmium toxicity by stimulating antioxidants. Environ Pollut 151:60–66

Hayat S, Mori M, Fariduddin Q, Bajguz A, Ahmad A (2010) Physiological role of brassinosteroids: an update. Indian J Plant Physiol 15:99–109

Hayat S, Ali B, Hasan SA, Ahmad A (2007) Brassinosteroid enhanced the level of antioxidants under cadmium stress in *Brassica juncea*. Environ Exp Bot 60:33–41

Hsu YT, Kao CH (2003) Role of abscisic acid in cadmium tolerance of rice (*Oryza sativa* L.) seedlings. Plant Cell Environ 26:867–874

Li L, van Staden J, Jäger AK (1998) Effects of plant growth regulators on the antioxidant system in seedlings of two maize cultivars subjected to water stress. Plant Growth Regul 25:81–87

Mazorra LM, Núñez M, Hechavarria M, Coll F, Sánchez-Blanco MJ (2002) Influence of brassinosteroids on antioxidant enzymes activity in tomato under different temperatures. Biol Plant 45:593–596

Mittler R (2002) Oxidative stress, antioxidants and stress tolerance. Trends Plant Sci 7:405–410

Müssig C, Fischer S, Altmann T (2002) Brassinosteroid-regulated gene expression. Plant Physiol 129:1241–1251

Núñez M, Mazzafera P, Mazorra LM, Siqueira WJ, Zullo MAT (2003) Influence of a brassinsteroid analogue on antioxidant enzymes in rice grown in culture medium with NaCl. Biol Plant 47:67–70

Őzdemir F, Bor M, Demiral T, Türkan I (2004) Effects of 24-epibrassinolide on seed germination, seedling growth, lipid peroxidation, proline content and antioxidative system of rice (*Oryza sativa* L.) under salinity stress. Plant Growth Regul 42:203–211

Ramonell KM, Kuang A, Porterfield DM, Crispi ML, Xiao Y, McClure G, Musgrave ME (2001) Influence of atmospheric oxygen on leaf structure and starch deposition in *Arabidopsis thaliana*. Plant Cell Environ 24:419–428

Sharma SS, Kumar V (2002) Responses of wild type and abscisic acid mutants of *Arabidopsis thaliana* to cadmium. J Plant Physiol 159:1323–1327

Sharma SS, Dietz K-J (2009) The relationship between metal toxicity and cellular redox imbalance. Trends Plant Sci 14:43–50

Synková H, Semoradová S, Burketová I (2004) High content of endogenous cytokinins stimulates activity of enzymes and proteins involved in stress responses in *Nicotiana tabacum*. Plant Cell Tissue Organ Cult 79:39–44

Triantaphylidès C, Havaux M (2009) Singlet oxygen in plants: production, detoxification and signalling. Trends Plant Sci 14:219–228

Vardhini BV, Rao SSR (2003) Amelioration of osmotic stress by brassinosteroids on seed germination and seedling growth of three varieties of sorghum. Plant Growth Regul 41:25–31

Yokota T (1999) Brassinosteroids. In: Hooykaas PJJ, Hall MA, Libbenga KR (eds) Biochemistry and molecular biology of plant hormones. Elsevier Science, London

Chapter 9
Hormonal Intermediates in the Protective Action of Exogenous Phytohormones in Wheat Plants Under Salinity

Farida M. Shakirova, Azamat M. Avalbaev, Marina V. Bezrukova, Rimma A. Fatkhutdinova, Dilara R. Maslennikova, Ruslan A. Yuldashev, Chulpan R. Allagulova, and Oksana V. Lastochkina

Abstract Unfavorable environmental factors, such as salinity, are known to induce significant shifts in the state of endogenous hormonal system of plants, usually associated with the accumulation of stress hormone ABA and the decrease in the levels of growth-stimulating hormones—auxins and cytokinins. Treatment by phytohormones to improve plant stress resistance also leads to rearrangement of the hormonal balance of plants, and this indicates the involvement of endogenous hormones in the protective effect of exogenous hormones on plants. The aim of this study was to identify the hormonal intermediates in the action of 24-epibrassinolide (EBR), methyl jasmonate (Me-JA), and salicylic acid (SA) on wheat plants under sodium chloride salinity. We found that during pretreatment of seedlings with EBR for 24 h, rapid and stable double accumulation of cytokinins (CKs) was observed without any changes in the contents of IAA and ABA. Pretreatment of seedlings with EBR for 24 h significantly reduced the damaging effect of salinity on plant growth, considerably reduced the stress-induced ABA accumulation as well as the decrease in the IAA content, and also maintained the concentration of CKs at the level of control plants. Interestingly, treatment of seedlings with Me-JA also caused almost a twofold reversible (in contrast to EBR action) CK accumulation without changes in the contents of IAA and ABA. Pretreatment with Me-JA also reduced the stress-induced accumulation of ABA and diminished the decrease in the IAA content and prevented salinity-induced decline in the CK level. These data suggest an important role of cytokinins as intermediates in the manifestation of the protective effect of EBR and Me-JA on wheat plants under salt stress. Comparative analysis of the influence of EBR, Me-JA, and CKs on growth, the state of proantioxidant system, and the level of osmoprotectants in the seedlings under

F.M. Shakirova (✉) • A.M. Avalbaev • M.V. Bezrukova • R.A. Fatkhutdinova •
D.R. Maslennikova • R.A. Yuldashev • C.R. Allagulova • O.V. Lastochkina
Institute of Biochemistry and Genetics, Ufa Scientific Centre, Russian Academy of Sciences, pr. Octyabrya, 71, 450054 Ufa, Russia
e-mail: shakirova@anrb.ru

N.A. Khan et al. (eds.), *Phytohormones and Abiotic Stress Tolerance in Plants*,
DOI 10.1007/978-3-642-25829-9_9,

salinity yielded experimental arguments in favor of this assumption. We have previously suggested that ABA may serve as endogenous intermediate in the protective effect of SA on wheat plants. Comparative analysis of the influence of SA and SA in mixture with fluridone, being an effective inhibitor of ABA biosynthesis, on both *PR-1* and *TADHN* dehydrin gene expression and the activity of antioxidant enzymes, as well as deposition of lignin in the cell walls under sodium chloride salinity, revealed the key role of endogenous ABA in the manifestation of the protective effect of SA on wheat plants.

Keywords 24-Epibrassinolide • Abscisic acid • Methyl jasmonate • Salicylic acid • Cytokinins • Dehydrin • Indoleacetic acid • Proline • Prooxidant–antioxidant balance • Wheat germ agglutinin • Salinity • *Triticum aestivum*

9.1 Introduction

Permanently changing environment demands plant adaptation to the conditions of their growth, which implies the development of a complex network of protective reactions aimed on the struggle for existence in the stressful environment. On the basis of this, there is the integration of the effective systems of regulation of cell metabolic activity for switching genetic programs from norm to stress aimed on development of an adequate protection on the level of whole organism. Salinity caused by increased content of soluble salts in soil is one of the most widely spread abiotic stress factors resulting in significant inhibition of plant growth and decline in crop productivity, sodium chloride being the most detrimental (Munns and Tester 2008; Cambrolle et al. 2011). The decline in cell growth processes is due to dehydration resulting from osmotic effect of salts accumulating in the root zone and due to toxic effect of sodium and chloride accumulation in the plant tissues causing great damaging effect on the most important physiological processes and cell membrane integrity (Munns and Tester 2008). It is necessary to emphasize that plants are able to develop a broad spectrum of protective reactions aimed on diminishing the detrimental effects of salinity (Flowers 2004; Munns and Tester 2008). Stress hormone ABA, which fast and significant accumulation is a characteristic response to salinity, makes an important contribution to protective reactions (Rock et al. 2010; Shakirova et al. 2010a).

ABA is known to play a key role in regulation of stomatal closure (Wilkinson and Davies 2010), resulting in a decline in transpiration and reduction of transpiration losses. Stomatal closure is one of early plant responses to salinity caused by ABA-induced increase in Ca^{2+} concentration in cytoplasm, subsequent activation of ion channels in plasmalemma, and turgor losses by guard cells also linked with ABA-induced enhancement of H_2O_2, production serving as signal intermediate of ABA in stomatal closure (Kim et al. 2010). At the same time, ABA also plays a key role in the synthesis and accumulation of the most important cell osmoprotectants, such as proline and dehydrin proteins, participating in stabilization of cell membranes and protection of biopolymers against the damaging action of reactive oxygen species (ROS) produced in response to dehydration (Silva-Ortega et al. 2008; Szabados and Savoure 2009; Hara 2010), as well as in production of many other components of plant protection (Rock et al. 2010). Induction of expression of ABA-responsive genes is on the basis of realization of regulatory action of ABA on the development of plant stress resistance. Thus, about two thirds of 2,000 genes induced in response to drought progression in *Arabidopsis thaliana* are under ABA control (Huang et al. 2008). In total, more than 2,900 ABA-responsive genes have been identified in *A. thaliana* plants (Nemhauser et al. 2006). At the same time, each phytohormone is known to be involved in regulation of such integral physiological processes as growth, development, and differentiation of plants under normal conditions and environmental changes (Chow and McCourt 2004; Kuppusamy et al. 2008; Stamm and Kumar 2010). This is convincingly demonstrated in experiments using transgenic plants and mutants defective in hormonal synthesis or response as well as in experiments with plants treated with exogenous hormones or inhibitors of their biosynthesis (Peng et al. 2009).

This indicates existence of active hormonal cross-signaling in the whole system of hormonal regulation interacting closely with other signaling systems and coordinating the induction of adequate plant response to internal and external stimuli (Nemhauser et al. 2006; Tuteja and Sopory 2008; Bari and Jones 2009; Shakirova et al. 2010a; Tognetti et al. 2011). At the same time, interaction of phytohormones with each other allows understanding how regulation of the same physiological process is achieved by different phytohormones.

Hormonal system is very sensitive to the slightest changes in conditions for plant growth (Shakirova 2001; Davies et al. 2005; Wang et al. 2008; Rubio et al. 2009; Dobra et al. 2010; Granda et al. 2011). Stress adaptation is determined by the pattern of changes in the concentration ratio of several interacting hormones as well as by the ability of each of hormones to influence endogenous levels of the others (Reski 2006; Kuppusamy et al. 2008; Shakirova et al. 2010a). Drought, disturbance of temperature regime, and salinity are the most important environmental factors, which are critical for survival of plants and lead to significant losses of crop yield. Responses to these stresses are known to be interconnected, employing common signaling pathways, which allow cell adaptation and lead to similar changes in plants on morphological, physiological, biochemical, and molecular/or genetic levels (Verslues and Bray 2006; Nemhauser et al. 2006; Shinozaki and Yamaguchi-Shinozaki 2007; Huang et al. 2008; Potters et al. 2009; Arbona et al. 2010; Des Marais and Juenger 2010). Significant changes in hormonal balance are on the basis of this, manifested in the decline in concentration of phytohormones participating in activation of growth processes (auxins, cytokinins, gibberellins, brassinosteroids) as well as in accumulation of ABA and other hormones involved in the control of stress responses (ethylene, jasmonic, and salicylic acids) (Davies et al. 2005; Ghanem et al. 2011; Wang et al. 2008; Albacete et al. 2008; Shakirova et al. 2010a; Peleg and Blumwald 2011; Granda et al. 2011).

Existence of plants themselves is known to be due to their ability to struggle against extreme environment protecting their vital potential. However, realization of natural protective mechanisms taking place in plants, when conditions become worse, is well known to be accompanied by a decline in their productivity. This raises important question about regulation of stress resistance. The problem of stress resistance is most important for plant breeding and is under a steadfast attention of researchers all over the world. This is indeed the case, since information concerning the chain of reactions taking place in plants in response to extreme external conditions may really contribute to an increase in plant resistance and productivity.

These goals are achieved not only by means of selection of stress tolerant cultivars but also through the purposeful manipulation of adaptation with the help of natural plant growth regulators (PGR) (Ashraf et al. 2008). Due to this, the interest of researchers to growth regulators is not casual. Thus, there is a lot of information concerning effective application of ABA resulting in an increase in plant resistance to stress factors leading to dehydration (Amzallag et al. 1990; Mantyla et al. 1995; Khadri et al. 2007). ABA treatment induces expression of genes coding for proteins involved in plant responses to the cold, drought, and salinity, contributing to preadaptation of plants to the forthcoming action of these stress factors, which in total is reflected in the reduction of the degree of their damaging effect on the growth processes (Tuteja and Sopory 2008). Most of these genes may be divided into two groups. The functional role of the first of them is in protection of cells against stress-induced damage (genes for LEA (late embryogenesis abundant) proteins including those coding for dehydrins widely spread among plants, genes for enzymes catalyzing biosynthesis of osmolytes and enzymes for detoxification), while the functional importance of the other group is linked to ABA signaling (Rock et al. 2010).

However, of special interest are those phytohormones which combine the properties of growth activators and inductors of unspecific resistance which reveal a perspective for their practical application in plant growth. And since plant breeding demands an intensification of not only plant resistance, but in the first place of productivity, aimed application of regulators of plant growth and development capable of increasing both of them are of high priority. It is obvious that not any PGR may be used, but those which are characterized by a wide spectrum of their protective action.

Until now, quite a lot of data have been accumulated in favor of effectiveness of application of phytohormones with the aim of increasing resistance of different cultivars to water deficit caused by drought, salinity, and cold providing the crop yield increase (Wu et al. 2007; Horvath et al. 2007; Walia et al. 2007; Argueso et al. 2009; Shakirova et al. 2010a; Krouk et al. 2011). However, effectiveness of application of PGRs depends strongly on their acting concentration (the use of excessively high concentrations may cause an opposite result). Effective concentrations are different with different cultivars, phases of plant development, and means of treatment. It is important to emphasize that purposeful use of phytohormones for the control of plant resistance and productivity demands a detailed study of the spectrum of reactions taking place in response to a treatment with phytohormones both under normal and stressful conditions for plant growth. Those may be expected to be to a great extent due to the effect of exogenous hormones on rearrangement of hormonal system being effective endogenous system of regulation of cell metabolism.

The present work is dedicated to purposeful revelation of the role of hormonal intermediates in the control of physiological action of brassinosteroids (BRs), salicylic acid (SA), and jasmonic acid (JA) on wheat plants under normal conditions and salinity.

9.2 Brassinosteroids as Inductors of Plant Resistance to Abiotic Stress Factors

Brassinosteroids as well as other phytohormones are involved in the control of a wide spectrum of physiological processes developing in the process of plant ontogenesis, while their belonging to this unique class of steroid phytohormones was proved by experiments using inhibitors of BRs synthesis as well as by insensitivity to BRs of mutants for their synthesis (Altmann 1999; Kim and Wang 2010; Li 2010). Moreover, taking into account the diversity of physiological effects of nmol concentrations of BRs, it is even considered to be the leader among phytohormones (Khripach et al. 2000). BRs are widely spread in the plant kingdom from lower to higher plants and include more than 60 compounds discovered in almost of all plant parts in free forms conjugated with carbohydrates and fatty acids (Clouse and Sasse 1998; Bajguz and Tretyn 2003). Their strikingly pronounced growth-promoting effect on plants achieved in extremely low concentrations is their distinguished property manifested in activation of germination of seeds, cell division and extension, differentiation of vascular system, formation of root system, formation of reproductive organs, and participation of BRs in the control of photomorphogenesis, flowering, and senescence (Clouse and Sasse 1998; Müssig and Altmann 2003; Yu et al. 2011).

Soon after identification of brassinolide, being the first representative of the new class of phytohormones brassinosteroids in plants, laboratory and field experiments revealed not only their growth-stimulating activity but also protective action in plants of different species experiencing low or high temperature, drought, salinity, toxic effect of heavy metals, oxidative stress, and attack of pathogens (Khripach et al. 2000; Kagale et al. 2007; Xia et al. 2009). All these effects lead to increased productivity of BR-treated plants (Cortes et al. 2003; Ali et al. 2008; Saygideger and Deniz 2008; Janeczko et al. 2010). Diverse genetic studies allowed identification of genes involved in BRs biosynthesis and signaling, as well as the great spectrum of genes implicated in plant responses to BRs under normal and stress conditions of growing and interaction of BRs with the signal systems of other hormones (Xia et al. 2009; Zhang et al. 2009; Kim and Wang 2010; Li 2010; Divi et al. 2010; Yu et al. 2011). Indeed, hundreds of genes showing sensitivity to BRs are simultaneously under positive or negative control of auxins, gibberellins, cytokinins, ethylene, ABA, JA, and SA, indicating the existence of cross-talk of

BRs with these hormones in the control of plant growth and development under normal and stress conditions (Arteca and Arteca 2008; Vert et al. 2008; Kurepin et al. 2008; Gendron et al. 2008; Wang et al. 2009; Zhang et al. 2009; Hansen et al. 2009; Choudhary et al. 2010; Divi et al. 2010; Zhang et al. 2011), which, in turn, serves as an evidence in favor of the regulation of BRs signaling by different hormones both under normal and stress conditions.

At the same time, the discussion of the broad spectrum of protective action of BRs raises a justified question concerning the role of endogenous BRs in the development of plant stress resistance (Jager et al. 2008; Xia et al. 2009; Divi et al. 2010) as well as the influence of exogenous BRs on the level of endogenous BRs (Janeczko and Swaczynova 2010). As mentioned above, the key role in induction of protective reactions in response to abiotic stress factors is attributed to endogenous ABA, which concentration is increased sharply under these conditions (Rock et al. 2010). The data concerning the effect of BRs on ABA content in plants under normal conditions and its role for realization of protective effect of BRs on different plant objects are rather contradictory. Thus, there are data showing BR-induced decline in ABA content (Eun et al. 1989) and, opposite to this, accumulation of ABA (Zhang et al. 2011) or no significant changes in the level of this hormone in BR-treated plants (Shakirova et al. 2002; Kurepin et al. 2008; Choudhary et al. 2010) in plants under normal conditions for plant growth. Moreover, there are data both about a decline in the level of stress-induced accumulation of ABA in BR-pretreated plants exposed to salinity (Shakirova and Bezrukova 1998; Avalbaev et al. 2010a) and, on the contrary, about an extra increase in ABA concentration in BR-treated plants under conditions of a short-term heat stress (Kurepin et al. 2008), hypothermia (Liu et al. 2011), and water stress (Yuan et al. 2010). It is of interest that treatment with different concentrations of 24-epibrassinolide (EBR, a synthetic analogue of brassinolide) resulted not only in additional accumulation of ABA in plants of *Raphanus sativus* L. under copper stress but also in an increase in the level of IAA (Choudhary et al. 2010). Discrepancy in the data concerning the role of endogenous ABA in realization of protective action of BRs may be due to specificity of responses of different plants species to BRs (Yuan et al. 2010).

Similar pattern is observed in case of analysis of the role of endogenous BRs in the development of plant stress resistance. Thus, in drought-stressed pea plants, manyfold accumulation of ABA was observed, while no significant changes in the level of endogenous castasterone have been registered, allowing the authors to conclude that endogenous BRs is not involved in plant responses to water stress (Jager et al. 2008). At the same time, development of stress resistance in cucumber plants was reported to correlate positively with the level of endogenous BRs (Xia et al. 2009). Indirect evidence of the involvement of BRs in the development of stress resistance may be provided by the data showing that *siz1-3* mutant of *Arabidopsis* with a reduced expression of genes responsible for BRs biosynthesis and signaling is characterized by significantly lower drought tolerance (Catala et al. 2007) as well as by the reports on a positive effect of BRs on the basic thermo and salt tolerance in a defective in ABA synthesis *aba1-1* mutant of *Arabidopsis*, indicating that in wild-type plants, ABA masks the effects of BRs in the plant stress responses (Divi et al. 2010).

Thus, further experiments are needed aimed on revealing the role of endogenous BRs in induction of plant stress resistance, which should open new perspectives in plant protection based on application of trifling concentrations of BRs in agriculture and encourage the study of BRs signaling involved in regulation of stress resistance of cultivated plants.

At the same time, simultaneous analysis of different groups of phytohormones in plants is of special interest, allowing to get a complex picture of rearrangements in the state of plant hormonal system in response to the treatment with exogenous hormones under normal and stress conditions and to reveal endogenous hormonal intermediates in realization of their biological action.

9.2.1 Role of Endogenous Cytokinins in the Manifestation of the Physiological Effects of 24-Epibrassinolide on Wheat Plants

9.2.1.1 Normal Growth Conditions

Two optimal concentrations of 24-epibrassinolide (EBR) (0.4 nM and 400 nM) stimulating the growth of wheat seedlings have been revealed by us (Shakirova et al. 2002) (Fig. 9.1).

Immunoassay of quantitative level of different groups of phytohormones in the same plants allowed revealing that growth-promoting effect of EBR is, first of all, due to EBR-induced accumulation of cytokinins in plants (Avalbaev et al. 2003) since no significant changes in IAA or ABA have been observed in EBR-treated plants (Fig. 9.2). The presence in the literature of rather contradictory data concerning the effect of synthetic analogues of brassinosteroids on concentration of cytokinins in different plants (both stimulating and inhibiting) should be emphasized (Gaudinova et al. 1995; Upreti and Murti 2004; Vlasankova et al. 2009; El-Khallal et al. 2009). Detailed analysis of the content of cytokinins (CKs) immunoreactive to the antibodies raised against zeatin riboside showed that EBR treatment resulted in a fast and persistent twofold accumulation of cytokinins, while withdrawal of EBR from the cultural media led to a gradual return of cytokinins to the control level (Fig. 9.3a, b). These results indicate the ability of EBR to regulate metabolism of endogenous CK.

Cytokinin oxidase/dehydrogenase (CKX) being a key enzyme in the CKs decay is known to contribute significantly to the control of cytokinin concentration (Galuszka et al. 2004; Avalbaev et al. 2006; Vysotskaya et al. 2010). In connection with this, we have carried out the experiments estimating the character of EBR effect on transcription level of the gene for CKX and activity of CKX enzyme in wheat seedlings. The analysis of transcription activity of CKX gene in wheat plants

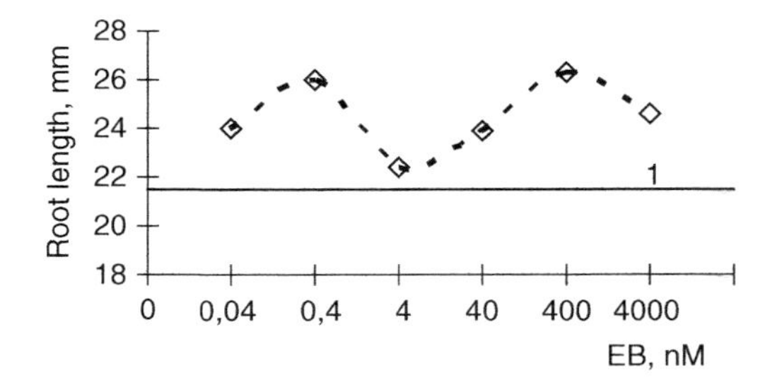

Fig. 9.1 The effect of presowing seed treatment with EBR on root length of 4-day-old seedlings. Mean data of three independent replicates and their SEs are presented

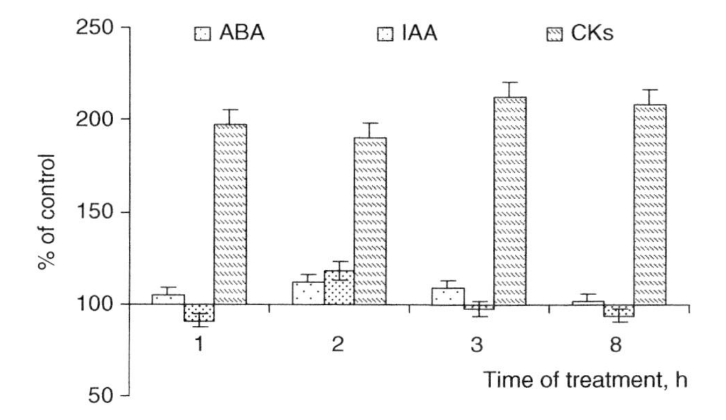

Fig. 9.2 The effect of treatment of 4-day-old seedlings with 400 nM EBR on phytohormone content. Mean data of three independent replicates and their SEs are presented

treated and untreated with EBR has been carried out using the primers chosen by us previously (Avalbaev et al. 2004).

The comparative analysis of the transcription level of CKX gene in plants treated and untreated with EBR showed that EBR caused inhibition of CKX gene expression, while gradual increase in the level of CKX gene expression was observed after withdrawal of EBR from the incubation medium (Fig. 9.3c, d), suggesting regulation of CK concentration by EBR at the expense of its effect on CKX enzyme activity. EBR did cause fast decline in enzyme activity of CKX in roots, while EBR withdrawal from the incubation medium resulted in fast return of CKX activity to the control level (Fig. 9.3e, f).

Thus, the sum of the data obtained by us suggests the ability of EBR to influence the metabolism of CKs and indicates in favor of the important role of endogenous cytokinins in realization of the physiological effects of EBR on wheat plants.

The broad spectrum of cytokinin effects is known to include regulation of growth, differentiation of leaves and roots, formation of vascular system, inhibition of apical dominance, biogenesis of chloroplasts, photomorphogenesis, transpiration, photosynthesis, nitrogen metabolism, fertility, seed development, and germination as well as adaptation of plants to stress factors of different nature (Thomas et al. 1992; Hare et al. 1997; Veselov et al. 2007; Srivastava et al. 2007; Zubo et al. 2008;

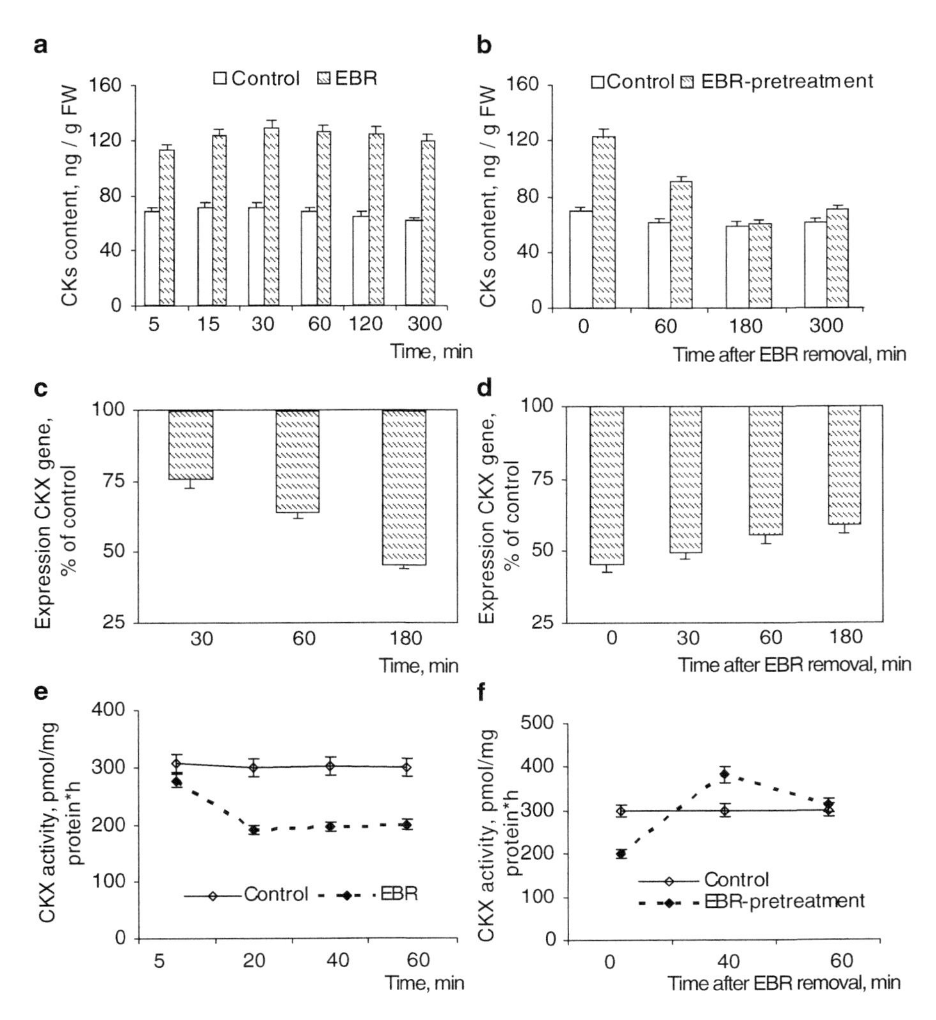

Fig. 9.3 Dynamics of the level of CK content (**a**, **b**), CKX gene expression (**c**, **d**), and enzyme CKX activity (**e**, **f**) in 4-day-old wheat seedlings in the course of 400 nM EBR treatment and immediately after the removal of EBR from the medium of incubation of seedlings. Mean data of three independent replicates and their SEs are presented

Argueso et al. 2009; Chernyad'ev 2009; Rivero et al. 2009; Kudo et al. 2010). Diversity of CK physiological effects is coupled to expression of a broad spectrum of genes sensitive to them (Zubo et al. 2008; Argueso et al. 2009; Xu et al. 2010).

Comparison between functional activity of BRs and CKs discovers a great similarity in their regulatory action on different physiological processes in the course of ontogenesis (Hu et al. 2000; Fatkhutdinova et al. 2002; Fu et al. 2008; Hansen et al. 2009; Avalbaev et al. 2010a, b).

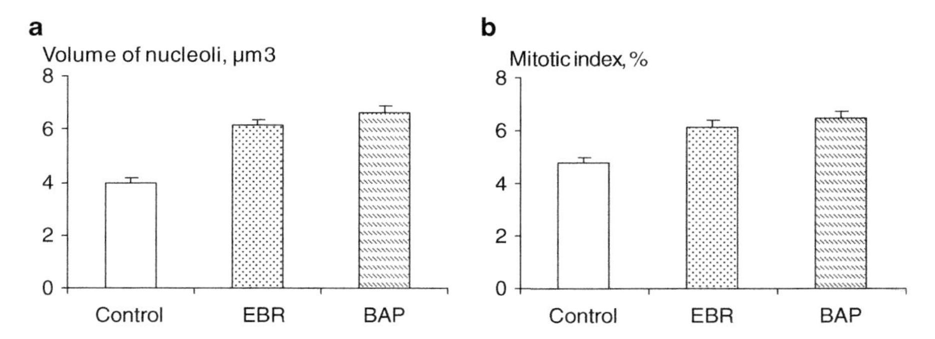

Fig. 9.4 The effect of treatment of 4-day-old wheat seedlings with 400 nM EBR or 40 nM BAP for 24 h on (**a**) the volumes of nucleoli in interphase nuclei and (**b**) the mitotic index (% of dividing cells in 3,000 root meristem cells)

Thus, comparative analysis of the effect of EBR and synthetic analogue of transzeatin 6-benzylaminopurin (BAP) on the size of nucleolar organizing regions (NOR), being an indicator of transcription activity of rDNA genes, and mitotic activity of the apical root meristem showed similarity in the level of stimulating effect of both hormones on these characteristics (Fig. 9.4). This indicates that the increase in the volume of translation apparatus induced by both of these hormones correlating with the total activity of protein synthesis contributes significantly to their growth-promotive action (Shakirova et al. 1982).

Ability of BRs and cytokinins to diminish the damaging effect of water stress on plant growth and productivity due to induction of different protective reactions involved in plant resistance is of special interest in connection with possibility of their application with the purpose of increasing stress resistance and productivity of different cultivated plants (Thomas et al. 1992; Khripach et al. 2000; Iqbal and Ashraf 2005; Srivastava et al. 2007; Veselov et al. 2007; Kagale et al. 2007; Shakirova et al. 2010a; Divi et al. 2010).

Since dehydration results in oxidative burst, associated with a sharp increase in ROS production resulting in imbalance of prooxidants/antioxidants, development of lipid peroxide oxidation (LPO), disturbance of integrity of membrane structures, and growth inhibition (Miller et al. 2010; Gill and Tuteja 2010); EBR- and CK-induced activation of the system of antioxidant defense may be expected to contribute significantly to their protective effect in stressed plants. There are actually a lot of data showing active action of exogenous BRs and CKs on the state of prooxidants/antioxidants systems both under normal and stress conditions (Zavaleta-Mancera et al. 2007; Zhang et al. 2008; Stoparih and Maksimovih 2008; Choudhary et al. 2010; Gangwar et al. 2010; Rubio-Wilhelmi et al. 2011). The study of different plants showed that the treatment with BRs and CKs itself leads to a reversible accumulation of ROS, which in turn results in activation of expression of the genes for antioxidant enzymes (in particular of those for superoxide dismutase (SOD), catalase and ascorbate peroxidase) as well as in an increase in their activity (Zhang et al. 2008; Xia et al. 2009; El-Khallal et al. 2009).

Since cytokinins are known to have a protective effect on plants in a broad range of their concentration (Chernyad'ev 2009), in preliminary experiments, concentration of BAP was chosen leading (similar to EBR) to twofold accumulation of endogenous cytokinins immunoreactive to the antibodies raised against zeatin riboside with the aim of proper comparison of effectiveness of realization of antistress effects of EBR and BAP on wheat plants. Figure 9.5 shows that incubation of wheat seedlings, on the medium containing 40 nM BAP, leads to a fast and stable twofold accumulation of active forms of cytokinins immunoreactive with corresponding antibodies, while withdrawal of BAP from the incubation medium results in gradual return of CK concentration to the control level.

In connection with this, we used 400 nM EBR and 40 nM BAP in the study of the effect of exogenous hormones on different physiological characteristics under normal and stress conditions. Thus, estimation of ROS content and activity of antioxidant enzymes in wheat plants in the course of EBR and CK treatment showed (Fig. 9.6) that incubation of seedling with their roots in the medium

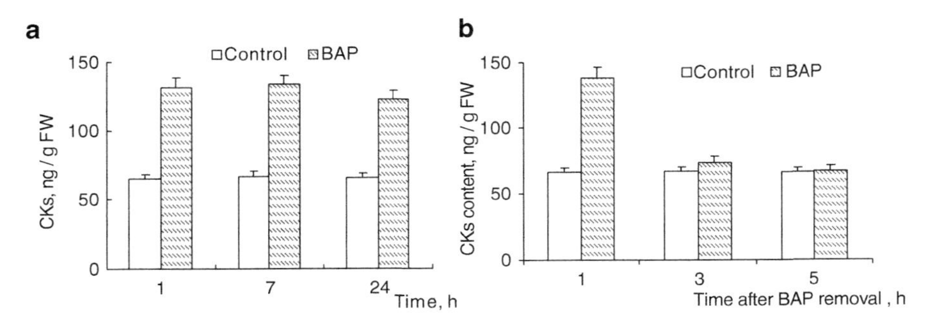

Fig. 9.5 The level of CK content in 4-day-old wheat seedlings (**a**) in the course of 40 nM BAP treatment and (**b**) immediately after removal of BAP from the medium of incubation of seedlings. Mean data of three independent replicates and their SEs are presented

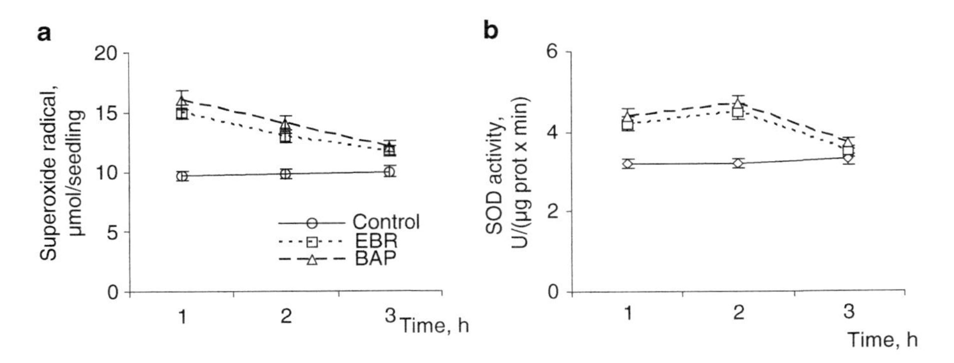

Fig. 9.6 The effects of 400 nM EBR and 40 nM BAP treatments on (**a**) $O_2^{\bullet -}$ generation and (**b**) SOD activity in 4-day-old wheat seedlings. Mean data of three independent replicates and their SEs are presented

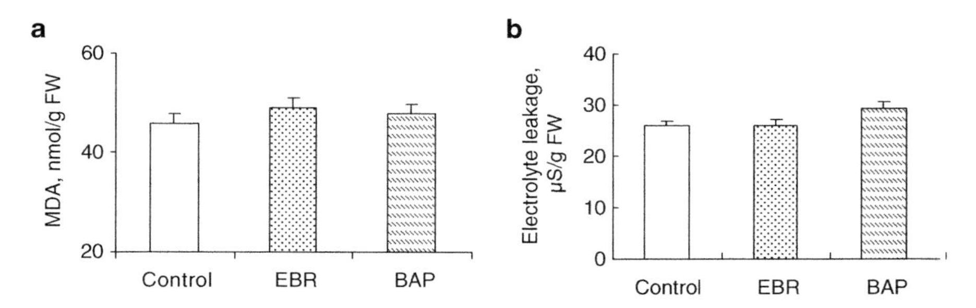

Fig. 9.7 The effects of treatments of 3-day-old wheat seedlings with 400 nM EBR and 40 nM BAP for 24 h on (**a**) MDA content and (**b**) electrolyte leakage. Mean data of three independent replicates and their SEs are presented

containing EBR or BAP results in a slight reversible acceleration of $O_2^{\bullet-}$ production and simultaneous increase in activity of SOD catalyzing $O_2^{\bullet-}$ conversion into H_2O_2.

This contributes to preservation of cell membrane structure integrity due to prevention of the processes of LPO (Gill and Tuteja 2010). The obtained data are likely to result from perception of hormones by plants as certain chemical agents causing changes in the state of prooxidant/antioxidant system, which is balanced in total and does not lead to a damaging effect on the seedlings. This notion is supported by the absence of the changes in the main product of LPO malondialdehyde (MDA) and of electrolyte leakage from the tissues of EBR- and BAP-treated seedlings (Fig. 9.7), as well as by the strictly manifested growth-stimulating effect of these hormones (Fig. 9.4).

Alongside with this, a slight increase in the level of ROS production may be of great importance for the action of EBR and BAP on seedlings, preadapting them to the forthcoming stresses, since ROS are known to act as signal molecules inducing a cascade of protective reactions in plants (Jaspers and Kangasjarvi 2010; Miller et al. 2010).

Until now, convincing data have been obtained in favor of proline functioning as not only osmoprotectant. Proline is also a protein and membrane stabilizer, a metal chelator and hydroxyl radical scavenger, and also an effective quencher of ROS formed in plants under stress-induced dehydration (Szabados and Savoure 2009; Gill and Tuteja 2010). Among ABA-induced proteins, important role belongs to dehydrins. These proteins are peculiar to all taxonomic groups of the plant kingdom. Their massive accumulation is observed in plant seed embryos during their dehydration. However, sharp increase in expression of dehydrin genes and accumulation of their protein products is registered in vegetative plant tissues subjected to dehydration. Dehydrins are the most abundant among stress proteins induced under these conditions (Allagulova et al. 2003). Physicochemical properties of dehydrins are determined by peculiarities of their structure, including the obligatory presence of 15-amino acid K-segment capable to form the secondary amphiphil α-spiral. These properties enable involvement of dehydrins in protecting against denaturation, preserving integrity of cellular structures, and stabilizing membranes under conditions of dehydration (Allagulova et al. 2003; Hara 2010). Moreover, some of dehydrins may be hydroxyl radical and peroxyl radical scavengers and inhibitors of lipid peroxidation (Hara 2010). Induction of expression of dehydrin genes and accumulation of their protein products also take place under normal conditions in ABA-treated plants (Close 1996; Allagulova et al. 2003; Shakirova et al. 2005). Sensitivity of the genes of dehydrins to ABA and different stress factors is determined by the presence in their promoters of different combinations of *cis*-regulated elements interacting with each other and with various ABA-induced *trans*-factors (Shinozaki and Yamaguchi-Shinozaki 2007; Chung and Parish 2008; Shakirova et al. 2009).

Thus, dehydrin and proline are characterized by their polyfunctionality (Szabados and Savoure 2009; Hara 2010), and both exogenous and endogenous ABA are involved in the regulation of their synthesis (Allagulova et al. 2003; Shakirova et al. 2005; Silva-Ortega et al. 2008). However, since application of BRs and CKs effectively increases plant resistance to stress factors, causing dehydration, it was of interest to check if such key osmoprotectants as proline and dehydrin are included into the spectrum of their antistress action on wheat plants.

We have shown a noticeable reversible enhancement of the expression of *TADHN* gene for dehydrin in wheat seedlings treated with EBR and BAP (Fig. 9.8a, b). At the same time, it is important to emphasize that the treatment with hormones themselves in the used concentrations did not cause significant changes in the content of endogenous ABA in wheat seedlings (Fig. 9.8c, d).

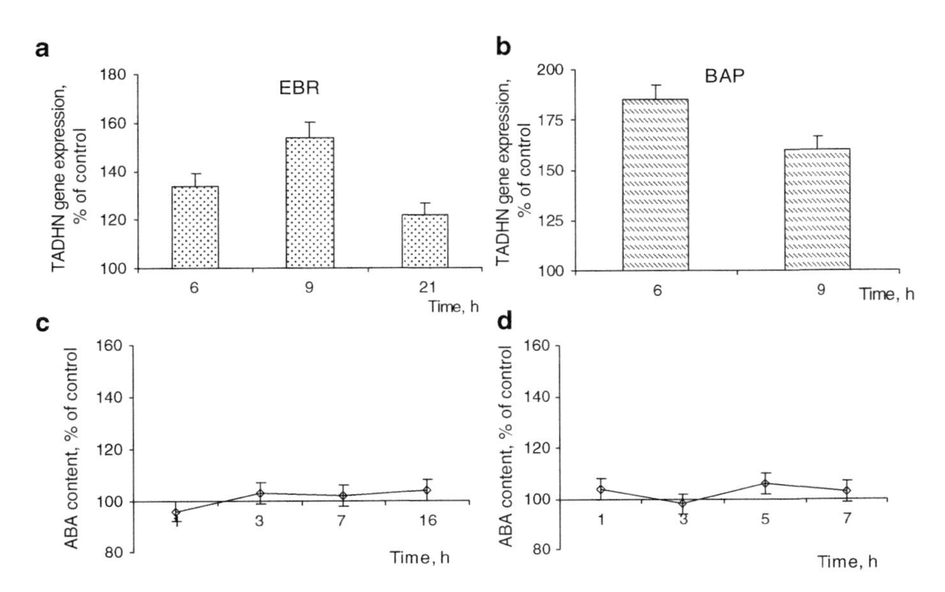

Fig. 9.8 The influence of 400 nM EBR (**a**, **c**) and 40 nM BAP (**b**, **d**) on relative level of dehydrin *TADHN* gene expression and ABA content in 4-day-old wheat seedlings. Mean data of three independent replicates and their SEs are presented

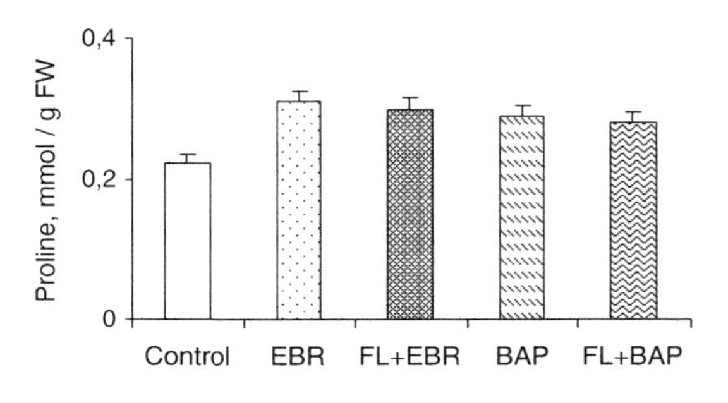

Fig. 9.9 The effect of treatments with 400 nM EBR and 40 nM BAP during 24 h with or without fluridone on proline content in 4-day-old wheat seedlings. Part of 3-day-old seedlings were pretreated by 5 mg/L fluridone for 24 h. Mean data of three independent replicates and their SEs are presented

This data present evidence in favor of and independent from ABA regulation of expression activity of *TADHN* gene for wheat dehydrin by exogenous EBR and CK.

This conclusion is also true in regard to the regulation of the content of proline, which is another key osmoprotectant. Since ABA is known to be involved in the regulation of proline accumulation in plants (Verslues and Bray 2006), we have carried out experiments estimating the effect of EBR and BAP on proline accumulation in wheat seedlings pretreated during 24 h with the inhibitor of ABA synthesis (5 mg/L fluridone) effective in preventing stress-induced accumulation of ABA (Shakirova et al. 2009).

Figure 9.9 shows that EBR and BAP treatments led to significant increase in the level of proline in the seedlings. At the same time, pretreatment with fluridone did not reduce the level of changes in this indicator induced by EBR and BAP in the applied concentrations.

The sum of obtained results demonstrates similarity in the level and rapidity of development of different indicators of cell metabolic activity of the seedlings in response to EBR and BAP, which serves as an argument in favor of possible involvement of endogenous cytokinins in realization of growth-promoting and preadaptive effects of EBR on wheat plants and possibility of the fulfillment by endogenous cytokinins of the role of hormonal intermediate in manifestation of the physiological effects of EBR on wheat plants.

9.2.1.2 Conditions of NaCl Salinity

Salinity belongs to one of the most important detrimental factors causing dramatic changes in activity of genes and proteins, disturbance of metabolic homeostasis, and the course of physiological processes (Jiang et al. 2007; Widodo et al. 2009; Janz et al. 2010; Cambrolle et al. 2011). All this is reflected in significant inhibition of growth processes and crop losses (Wu et al. 2007; Munns and Tester 2008). Alongside with this, participation of BRs and CKs in increasing resistance and productivity of different cultivars to conditions of sodium chloride salinity has been demonstrated (Thomas et al. 1992; Khan et al. 2004; Iqbal and Ashraf 2005; Veselov et al. 2007; Srivastava et al. 2007; Avalbaev et al. 2010a, b; Divi et al. 2010;

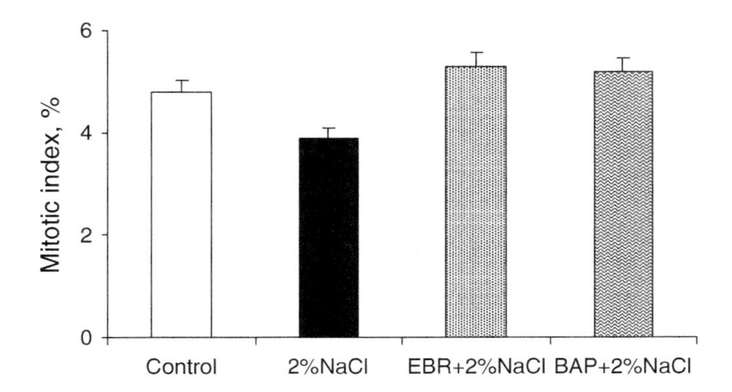

Fig. 9.10 The effect of 2% NaCl (7 h) on mitotic index of roots of 4-day-old wheat seedlings pretreated with 400 nM EBR or 40 nM BAP. Mean data of three independent replicates and their SEs are presented

Ghanem et al. 2011). Comparative analysis of the effect of pretreatment of seedlings with EBR and BAP on different indicators of cell metabolic activity under conditions of salt stress has been carried out with the aim of getting an answer to the question whether endogenous cytokinins are involved in realization of protective action of EBR on plants.

Pretreatment with EBR and BAP for 24 h did not prevent the damaging effect of 2% NaCl on plant growth as was evident from the results of the measurement of mitotic activity of root apical meristem, but significantly reduced its degree manifested in maintaining mitotic index (MI) at least on the control level in roots of EBR- and BAP-treated plants (Fig. 9.10). It is of interest that EBR and BAP exerted protective influence on this growth indicator to the similar degree. The protective effect of EBR is manifested in significantly lower stress-induced shifts in hormonal balance in wheat seedlings.

Actually, dramatic changes in hormonal system associated with accumulation of ABA and gradual decline in the level of auxins and cytokinins may be attributed to the typical plant responses to salinity (Shakirova et al. 2003, 2010a; Davies et al. 2005; Albacete et al. 2008; El-Khallal et al. 2009; Avalbaev et al. 2010a, b). At the same time, plants pretreated with EBR are characterized by significantly smaller stress-induced changes in concentration of ABA and IAA and maintenance of cytokinins at the control level (Fig. 9.11).

Attention should be focused on significant increase in cytokinin content in EBR-pretreated seedlings registered during first 2 h of 2% NaCl action being the consequence of the increase in cytokinin level in the course of pretreatment of plants with EBR (Fig. 9.3a). At the same time, the prevention of the drop of cytokinins during the subsequent salinity action is linked to the inhibition of transcription activity of the CKX gene in EBR-pretreated seedlings under stress (Fig. 9.11d). Thus, under salinity, EBR-pretreated plants are characterized by decreased level of CKX gene expression, which in turn is reflected in maintaining CK concentration at the level close to the control and preventing the inhibitory

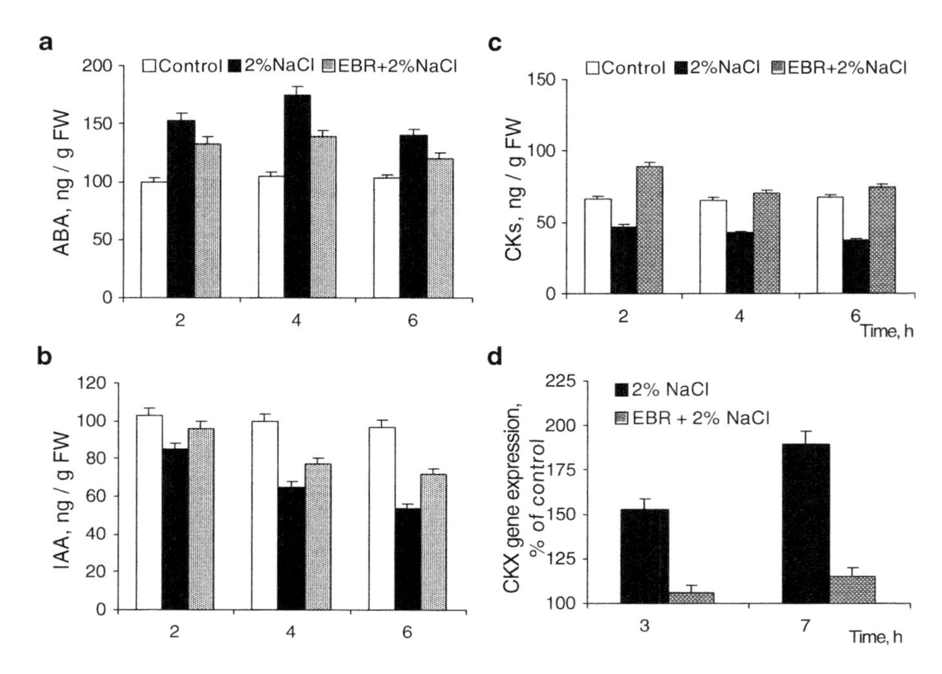

Fig. 9.11 The effect of pretreatment with 400 nM EBR on (**a**) ABA, (**b**) IAA, (**c**) CK content, and (**d**) level of CKX gene expression in wheat seedlings under salinity. Mean data of three independent replicates and their SEs are presented

action of salinity on these growth indicators (Fig. 9.10). Actually, there are a lot of literature data showing that the decline in the level of CKX gene expression and in the level of activity of this enzyme participating in the degradation of free cytokinin forms correlates directly with the increased CK content and plant productivity (Ashikari et al. 2005; Zalewski et al. 2010).

Since EBR and BAP treatment itself exerted small balanced reversible increase in ROS production and activity of SOD and peroxidase in wheat seedlings, it was of interest to compare the dynamics of ROS and activity of antioxidant enzymes in EBR- and BAP-pretreated plants exposed to salinity. The results are presented in Fig. 9.12.

Already within 3 h, salinity resulted in twofold increase in superoxide radical production and subsequent increase in SOD and then in peroxidase activity in the seedlings, being a typical plant stress response (Miller et al. 2010). However, in plants untreated with the hormones, there was an imbalance between ROS and activity of the antioxidant enzymes manifested in preferential ROS accumulation. This exerted an evident damaging effect on plants evident from the decline in mitotic activity of root apical meristem (Fig. 9.9) as well as from the fast (already within 3 h) increase in the level of peroxide oxidation of lipids and electrolyte leakage from the tissues (Fig. 9.13).

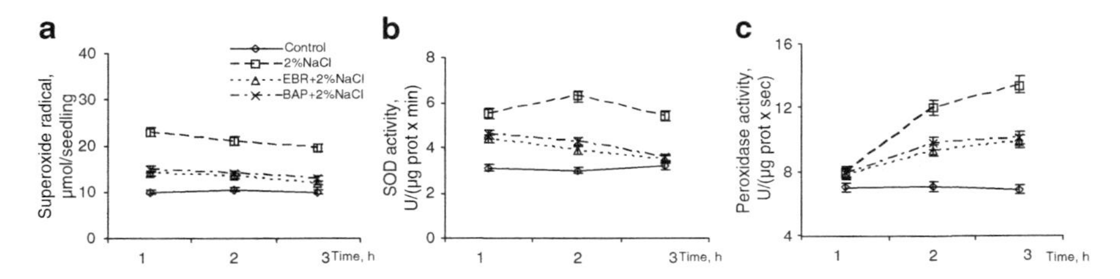

Fig. 9.12 The effects of pretreatments of 3-day-old wheat seedlings with 400 nM EBR and 40 nM BAP for 24 h on (**a**) $O_2^{\bullet-}$ generation, (**b**) SOD, and (**c**) peroxidase activities in roots of 4-day-old wheat seedling under salinity stress. Mean data of three independent replicates and their SEs are presented

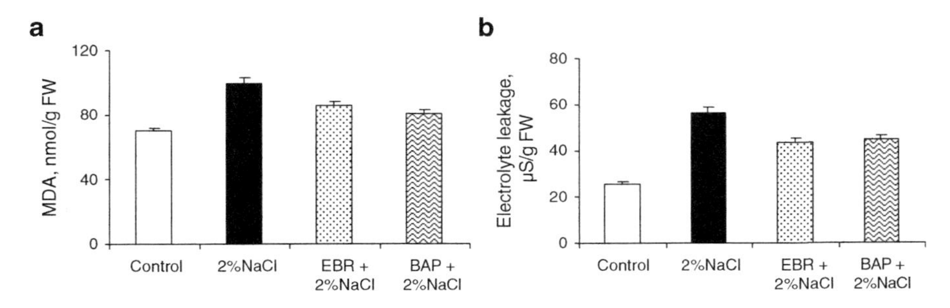

Fig. 9.13 The effects of pretreatments of 3-day-old wheat seedlings with 400 nM EBR and 40 nM BAP for 24 h on (**a**) MDA content and (**b**) electrolyte leakage under 3-h-NaCl salinity. Mean data of three independent replicates and their SEs are presented

At the same time, EBR- and BAP-pretreated seedlings were characterized by significantly lower level of stress-induced accumulation of ROS and smaller activation of SOD and peroxidase (Fig. 9.12) and, correspondingly, by the lower level of the damaging effect of salinity on integrity of membrane structures (Fig. 9.13). These data serve as evidence in favor of the notion that EBR and BAP pretreatment under normal conditions results in slight increase in production of nontoxic levels of ROS, inducing components of antioxidant protection in a balanced way, which in turn contributes to preventing the development of toxic NaCl-induced oxidative burst, so that EBR- and BAP-pretreated seedlings were characterized under salinity by about 50% lower level of ROS accumulation and correspondingly lower activity of SOD and peroxidase. Thus, EBR and BAP were likely to prepare plants to the possible forthcoming stresses by inducing the network of protective reactions those involving ROS signal molecules being one of them.

More than twofold accumulation of proline is observed in response to salinity (Fig. 9.14), which contributes significantly to the development of plant resistance to the stress factor and, as mentioned above, is involved in cell protection against stress-induced oxidative burst (Szabados and Savoure 2009). About 50% reduction of proline accumulation was also observed in EBR- and BAP-pretreated plants

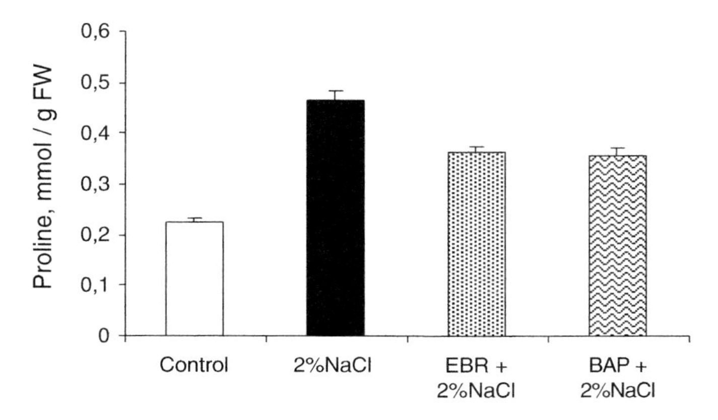

Fig. 9.14 The effects of pretreatments of 3-day-old wheat seedlings with 400 nM EBR and 40 nM BAP for 24 h on proline accumulation in plants during 3-h-NaCl salinity. Mean data of three independent replicates and their SEs are presented

exposed to salinity. These results may be considered as an indicator of the smaller degree of the damaging effect of salinity in plants pretreated with the hormones.

While discussing the results obtained by us, it is necessary to mention the literature data showing that under stress conditions, plants treated with BRs and CKs are characterized by lowered level of stress-induced accumulation of ROS as compared to untreated plants, which is, however, accompanied by additional activation of the antioxidant system possibly due to the increased level of ABA induced by these hormones (Zhang et al. 2008; Yuan et al. 2010; Gangwar et al. 2010; Liu et al. 2011). The discrepancy between our data and literature data may be due to the difference in the plant objects, duration of hormonal treatment (in some experiments plants were exposed to hormones and stress factors simultaneously), the use of different concentrations, which were lower in our experiments, and other reasons.

The sum of the data obtained by us demonstrates similar level of protective effect of both EBR and BAP pretreatments on wheat plants manifested in the decline in the amplitude of stress-induced rearrangements of the state on hormonal and pro-/antioxidant systems as well as in the level of stress-induced accumulation of proline. In our opinion, this is the consequence of a preadaptive effect of hormones causing a slight increase in ROS production in the course of their pretreatment and exerting no negative effect on plants as evident from the strict growth-promoting effect of EBR and BAP on plants, although contributing to activation of antioxidant system of protection as well as to development at the slight activation of the system of osmoprotection (accumulation of proline, TADHN dehydrin).

Taking into account the prevention of stress-induced drop of cytokinins in wheat seedling pretreated with EBR, linked to the decline in the level of activation of the expression of the gene for CKX as a key enzyme of CK degradation, it may be concluded that endogenous cytokinins may be considered as hormonal intermediates playing an important role in realization of the physiological effect of EBR in wheat plants manifested in stimulation of growth processes under normal conditions as well as in a clearly pronounced antistress effect of EBR pretreatment on the wheat plants under salinity. This is evident from the data showing a similar effect of EBR and BAP on wheat seedlings both under normal and stress conditions, when the chosen 0.04 μM BAP concentration exerted the same as EBR twofold accumulation of endogenous cytokinins on the background of the absence of any changes in ABA content.

9.3 Salicylic Acid as Inductor of Unspecific Plant Resistance

Salicylic acid (SA) being an endogenous regulator of growth and development of phenolic nature draws a lot of attention due to its practical importance for increasing plant resistance and productivity. The great attention to SA was initiated by discovery of its key role in induction of systemic acquired resistance (SAR), on the basis of which is the expression of SA-sensitive genes for PR (pathogenesis-related) proteins (Metraux 2002; Vlot et al. 2009). The knowledge about SA signaling is still limited, which is due to the absence of information about its receptors, although a range of SA-binding proteins have been discovered in tobacco: SABP2, having the greatest affinity to SA, chloroplast carboanhydrase, as well as catalase and cytoplasmic ascorbate peroxidase—which indicates the important role of H_2O_2 in the development of SA-induced SAR (Vlot et al. 2009). Information about other components particularly necessary for development of SAR is available due to the use of a set of mutants and transgenic plants (Bari and Jones 2009; Zhang et al. 2010; An and Mou 2011). Transduction of SA signal demands the presence of a regulatory protein NPR1 (nonexpressor of PR genes1) which contains an ankyrin-repeat motif and a BTB/POZ domain, enabling protein–protein interactions (Vlot et al. 2009). Interaction of nuclear localized NPR1 with the trans-factors of TGA family resulted in activation of PR proteins sensitive to SA. The promoter region of the *NPR1* gene also contains W-box sequences, which are binding sites of WRKY family protein, suggesting that WRKY transcription factor plays an important role in mediating signaling between SA and NPR1 (Koornneef and Pieterse 2008; An and Mou 2011). *PR1* is the best-studied gene containing in its promoter *as-1* motive necessary for binding of TGA factors of transcription and gene expression and serving as a marker of SAR (Vlot et al. 2009).

Alongside with this, SA participates in the regulation of different physiological processes in the course of plant ontogenesis under normal conditions of growth (Raskin 1992; Shakirova 2001). Moreover, by now, convincing experimental evidences of the participation of exogenous and endogenous SA in the regulation of plant resistance to different abiotic stress factors (drought, salinity, hypo- and hyperthermia, action of heavy metals, ozone, UV radiation) have been obtained (Shakirova and Bezrukova 1997; Shakirova et al. 2003; Yang et al. 2004; Liu et al. 2006; Horvath et al. 2007; Shakirova 2007; Gemes et al. 2011; Nazar et al. 2011; Syeed et al. 2011).

Thus, SA fulfills an important role in the regulation of unspecific plant resistance due to induction under its influence of a broad spectrum of protective reactions, which allows to consider this growth regulator as a plant stress hormone (Shakirova 2001, 2007).

9.3.1 The Role of Endogenous ABA in the Regulation of Preadaptive Effect of Salicylic Acid on Wheat Plants to Dehydration Conditions

Previously, it was shown by us that SA treatment causes in wheat plants fast shift in the state of hormonal system, associated with parallel reversible accumulation of ABA and IAA on the background of the absence of changes in CK level. This allowed us to suggest that endogenous ABA may serve as a hormonal intermediate in the realization of SA-induced preadaptation of plant to the forthcoming stress (Shakirova et al. 2003; Shakirova 2007). To test this possibility, we have carried out experiments when wheat plants were treated with the inhibitor of ABA synthesis fluridone with subsequent SA treatment and measurement of different indicators of cell metabolic activities.

Figure 9.15 shows that pretreatment with fluridone in selected concentration completely prevented SA-induced accumulation of ABA in wheat seedlings. Consequently, ABA accumulated in SA-treated plants was newborn, indicating participation of SA in the control of de novo ABA synthesis in wheat plants. In connection with this, it was important to estimate the importance of SA-induced accumulation of ABA in the regulation of protective reactions of wheat seedlings developing in plants in response to SA treatment.

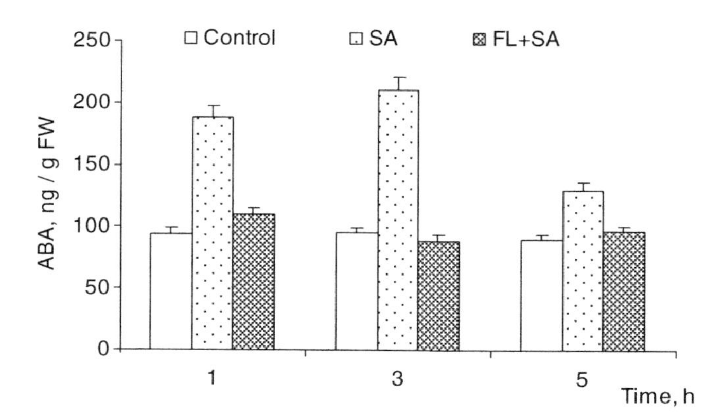

Fig. 9.15 The effect of 0.05 mM SA on ABA content in 4-day-old wheat seedlings pretreated or not treated with 5 mg/L fluridone for 24 h. Mean data of three independent replicates and their SEs are presented

Enhancement of the expression of *PR-1* gene may be attributed to the most typical responses to SA treatment, which may be a marker of SA-induced SAR (Metraux 2002). Alongside with this, by now, there appeared a lot of information about involvement of PR-proteins, PR-1 being among them, in diversity of developmental processes, plant responses to different abiotic stress factors and about sensitivity of different genes coding for PR-proteins not only to infection, treatment with SA, JA, and ethylene being indicators of systemic resistance but with ABA as well as BRs and cytokinins (Liu and Ekramoddoullah 2006; Seo et al. 2008; Lee et al. 2008).

Figure 9.16 shows that the treatment of wheat seedling with SA results in reversible enhancement of expression of *PR-1* gene with a maximum at 9 h. Since SA causes fast accumulation of ABA in the seedlings preceding activation of *PR-1* gene, it could be expected that this gene is sensible to ABA treatment too. Actually, ABA leads to fast (with the maximum at 3 h) enhancement of *PR-1* gene transcription (Fig. 9.16), which may serve as an indirect evidence in favor of the involvement of ABA in the regulation of protective reactions induced in response to SA and contributing to possible stresses of different nature.

Deposition of lignin and suberin in the root cell walls resulting in strengthening of their barrier functions is known to contribute significantly to the development of plant resistance to the stress factors leading to dehydration (Boudet et al. 1995; Moura et al. 2010). Biosynthesis of those biopolymers is realized with the help of H_2O_2 as well as enzymes phenylalanine ammonia-lyase (PAL) and anion peroxidase, ABA, and SA being involved in the control of expression of their genes and enzyme activity (Thulke and Conrath 1998; Hiraga et al. 2001; Fernandes et al. 2006; Moura et al. 2010). In connection with this, it was of interest to compare the effect of SA on dynamics of H_2O_2, activity of PAL and peroxidase, as well as that of lignin deposition in the cell walls of the basal part of roots of wheat seedlings, and to estimate the role of ABA in the control of these processes.

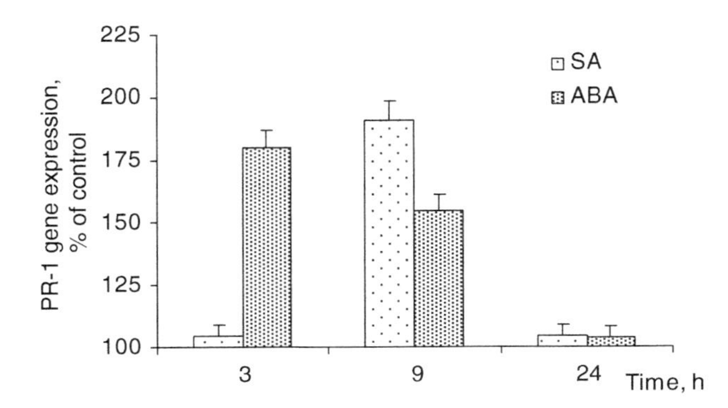

Fig. 9.16 The effect of treatment with 0.05 mM SA and 4 μM ABA on the level of *PR-1* gene expression in 4-day-old wheat seedlings. Mean data of three independent replicates and their SEs are presented

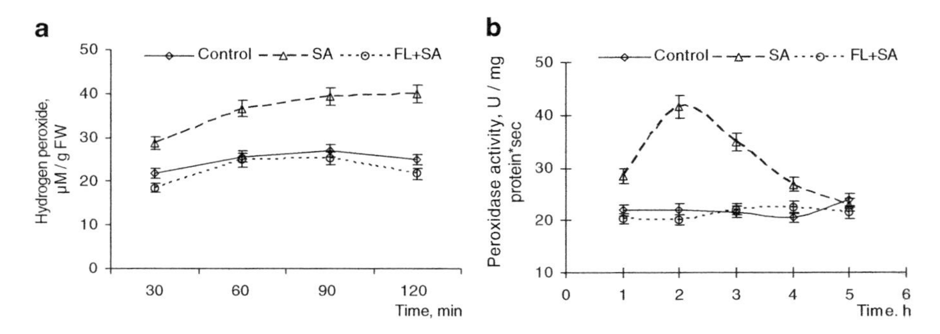

Fig. 9.17 The effect of 0.05 mM SA on (**a**) H_2O_2 production and (**b**) peroxidase activity in 4-day-old wheat seedlings pretreated or not treated with 5 mg/L fluridone for 24 h. Mean data of three independent replicates and their SEs are presented

The treatment with SA itself has been discovered by us earlier to result not only in the transitory accumulation of ABA but also in the enhancement of $O_2^{\bullet-}$ and H_2O_2 in wheat seedlings, accompanied by activation of SOD and peroxidase, which was beneficial for plants as judged from the growth-promoting effect of 0.05 mM SA (Fatkhutdinova et al. 2004; Shakirova 2007). In connection with this, it was important to carry out further a detailed analysis of the effect of SA on H_2O_2 concentration and activity of peroxidase in the seedlings untreated and pretreated with fluridone for 24 h under normal conditions for plant growth.

Figure 9.17 shows that pretreatment of wheat seedlings with fluridone, which prevented SA-induced accumulation of ABA, completely inhibited the enhancement of H_2O_2 production and peroxidase activation exerted by SA alone. These results demonstrate that the balanced increase in the generation of H_2O_2 and activation of peroxidase observed in plants in response to SA treatment is mediated by SA-induced synthesis of ABA. The obtained data indicate the key role of ABA in regulation of H_2O_2 production and confirm the data showing that under different external influences leading to ABA accumulation, pretreatment of plants with inhibitors of ABA synthesis inhibits both H_2O_2 production and activation of antioxidant enzymes (Jiang and Zhang 2002; Ye et al. 2011).

Similar pattern have been also revealed in the course of analyzing the dynamics of PAL enzyme activity presented on Fig. 9.18. It shows that the SA treatment itself exerted a significant increase in PAL activity in the seedlings, while fluridone pretreatment completely prevented SA-induced activation of PAL. These results also indicate that under normal conditions, activation of PAL in plants in response to SA treatment is due to SA-induced production of ABA de novo.

We further carried out the analysis of dynamics of lignin deposition in the central cylinder of the basal part of the control and treated wheat plants (Table 9.1). It shows that treatment of seedlings with SA for 24 h contributes to acceleration of lignification in the cell walls of root xylem vessels as compared to the control: staining of cell walls with phloroglucinol was clearly revealed in the roots of 5-day-old seedlings, while in 6-day-old seedlings, it was additionally enhanced. At the

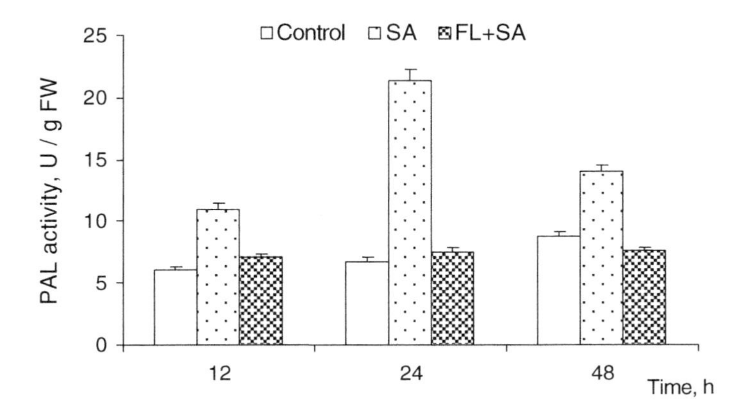

Fig. 9.18 The effect of 0.05 mM SA on PAL activity in 4-day-old wheat seedlings pretreated or not treated with 5 mg/L fluridone for 24 h. Mean data of three independent replicates and their SEs are presented

Table 9.1 Qualitative assay of specific lignin staining with phloroglucinol in the basal part of roots of wheat seedlings pretreated with SA during 24 h in the presence or absence of fluridone (5th day) and then 24 h after exposure of seedlings on 4 μM ABA (6th day)

Variant	5th day	6th day
Control	–	+
SA	+	++
FL + SA	–	–
FL + SA + ABA		++

"-" indicates absence of staining; number of "+" reflects the extent of staining intensity

same time, in the control variant lignin deposition in the cell walls was discovered later on, starting from the 6th day (Table 9.1).

Alongside with this, pretreatment of 3-day-old seedlings with fluridone for 24 h inhibited deposition of lignin in cell wall of roots not only in 5-day-old but also in 6-day-old seedlings (Table 9.1). Consequently, pretreatment with fluridone preventing SA-induced accumulation of endogenous ABA inhibited the enhancement of H_2O_2 production and activation of PAL and peroxidase being the key enzymes in lignin biosynthesis and, as a consequence, inhibited lignin deposition in the cell walls of the root central cylinder. The obtained results convincingly indicate the regulatory role of endogenous ABA in acceleration of lignin deposition in the cell walls of roots of SA-treated seedlings as also evident from the data showing a complete recovery of this process during the subsequent action of exogenous ABA for 24 h on fluridone- and SA-pretreated seedlings (Table 9.1).

The sum of obtained data indicates the important role of SA-induced synthesis of ABA de novo in the regulation of SA-induced activation of the key components of lignin biosynthesis significantly contributing to the strengthening of barrier properties of root cell walls and preadaptation of plants to possible forthcoming action of environmental stress factors.

9.3.2 *The Role of Endogenous ABA in SA-Induced Activation of Protective Reactions in Salinity Stressed Plants*

At the beginning of experiments, it was important to compare the effects of treating the plants with SA and a mixture of SA and fluridone on ABA content in the seedlings exposed to salinity stress. Figure 9.19 shows that salinity led to significant accumulation of ABA in untreated wheat seedlings, which is not surprising since this response is a typical stress reaction (Shakirova 2001). SA-pretreated seedlings were characterized by significantly lower level of salt-induced accumulation of ABA, which may be an indicator of lower degree of the damaging stress effect in these plants due to preadaptive action of SA in the course of pretreatment. At the same time, a pretreatment of seedlings initially with fluridone and then with a mixture of fluridone and SA during 24 h completely prevented the salt-induced increase in the level of ABA (Fig. 9.19a).

This suggests inhibition of ABA-mediated protective reactions and consequent prevention of protective action of SA on plants under salinity. Actually, pretreatment of SA significantly decreased the degree of the negative stress effect of the seedlings, while the treatment with the mixture of SA with fluridone prevented the protective action of SA on growth (Fig. 9.19b), indicating the important role of endogenous ABA in realization of SA protective action in stressed plants.

Salinity induces a fast increase initially in activity of SOD followed by peroxidase in wheat seedlings (Fig. 9.20), which is a typical plant response to stress-induced enhancement of ROS production (Jaspers and Kangasjarvi 2010). Stressed plants pretreated with SA are characterized by a significantly lower activity of

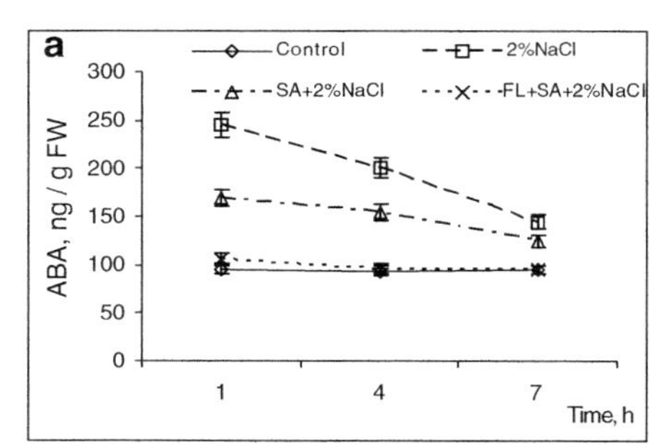

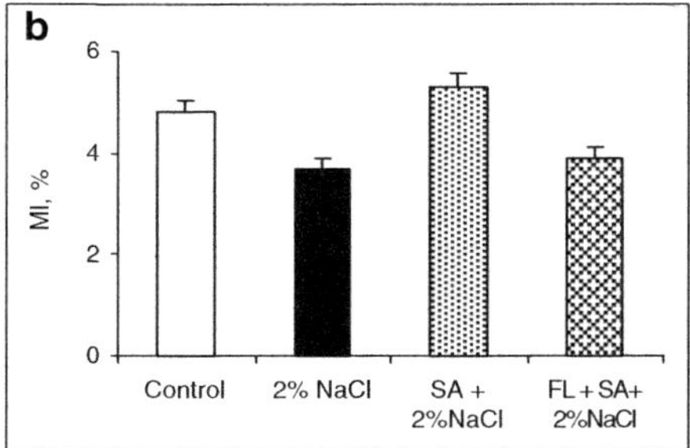

Fig. 9.19 The effect of pretreatment of 3-day-old seedlings initially with 5 mg/L fluridone alone during 3 h and then mix FL with 0.05 mM SA for 24 h on (**a**) ABA content in 4-day-old wheat plants under salinity and (**b**) mitotic index (%) of root meristem cells after 7 h exposure on 2% NaCl. Mean data of three independent replicates and their SEs are presented

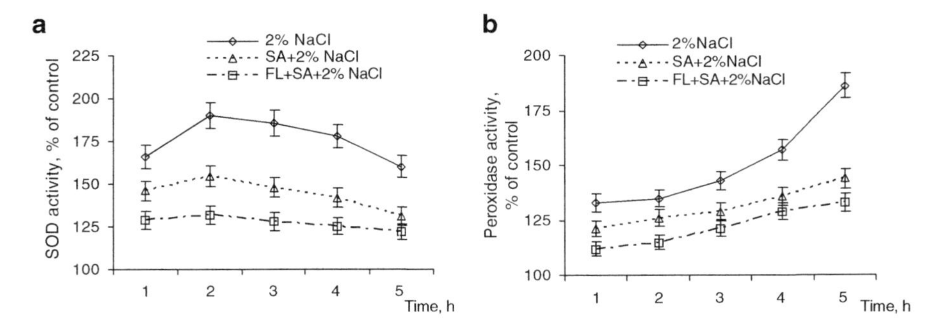

Fig. 9.20 The effect of pretreatment of 3-day-old seedlings initially with 5 mg/L fluridone alone during 3 h and then with the mixture of FL with 0.05 mM SA for 24 h on activity of (**a**) SOD and (**b**) peroxidase in 4-day-old wheat plants under salinity. Mean data of three independent replicates and their SEs are presented

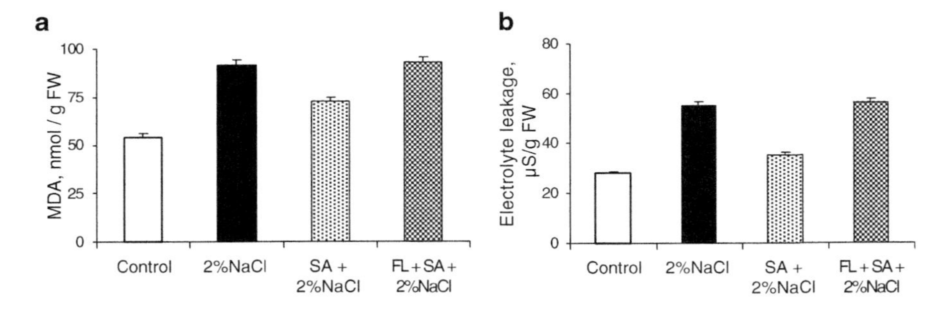

Fig. 9.21 The effect of pretreatment of 3-day-old seedlings initially with 5 mg/L fluridone alone during 3 h and then with the mixture of FL with 0.05 mM SA for 24 h on (**a**) MDA content and (**b**) electrolyte leakage in wheat plants after 24-h salinity. Mean data of three independent replicates and their SEs are presented

antioxidant enzymes as compared to untreated plants (Fig. 9.20), which is likely to be due to the lower production of ROS in the plants (Shakirova 2007). This is associated with the fact that SA treatment itself (before stress) causes a balanced increase in production of ROS and activity of antioxidant enzymes (Fig. 9.17), sufficient for the strengthening of barrier properties of cell walls and enabling effective neutralization of salt-induced oxidative burst (Shakirova 2007).

This notion is supported by 50% decline in the level of MDA in comparison with untreated plants under salt stress (Fig. 9.21a) and reduction of electrolyte leakage from the tissues (Fig. 9.21b). At the same time, treatment with SA in the mixture with the inhibitor of ABA synthesis fluridone led to significant inhibition of stress-induced activation of SOD and peroxidase in the seedlings in comparison to those untreated with SA (Fig. 9.20). This was reflected in that in the variant of treatment with the mixture of fluridone and SA, plants were characterized by the same level of

Table 9.2 Qualitative assay of specific lignin staining with phloroglucinol in the basal part of roots of 5-day-old wheat seedlings pretreated with 0.05 mM SA during 24 h in the presence or absence of 5 mg/L fluridone and then 24 h after exposure of seedlings on 2% NaCl

Variant	5th day
Control	–
2% NaCl	+
SA + 2% NaCl	++
FL + SA + 2% NaCl	–/+

"–" indicates absence of staining; "–/+" indicates very weak staining; number of "+" reflects the extent of staining intensity, with "++" being very strong staining

MDA accumulation and electrolyte leakage as in the salt-stressed plants untreated with SA (Fig. 9.21).

The obtained data illustrate the importance of ABA in realization of preadaptive effect of SA on wheat plants manifested in maintaining enzyme activity of SOD and peroxidase in SA-pretreated salt-stressed plants at the level sufficient for their protective effect and in significant decline in the level of the damaging effect of salinity on the integrity of cell membranes and their permeability in SA-pretreated plants under NaCl salinity.

At additional argument in favor of involvement of ABA in the regulation of SA-induced protective reactions aimed on the decline in the damaging effect of salinity on wheat plants is in the data showing significant decline in lignin accumulation in the cell walls of the basal part of seedling roots pretreated with the mixture of fluridone with SA (Table 9.2). This is likely to be due to prevention by fluridone of the SA-induced production of H_2O_2 and activation of the key enzymes of lignin biosynthesis (PAL and peroxidase) being under the control of ABA (Moura et al. 2010).

The sum of data serves as the evidence in favor of fulfillment by ABA of the role of hormonal intermediate in triggering SA-induced net of protective reactions contributing to the development of SA-induced resistance to abiotic stress factors.

At the same time, the analyzed data concerning combination in SA of the properties of induction of SAR and resistance to a broad range of abiotic stress factors draw attention to the fact that another growth regulator jasmonic acid (JA) and its active metabolites joined in general group of jasmonates, which similar to SA is now attributed to the class of stress hormones, also play a key role in triggering induced systemic resistance (ISR) to different causative agents of plants diseases and are able to increase resistance to different abiotic stresses.

9.4 Importance of Endogenous Cytokinins in Manifestation of Physiological Effects of Jasmonates on Wheat Plants

Jasmonic acid (JA) and its active derivatives are joined into the group of endogenous regulators of growth and development participating in regulation of integral physiological processes like seed germination, root growth, formation of generative organs, flowering, transport of assimilated, nutrient storage, tuber formation, fruit ripening, and senescence (Creelman and Mullet 1995; Wasternack 2007; Balbi and Devoto 2008; Sun et al. 2009; Shimizu et al. 2011). However, special attention to jasmonates was initiated by discovery of their key role in cooperation with ethylene in triggering ISR in response to wounding, affection with pests, necrotroph pathogens, associated with expression of *PR*-genes, and other protective proteins (Wasternack 2007; Van der Ent et al. 2009; Bari and Jones 2009; Fujimoto et al. 2011; Ballare 2011). Alongside with this, information about the increase in biosynthesis of jasmonates in different cultivars under conditions of drought, salinity, extreme temperatures, heavy metals, ozone, UV radiance as well as about the decline in the damaging effect of these stresses on plants exerted by exogenous jasmonates serves as the evidence of the involvement of jasmonates in the control of different protective reactions laying on the basis of development of plant resistance to different adverse abiotic factors (Bandurska et al. 2003; Kang et al. 2005; Walia et al. 2007; Tamaoki et al. 2008; Battal et al. 2008; Clarke et al. 2009; Shakirova et al. 2010b; Shan and Liang 2010). Thus, the revealed diversity of regulatory action of jasmonates in plants allowed attributing them to the main classes of plant hormones (Goda et al. 2008; Katsir et al. 2008; Browse 2009).

Discovery of the functional importance of jasmonates was to a great extent due to obtaining of *Arabidopsis coronatine-insensitive 1* (*coi1*) mutant. *Coi1* locus codes for F-box protein, being the subunit of SCF ubiquitin ligase E3 complex, in the presence of hormone SCF^{COI1} binds to JAZ (jasmonate ZIM-domain) proteins of fast response to JA, being repressors of JA signaling, which leads to the degradation of JAZ proteins with the help of 26S proteasomes and switching on the expression of JA-sensitive genes (Katsir et al. 2008; Browse 2009; Ballare 2011). In the absence of JA, JAZ proteins interact with MYC2 and other JA-regulated trans-factors and suppress the response genes to jasmonates (Katsir et al. 2008; Browse 2009; Galis et al. 2009). By now, a lot of data have been obtained showing that diversity of physiological effects of jasmonates is due to their effect on expression state of hundreds of genes (Wasternack 2007; Walia et al. 2007; Ziosi et al. 2008; Goda et al. 2008; Browse 2009; Pauwels et al. 2009; Chen et al. 2011), many of which are also under the control of other hormones (Fujita et al. 2004; Goda et al. 2008; Peng et al. 2009), indicating the integration of jasmonates into the whole hormonal system regulating metabolism of plant organism in the course of ontogenesis under normal and stress conditions in interaction with Ca^{2+}, ROS and NO signaling (Sasaki-Sekimoto et al. 2005; Hung et al. 2006; Tuteja and Sopory 2008; Pauwels et al. 2009; Clarke et al. 2009; Sun et al. 2009; Yang et al. 2011; Peng et al. 2011).

In the literature, there is information about effectiveness of application of jasmonates on different cultivated plants with the aim of increasing their resistance to abiotic stress factors. Thus, recently, some data have been obtained showing that JA pretreatment contributes to an increase in resistance of barley to sodium chloride salinity (Walia et al. 2007). This is evident from the data showing a decline in accumulation of sodium ions in the shoot tissues and significant amelioration of indicators of photosynthesis as compared to untreated with growth regulators stressed plants, which is likely to be due the increase in expression of genes involved in adaptation of plants to salt stress in plants pretreated with JA (Walia et al. 2007). At the same time, it is necessary to mention that barley is characterized by elevated salt tolerance among other cultivated plants (Munns and Tester 2008). By the beginning of our experiments, we knew nothing about protective action of jasmonates and active natural JA derivative methyl jasmonate (Me-JA), in particular, on wheat plants under sodium chloride salinity.

Initially, it was necessary to chose acting concentration of Me-JA efficient for wheat seedlings. Beforehand, experiments were carried out with presowing treatment of seeds during 3 h with Me-JA in the range of 0.001–1,000 nM. Analysis of the linear size, raw, and dry mass of seedlings revealed two most effective concentrations of Me-JA (0.001 and 100 nM) (Shakirova et al. 2010b), which were chosen for further experiments with the treatment of seedlings. The experiments showed the advantage of the use of pretreatment of wheat seedlings with 100 nM methyl jasmonate, and this concentration was applied in further testing (Fig. 9.22).

Further, it was of interest to study the character of the changes in hormonal balance of wheat seedlings in the course of Me-JA treatment. This analysis was needed since we had no data about the effect of jasmonates on the content of the main classes of hormones in the same wheat plants. The experiments showed that incubation of seedlings on the medium with Me-JA led to a fast, about twofold, transitory accumulation of cytokinins (Fig. 9.23c) without significant changes in the content of ABA and IAA (Fig. 9.23a, b).

It is necessary to emphasize the literature data showing that the diversity of regulatory action of jasmonates is due to their positive and negative interaction with other phytohormones, such as ethylene and SA (Tamaoki et al. 2008; Bari and Jones 2009; Clarke et al. 2009; Yang et al. 2011), ABA (Hays et al. 1999; Si et al. 2009), auxins (Wong et al. 2009; Sun et al. 2009), gibberellins (Traw and Bergelson 2003), brassinosteroids (Campos et al. 2009), and cytokinins (Ananieva et al. 2004); however, information about the effect of jasmonates on the content of hormones from different classes obtained in the same experiment is rather limited (Bandurska

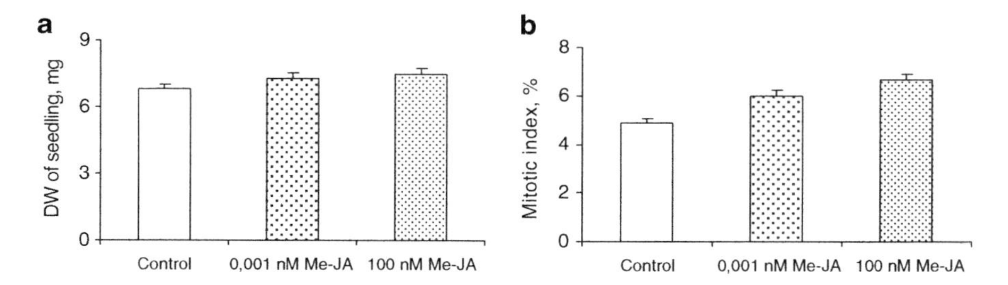

Fig. 9.22 The effect of treatment of 3-day-old seedlings with 0.001 or 100 nM Me-JA during 24 h on (**a**) seedling dry weight and (**b**) mitotic index of roots of 4-day-old seedlings. Mean data of three independent replicates and their SEs are presented

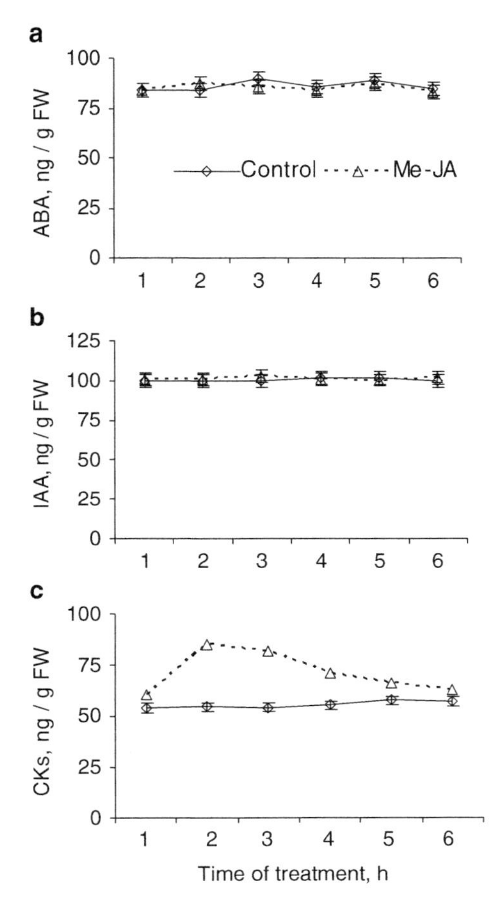

Fig. 9.23 The effect of 100 nM Me-JA treatment on phytohormone content in 4-day-old wheat seedlings. Mean data of three independent replicates and their SEs are presented

et al. 2003; Kang et al. 2005; Battal et al. 2008; Sun et al. 2009; Shakirova et al. 2010a; Yang et al. 2011). The data concerning changes in the content of different forms of cytokinins in response to JA treatment are rather contradictory (Dermastia et al. 1994; Ananieva et al. 2004; Battal et al. 2008; Shakirova et al. 2010b), which may be due to the use of different model objects and application of rather broad range of jasmonates concentrations. Thus, recently, there appeared data showing a dose-dependent increase in concentration of cytokinins in maize plants pretreated with JA and exposed to the action of hypothermia (Battal et al. 2008). The data obtained by us allow to suggest an important role of cytokinins not only in the growth-promoting action of Me-JA on wheat plant but also in preadaptation of plants to possible forthcoming stresses, since cytokinins are known to be implicated in the control of a broad spectrum of metabolic processes under normal conditions, some of which contribute significantly to plant stress adaptation (Hare et al. 1997; Veselov et al. 2007; Chernyad'ev 2009).

9.4.1 Effect of Me-JA on Wheat Resistance to Sodium Chloride Salinity

Application in the practice of plant breeding of different ways of plant treatment with growth regulators has been discussed above. Testing their effect on seed germination is used for the analysis of the effectiveness of the application of growth regulators for protection of plants against adverse environmental factors. We have studied the effect on germination of wheat seeds of their presowing treatment with the solution of 100 nM Me-JA during 3 h. Figure 9.24 shows that presowing treatment of seeds with Me-JA did not prevent but significantly reduced the extent of the negative effect of sodium chloride on germination capacity of seeds, estimated 36 h after the start of germination and on accumulation of seedling mass registered after 3 days of incubation on 2% NaCl. This indicates protective effect of Me-JA on wheat resulting from presowing treatment.

Alongside with the soaking of seeds, the treatment of vegetative plants is also frequently applied. Thus, it was important to find out further if 100 nM Me-JA exerts a protective effect on wheat plants during pretreatment for 24 h before stress. In connection with the fact that roots are the immediate target exposed to salty environment, we used the analysis of mitotic activity of root apical meristem as an indicator of the effect of Me-JA pretreatment on wheat resistance to the given stress.

Figure 9.25 showed that seedlings pretreated with Me-JA during 24 h were characterized by increased root MI (in comparison with the control variant), being an indicator of growth promoting effect of 100 nM Me-JA on wheat plants. The action of 2% NaCl during 7 h led to a noticeable decline in MI of cells of root tips of plants both treated and untreated with Me-JA. However, it is important to emphasize that pretreatment with Me-JA contributed to maintaining root MI of the seedlings at the control level during the poststress period (Fig. 9.25). The results demonstrate protective effect of Me-JA pretreatment on wheat plants under salinity.

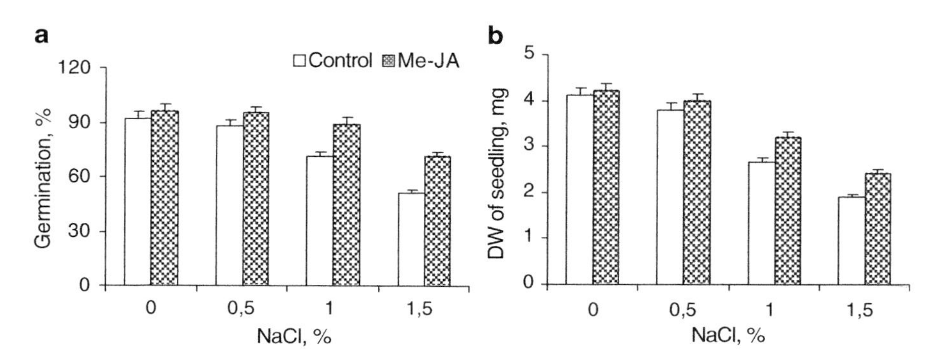

Fig. 9.24 The effect of presowing seed treatment with 100 nM Me-JA for 3 h on (**a**) seed germination (after 36 h) and (**b**) dry weight of seedlings (after 3 days) under salinity. Mean data of three independent replicates and their SEs are presented

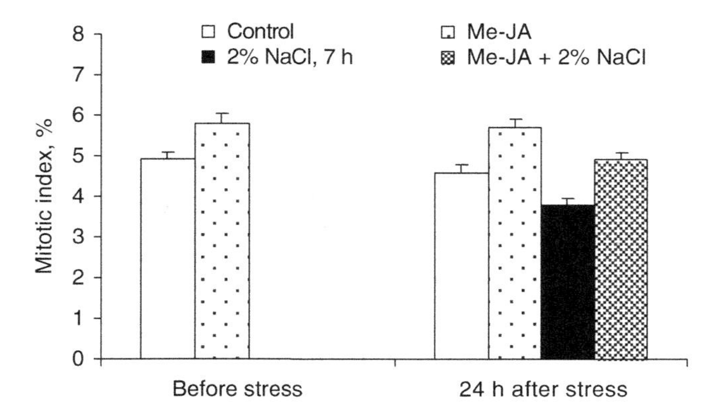

Fig. 9.25 The effect of 24-h pretreatment of 3-day-old wheat seedlings with 100 nM Me-JA on mitotic index of root meristem cells before and after 24 h of 7-h treatment with 2% NaCl. Mean data of three independent replicates and their SEs are presented

The key role is attributed to hormonal system in the regulation of whole network of protective reactions laying on the basis of the development of plant resistance to stress environmental factors (Shakirova 2001). Effectiveness of development of stress adaptation is mostly determined by the character of changes in concentration of phytohormones in relation to each other, interacting in united regulatory system, as well as by the ability of each hormone to influence the level of other endogenous hormones (Reski 2006).

Analysis of hormonal status of wheat seedlings exposed to salinity showed (Fig. 9.26) that Me-JA pretreatment exerts in total a stabilizing effect on the phytohormonal balance associated with the decline in the amplitude of stress-induced accumulation of ABA and in the extent of decline in IAA, as well as in prevention of salt-induced drop of cytokinin content.

The obtained results allow to suggest that maintaining cytokinin concentration in Me-JA-pretreated stressed plants at the control level is an important regulatory component of the realization of protective action of Me-JA on wheat, since cytokinins used in different ways (including presowing of seeds) also contribute to the increase in resistance of plants to stresses caused by disturbance of water relations (Chernyad'ev 2009).

Since salinity causes a sharp enhancement of ROS generation (Munns and Tester 2008), leading to destructive consequences manifested in the damage of the integrity of membrane cell structures and their permeability, it is important to emphasize the ability of jasmonates to increase activity of catalase, cell-wall peroxidase, ascorbate peroxidase, NADH peroxidase as well as to increase the content of ascorbate and glutathione in plant tissues (Sasaki-Sekimoto et al. 2005; Maksymiec and Krupa 2006; Xue et al. 2008; Piotrowska et al. 2009; Shan and Liang 2010). This allows suggesting that in the course of pretreatment (prior to stress) Me-JA contributes to the development of an efficient system of antioxidant defense, enabling neutralization of ROS production, which, in turn, should be reflected in

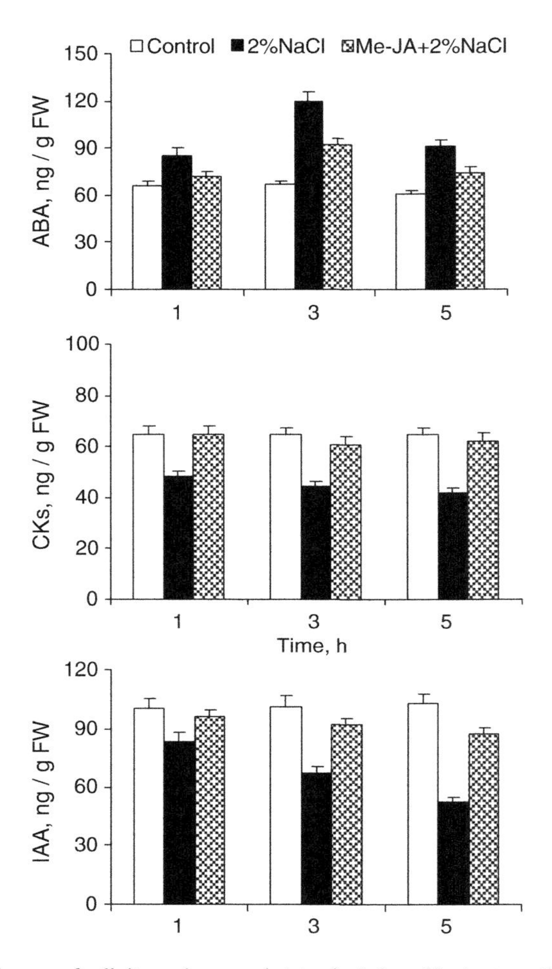

Fig. 9.26 The influence of salinity on hormonal status in 4-day-old wheat seedlings untreated and pretreated for 24 h with 100 nM Me-JA. Mean data of three independent replicates and their SEs are presented

weakening the development of salt-induced oxidative burst and reducing the degree of the damaging effect of salinity on cell membrane structures.

Actually, Me-JA-pretreated wheat seedlings exposed to salinity are characterized by the lower level of peroxide lipid oxidation (Fig. 9.27a) and reduced indicators of electrolyte leakage (Fig. 9.27b). The obtained results are in agreement with literature data showing the decline in the level of damaging effect of water deficit on the cell membrane integrity in plants pretreated with JA (Bandurska et al. 2003; Shan and Liang 2010). At the same time, the treatment with Me-JA itself exerted only slight changes in the level of these indicators which serve as an evidence of a quite beneficial effect of Me-JA used in this concentration (Fig. 9.27).

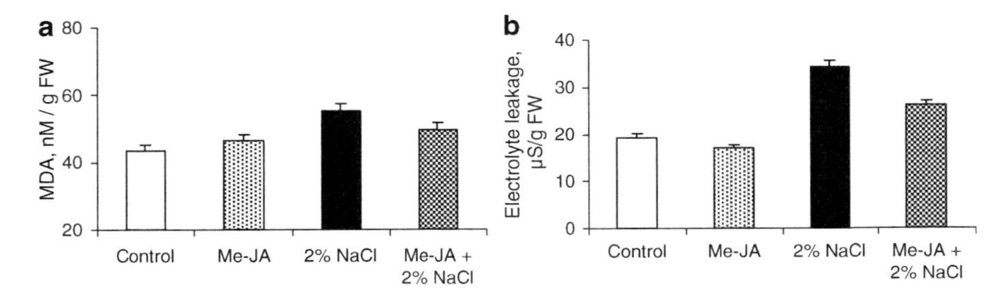

Fig. 9.27 The effect of pretreatment of 3-day-old seedlings with 100 nM Me-JA on (**a**) MDA content and (**b**) electrolyte leakage in wheat plants after 24 h salinity. Mean data of three independent replicates and their SEs are presented

Table 9.3 Qualitative assay of lignin accumulation in the basal part of roots of 5-day-old wheat seedlings pretreated and untreated with Me-JA or BAP for 24 h and after exposure to 2% NaCl for 24 h[a]

Variant	5th day
Control	–
100 nM Me-JA	+
40 nM BAP	+
2% NaCl	+
100 nM Me-JA + 2% NaCl	++
40 nM BAP + 2% NaCl	++

[a]Three-day-old seedlings pretreated with Me-JA or BAP for 24 h were transferred to the solution of NaCl for 24 h

Mark "–" indicates absence of staining by phloroglucinol; mark "+" reflects the extent of color intensity

Enhancement of lignin synthesis and its deposition in the cell walls contributes significantly to the development of plants resistance to the toxic effect of sodium chloride (Moura et al. 2010). In the literature, there are data concerning the jasmonate-induced enhancement of the expression of the gene for phenylalanine ammonia-lyase, increase in activity of apoplastic peroxidase and the level of H_2O_2, being the main players in lignin synthesis as well as its accumulation in different cultivated plants under normal and stress conditions (Maksymiec and Krupa 2006; Xue et al. 2008; Pauwels et al. 2009). Analysis of the effect of Me-JA on lignin deposition in the walls of the central cylinder of the basal part of the roots of wheat seedlings showed that similar to salicylate, Me-JA contributes to the increase in lignification of cell walls in the course of pretreatment as compared to the control (Table 9.3). Alongside with this, let us recall that in case of SA treatment, regulation of this process was realized through SA-induced accumulation of endogenous ABA (Table 9.1). At the same time, treatment with Me-JA in the concentration used by us did not cause any shifts in ABA content (Fig. 9.23a), showing the ability of Me-JA to regulate acceleration of lignin deposition in the cell walls independently of endogenous ABA.

Alongside with this, endogenous and exogenous cytokinins are known to be involved in the regulation of biosynthesis and accumulation of lignin (Funk and Brodelius 1990; Guo et al. 2005). Relying on the revealed fact of the fast reversible accumulation of cytokinins in wheat seedlings in response to the treatment with

Me-JA without any visible changes in the level of ABA, analysis of the effects of BAP treatment in 40 nM concentration, which did not cause changes in the level of ABA, was further carried out. Table 9.3 shows acceleration over the control of lignin deposition in the basal root parts of BAP-pretreated seedlings similar to the effect of Me-JA on this process. Moreover, 2% NaCl also accelerated lignin accumulation in the cell walls within 24 h in comparison to the control, while in the roots of seedlings pretreated with Me-JA and BAP significant additive accumulation of lignin was observed in the cells walls of roots (Table 9.3).

The obtained data indicate involvement of Me-JA in the control of lignin accumulation in the root cell walls contributing significantly to the strengthening of barrier properties and reduction of the degree of the damaging effect of toxic ions on the wheat plants. It may be suggested that the revealed effect of Me-JA pretreatment on lignification of root cell walls may be associated with Me-JA-induced increase in the content of cytokinins in the seedlings in the course of hormonal treatment and maintaining the concentration of cytokinins on the control level in the salt-stressed wheat plants pretreated with Me-JA.

Thus, the effect of Me-JA on accumulation of cytokinins under normal conditions and prevention of the stress-induced decline in cytokinins in the salt-stressed seedlings pretreated with Me-JA is likely to play the key role in manifestation of the preadaptive and protective effects of Me-JA on wheat plants.

9.5 Conclusion

Phytohormones are widely used in plant breeding to increase their productivity; however, the greatest effectiveness may be expected from the use of those combining simultaneously the properties of growth regulators and inducers of plants resistance. The present work is dedicated to revealing the mechanisms of regulatory action of 24-epibrassinolide, salicylic acid, and methyl jasmonate used in concentration concentrations optimal for plant growth under normal growth conditions and sodium chloride salinity. Since the plant growth is under the control of hormonal system, the effect of exogenous phytohormones inevitably leads to the shifts in its state, which may be discovered in the same plants with the help of immunoassay of hormones belonging to different classes.

Basing on the previously reported suggestion about possible fulfillment by ABA of the role of intermediate in SA-induced resistance of wheat to abiotic stresses, we have carried out experiments with inhibitor of ABA synthesis fluridone. They showed that pretreatment with fluridone completely inhibited the SA-induced accumulation of newborn ABA and development of preadaptive effect of SA on wheat plants under normal conditions indicated by the ROS production, activity of peroxidase and PAL, and the rate of lignin deposition in the root cell walls. Moreover, it was discovered that prevention by fluridone of maintenance of increased ABA content in SA-pretreated salt-stressed plants resulted in a complete loss of SA-induced resistance in wheat plants as judged by the level of peroxide oxidation of lipids and electrolyte leakage from tissues. Comparative analysis of the net of protective reactions in SA-treated plants in the presence and absence of fluridone allowed to receive the arguments in favor of the notion that ABA plays the role of intermediate in the regulation of preadaptive and protective action of SA on the wheat plants under conditions of sodium chloride salinity.

It was discovered that in response to the treatment with EBR and Me-JA, an almost twofold increase in cytokinins is observed in plants without significant changes in the level of ABA and IAA despite the difference in the dynamics of CKs: EBR caused a fast and stable increase in the level of CKs in plants, maintained in the course of the whole period of incubation of EBR, while Me-JA treatment led to a reversible accumulation of CKs in wheat seedlings with the maximum at 2–3 h. At the same time, plants pretreated with both EBR and Me-JA are characterized by smaller amplitude of NaCl-induced imbalance of ABA and IAA and maintenance of CKs under stress at the control level, indicating the important role of endogenous cytokinins in realization of the growth-promoting and preadaptive effects of EBR and Me-JA on wheat plants. The data on the independent endogenous ABA ability of EBR and Me-JA to cause, in the course of pretreatment, an increase in cytokinin content, balanced activation of proantioxidant system, accumulation of osmoprotectants, and accelerated lignification of cell walls contributing significantly to preadaptation to the forthcoming salinity and to the decrease in the level of the damaging effect of stress on the growth serve as the evidence in favor of possible fulfillment by the endogenous cytokinins of the role of intermediates in the physiological action of EBR and Me-JA on wheat plants. This conclusion is supported by the data showing ability of exogenous cytokinin BAP, the concentration of which was chosen as inducing the same twofold increase in the content of cytokinins immunoreactive toward antiserum against zeatin riboside on the background of the absence of any changes in ABA content, to exert similar to EBR and Me-JA preadaptive and protective effects on wheat plants.

The sum of the obtained data indicates the similar effect of EBR, SA, and Me-JA on wheat plants under conditions of salinity and is realized with the help of hormonal intermediates. Their role is fulfilled by endogenous ABA in case of SA treatment, while in case of EBR и Me-JA, by endogenous cytokinins.

Acknowledgments This work was supported by the Russian Foundation for Basic Research—Povoljie (project no. 11-04-97051).

References

Albacete A, Ghanem ME, Martinez-Andujar C, Acosta M, Sanchez-Bravo J, Martinez V, Lutts S, Dodd IC, Perez-Alfocea F (2008) Hormonal changes in relation to biomass partitioning and shoot growth impairment in salinized tomato (*Solanum lycopersicum* L.) plants. J Exp Bot 59:4119–4131

Ali Q, Athar H, Ashraf M (2008) Modulation of growth, photosynthetic capacity and water relations in salt stressed wheat plants by exogenously applied 24-epibrassinolide. Plant Growth Regul 56:107–116

Allagulova ChR, Gimalov FR, Shakirova FM, Vakhitov VA (2003) The plant dehydrins: structure and putative functions. Biochemistry (Moscow) 68:945–951
Altmann T (1999) Molecular physiology of brassinosteroids revealed by the analysis of mutants. Planta 208:1–11
Amzallag GN, Lerner HR, Poljakoff-Mayber A (1990) Exogenous ABA as a modulator of the response of sorghum to high salinity. J Exp Bot 41:1529–1534
An C, Mou Z (2011) Salicylic acid and its function in plant immunity. J Integr Plant Biol 53:412–428
Ananieva K, Melbeck J, Kaminek M, Staden J (2004) Methyl jasmonate down-regulated cytokinin levels in cotyledons of *Cucurbita pepo* (zucchini) seedlings. Physiol Plant 122:496–503
Arbona V, Argamasilla R, Gomez-Cadenas A (2010) Common and divergent physiological, hormonal and metabolic responses of *Arabidopsis thaliana* and *Thellungiella halophila* to water and salt stress. J Plant Physiol 167:1342–1350
Argueso CT, Ferreira FJ, Kieber JJ (2009) Environmental perception avenues: the interaction of cytokinin and environmental response pathways. Plant Cell Environ 32:1147–1160
Arteca RN, Arteca JM (2008) Effects of brassinosteroid, auxin, and cytokinin on ethylene production in Arabidopsis thaliana plants. J Exp Bot 59:3019–3026
Ashikari M, Sakakibara H, Lin S, Yamamoto T, Takashi T, Nishimura A, Angeles ER, Qian Q, Kitano H, Matsuoka M (2005) Cytokinin oxidase regulates rice grain production. Science 309:741–745
Ashraf M, Athar HR, Harris PJC, Kwon TR (2008) Some prospective strategies for improving crop salt tolerance. Adv Agron 97:45–110
Avalbaev AM, Bezrukova MV, Shakirova FM (2003) Effect of brassinosteroids on the hormonal balance in wheat seedlings. Dokl Biol Sci 391:337–339
Avalbaev AM, Yuldashev RA, Shakirova FM (2004) *Triticum aestivum* cytokinin oxidase gene, partial cds. NCBI Accession No AY831399
Avalbaev AM, Yuldashev RA, Vysotskaya LB, Shakirova FM (2006) Regulation of gene expression and activity of cytokinin oxidase in the roots of wheat seedlings by 24-epibrassinolide. Dokl Biochem Biophys 410:317–319
Avalbaev AM, Yuldashev RA, Fatkhutdinova RA, Urusov FA, Safutdinova YuV, Shakirova FM (2010a) The influence of 24-epibrassidinolide on the hormonal status of wheat plants under sodium chloride. Appl Biochem Microbiol 46:99–102
Avalbaev AM, Yuldashev RA, Safutdinova YuV, Allagulova ChR, Fatkhutdinova RA, Shakirova FM (2010b) Effect of 6-benzylaminopurine on the growth and hormonal system of wheat seedlings under salinity conditions. Agrochimiya 9:60–65
Bajguz A, Tretyn A (2003) The chemical characteristic and distribution of brassinosteroids in plants. Phytochemistry 62:1027–1046
Balbi V, Devoto A (2008) Jasmonate signaling network in *Arabidopsis thaliana*: crucial regulatory nodes and new physiological scenarios. New Phytol 177:301–318
Ballare CL (2011) Jasmonate-induced defenses: a tale of intelligence, collaborators and rascals. Trends Plant Sci 16:249–257
Bandurska H, Stroinski A, Kubis J (2003) The effect of jasmonic acid on the accumulation of ABA, proline and spermidine and its influence on membrane injury under water deficit in two barley genotypes. Acta Physiol Plant 25:279–285
Bari R, Jones JD (2009) Role of plant hormones in plant defence responses. Plant Mol Biol 69:473–488
Battal P, Erez ME, Turker M, Berber I (2008) Molecular and physiological changes in maize (*Zea mays*) induced by exogenous NAA and MeJa during cold stress. Ann Bot Fenn 45:173–185
Boudet AM, Lapierre C, Grima-Pettenati J (1995) Biochemistry and molecular biology of lignification. New Phytol 129:203–236
Browse J (2009) Jasmonate passes muster: a receptor and targets for the defense hormone. Annu Rev Plant Biol 60:183–205

Cambrolle J, Redondo-Gomez S, Mateos-Naranjo E, Luque T, Figueroa ME (2011) Physiological responses to salinity in the yellow-horned poppy, *Glaucium flavum*. Plant Physiol Biochem 49:186–194

Campos ML, de Almeida M, Rossi ML, Martinelli AP, Litholdo JCG, Figueira A, Rampelotti-Ferreira FT, Vendramim JD, Benedito VA, Peres LEP (2009) Brassinosteroids interact negatively with jasmonates in the formation of anti-herbivory traits in tomato. J Exp Bot 60:4347–4361

Catala R, Ouyang J, Abreu IA, Hu Y, Seo H, Zhang X, Chua N-H (2007) The *Arabidopsis* E3 SUMO ligase SIZ1 regulates plant growth and drought responses. Plant Cell 19:2952–2966

Chen Y, Pang Q, Dai S, Wang Y, Chen S, Yan X (2011) Proteomic identification of differentially expressed proteins in Arabidopsis in response to methyl jasmonate. J Plant Physiol. doi:10.1016/j.jplph.2011.01.018

Chernyad'ev II (2009) The protective action of cytokinins on the photosynthetic machinery and productivity of plants under stress (review). Appl Biochem Microbiol 45:351–362

Choudhary SP, Bhardwaj R, Gupta BD, Dutt P, Gupta RK, Biondi S, Kanwar M (2010) Epibrassinolide induces changes in indole-3-acetic acid, abscisic acid and polyamine concentrations and enhances antioxidant potential of radish seedlings under copper stress. Physiol Plant 140:280–296

Chow B, McCourt P (2004) Hormone signalling from a developmental context. J Exp Bot 55:247–251

Chung S, Parish RW (2008) Combination interactions of multiple cis-elements regulating the induction of the Arabidopsis XERO2 dehydrin gene by abscisic acid and cold. Plant J 54:15–29

Clarke SM, Cristescu SM, Miersch O, Harren FJM, Wasternack C, Mur LAJ (2009) Jasmonates act with salicylic acid to confer basal thermotolerance in *Arabidopsis thaliana*. New Phytol 182:175–187

Close TJ (1996) Dehydrins: emergence of a biochemical role of a family of plant dehydration proteins. Physiol Plant 96:795–803

Clouse SD, Sasse JM (1998) Brassinosteroids: essential regulators of plant growth and development. Annu Rev Plant Physiol Plant Mol Biol 49:427–451

Cortes PA, Terrazas T, Leon TC, Larque-Saavedra A (2003) Brassinosteroid effect on the precocity and yield of cladodes of cactus pear (*Opuntia ficus-indica* (L) Mill.). Sci Hortic 97:65–73

Creelman RA, Mullet JE (1995) Jasmonic acid distribution and action in plants: regulation during development and response to biotic and abiotic stress. Proc Natl Acad Sci USA 92:4114–4119

Davies WJ, Kudoyarova GR, Hartung W (2005) Long-distance ABA signaling and its relation to other signaling pathways in the detection of soil drying and the mediation of the plant's response to drought. J Plant Growth Regul 24:285–295

Dermastia M, Ravnikar M, Vilhar B, Kovae M (1994) Increased level of cytokinins ribosides in jasmonic acid treated potato (*Solanum tuberosum* L.) stem node cultures. Physiol Plant 92:241–246

Des Marais DL, Juenger TE (2010) Pleiotropy, plasticity, and the evolution of plant abiotic stress tolerance. Ann NY Acad Sci 1206:56–79

Divi UK, Rahman T, Krishna P (2010) Brassinosteroid-mediated stress tolerance in Arabidopsis shows interactions with abscisic acid, ethylene and salicylic acid pathways. BMC Plant Biol 10:151–165

Dobra J, Motyka V, Dobrev P, Malbeck J, Prasil IT, Haisel D, Gaudinova A, Havlova M, Gubis J, Vankova R (2010) Comparison of hormonal responses to heat, drought and combined stress in tobacco plants with elevated proline content. J Plant Physiol 167:1360–1370

El-Khallal SM, Hathout TA, Ashour AA, Kerrit AA (2009) Brassinolide and salicylic acid induced antioxidant enzymes, hormonal balance and protein profile of maize plants grown under salt stress. Res J Agric Biol Sci 5:391–402

Eun J-S, Kuraishi S, Sakurai N (1989) Changes in levels of auxin and abscisic acid and the evolution of ethylene in squash hypocotyls after treatment with brassinolide. Plant Cell Physiol 30:807–810

Fatkhutdinova RA, Shakirova FM, Chemeris AV, Sabirzhanov BE, Vakhitov VA (2002) NOR activity in wheat species with different ploidy levels treated with phytohormones. Russ J Genet 38:1335–1338

Fatkhutdinova DR, Sakhabutdinova AR, Maksimov IV, Yarullina LG, Shakirova FM (2004) The effect of salicylic acid on antioxidant enzymes in wheat seedlings. Agrochimiya 8:27–31

Fernandes CF, Moraes VCP, Vasconcelos IM, Silveira JAG, Oliveira JTA (2006) Induction of an anionic peroxidase in cowpea leaves by exogenous salicylic acid. J Plant Physiol 163:1040–1048

Flowers TJ (2004) Improving crop salt tolerance. J Exp Bot 55:307–319

Fu FQ, Mao WH, Shi K, Zhou YH, Asami T, Yu JQ (2008) A role of brassinosteroids in early fruit development in cucumber. J Exp Bot 59:2299–2308

Fujimoto T, Tomitaka Y, Abe H, Tsuda S, Futai K, Mizukubo T (2011) Expression profile of jasmonic acid-induced genes and the induced resistance against the root-knot nematode (*Meloidogyne incognita*) in tomato plants (*Solanum lycopersicum*) after foliar treatment with methyl jasmonate. J Plant Physiol 168:1084–1097

Fujita M, Fujita Y, Maruyama K, Seki M, Hiratsu K, Ohme-Takagi M, Tran L-SP, Yamaguchi-Shinozaki K, Shinozaki K (2004) A dehydration-induced NAC protein, RD26, is involved in a novel ABA-dependent stress-signaling pathway. Plant J 39:863–876

Funk C, Brodelius P (1990) Influence of growth regulators and elicitor on phenylpropanoid metabolism in suspension culture of *Vanilla planifolia*. Phytochemistry 29:845–848

Galis I, Gaquerel E, Pandey SP, Baldwin IT (2009) Molecular mechanisms underlying plant memory in JA-mediated defence responses. Plant Cell Environ 32:617–627

Galuszka P, Frebortova J, Werner T, Yamada M, Strnad M, Schmulling T, Frebort I (2004) Cytokinin oxidase/dehydrogenase genes in barley and wheat. Cloning and heterologous expression. Eur J Biochem 271:3990–4002

Gangwar S, Singh VP, Prasad SM, Maurya JN (2010) Modulation of manganese toxicity in *Pisum sativum* L. seedlings by kinetin. Sci Hortic 126:467–474

Gaudinova A, Sussenbekova H, Vojtechova M, Kaminek M, Eder J, Kohout L (1995) Different effects of two brassinosteroids on growth, auxin and cytokinin content in tobacco callus tissue. Plant Growth Regul 17:121–126

Gemes K, Poor P, Horvath E, Kolbert Z, Szopko D, Szepesi A, Tari I (2011) Cross-talk between salicylic acid and NaCl-generated reactive oxygen species and nitric oxide in tomato during acclimation to high salinity. Physiol Plant 142:179–192

Gendron JM, Haque A, Gendron N, Chang T, Asami T, Wang Z-Y (2008) Chemical genetic dissection of brassinosteroids-ethylene interaction. Mol Plant 1:368–379

Ghanem ME, Albacete A, Smigocki AC, Frebort I, Pospisilova H, Martinez-Andujar C, Acosta M, Sanchez-Bravo J, Lutts S, Dodd IC, Perez-Alfocea F (2011) Root-synthesized cytokinins improve shoot growth and fruit yield in salinized tomato (*Solanum lycopersicum* L.) plants. J Exp Bot 62:125–140

Gill SS, Tuteja N (2010) Reactive oxygen species and antioxidant machinery in abiotic stress tolerance in crop plants. Plant Physiol Biochem 48:909–930

Goda H, Sasaki E, Akiyama K, Maruyama-Nakashita A, Nakabayashi K, Li W, Ogawa M, Yamauchi Y, Preston J, Aoki K, Kiba T, Takatsuto S, Fujioka S, Asami T, Nakano T, Kato H, Mizuno T, Sakakibara H, Yamaguchi S, Nambara E, Kamiya Y, Takahashi H, Hirai MY, Sakurai T, Shinozaki K, Saito K, Yoshida S, Shimada Y (2008) The AtGenExpress hormone and chemical treatment data set: experimental design, data evaluation, model data analysis and data access. Plant J 55:526–542

Granda V, Cuesta C, Alvarez R, Ordas R, Centeno ML, Rodriguez A, Majada JP, Fernandez B, Feito I (2011) Rapid responses of C14 clone of *Eucalyptus globulus* to root drought stress: time-course of hormonal and physiological signaling. J Plant Physiol 168:661–670

Guo J, Hu X, Duan R (2005) Interactive effects of cytokinins, light, and sucrose on the phenotypes and the syntheses of anthocyanins and lignins in cytokinin overproducing transgenic Arabidopsis. J Plant Growth Regul 24:93–101

Hansen M, Chae HS, Kieber JJ (2009) Regulation of ACS protein stability by cytokinin and brassinosteroids. Plant J 57:606–614

Hara M (2010) The multifunctionality of dehydrins. Plant Signal Behav 5:503–508

Hare PD, Cress WA, van Staden J (1997) The involvement of cytokinins in plant responses to environmental stress. Plant Growth Regul 23:79–103

Hays DB, Wilen RW, Sheng C, Moloney MM, Pharis RP (1999) Embryo-specific gene expression in microspore-derived embryos of *Brassica napus*. An interaction between abscisic acid and jasmonic acid. Plant Physiol 119:1065–1072

Hiraga S, Sasaki K, Ito H, Ohashi Y, Matsui H (2001) A large family of class III plant peroxidases. Plant Cell Physiol 42:462–468

Horvath E, Szalai G, Janda T (2007) Induction of abiotic stress tolerance by salicylic acid signaling. J Plant Growth Regul 26:290–300

Hu Y, Bao A, Li J (2000) Promotive effect of brassinosteroids on cell division involves a distinct CycD3-induction pathway in *Arabidopsis*. Plant J 24:693–701

Huang D, Wu W, Abrams SR, Cutler AJ (2008) The relationship of drought-related gene expression in *Arabidopsis thaliana* to hormonal and environmental factors. J Exp Bot 59:2991–3007

Hung KT, Hsu YT, Kao CH (2006) Hydrogen peroxide is involved in methyl jasmonate-induced senescence of rice leaves. Physiol Plant 127:293–303

Iqbal M, Ashraf M (2005) Presowing seed treatment with cytokinins and its effect on growth, photosynthetic rate, ionic levels and yield of two wheat cultivars differing in salt tolerance. J Integr Plant Biol 47:1315–1325

Jager CE, Symons GM, Ross JJ, Reid JB (2008) Do brassinosteroids mediate the water stress response? Physiol Plant 133:417–425

Janeczko A, Swaczynova J (2010) Endogenous brassinosteroids in wheat treated with 24-epibrassinolide. Biol Plant 54:477–482

Janeczko A, Biesaga-Koscielniak J, Oklest'kova J, Filek M, Dziurka M, Szarek-Lukaszewska G, Koscielniak J (2010) Role of 24-epibrassinolide in wheat production: physiological effects and uptake. J Agron Crop Sci 196:311–321

Janz D, Behnke K, Schnitzler J-P, Kanawati B, Schmitt-Kopplin P, Polle A (2010) Pathway analysis of the transcriptome and metabolome of salt sensitive and tolerant poplar species reveals evolutionary adaption of stress tolerance mechanisms. BMC Plant Biol 10:150–165

Jaspers P, Kangasjarvi J (2010) Reactive oxygen species in abiotic stress signaling. Physiol Plant 138:405–413

Jiang M, Zhang J (2002) Water stress-induced abscisic acid accumulation triggers the increased generation of reactive oxygen species and up-regulates the activities of antioxidant enzymes in maize leaves. J Exp Bot 53:2401–2410

Jiang Y, Yang B, Harris NS, Deyholos MK (2007) Comparative proteomic analysis of NaCl stress-responsive proteins in Arabidopsis roots. J Exp Bot 58:3591–3607

Kagale S, Divi UK, Krochko JE, Keller WA, Krishna P (2007) Brassinosteroid confers tolerance in *Arabidopsis thaliana* and *Brassica napus* to a range of abiotic stresses. Planta 225:353–364

Kang D-J, Seo Y-J, Lee J-D, Ishii R, Kim KU, Shin DH, Park SK, Jang SW, Lee I-J (2005) Jasmonic acid differentially affects growth, ion uptake and abscisic acid concentration in salt-tolerant and salt-sensitive rice cultivars. J Agron Crop Sci 191:273–282

Katsir L, Chung HS, Koo AJK, Howe GA (2008) Jasmonate signaling: a conserved mechanism of hormone sensing. Curr Opin Plant Biol 11:428–435

Khadri M, Tejera NA, Lluch C (2007) Sodium chloride–ABA interaction in two common bean (*Phaseolus vulgaris*) cultivars differing in salinity tolerance. Environ Exp Bot 60:211–218

Khan MA, Gul B, Weber DJ (2004) Action of plant growth regulators and salinity on seed germination of *Ceratoides lanata*. Can J Bot 82:37–42
Khripach V, Zhabinskii V, de Groot A (2000) Twenty years of brassinosteroids: steroidal plant hormones warrant better crops for the XXI century. Ann Bot 86:441–447
Kim T-W, Wang Z-Y (2010) Brassinosteroid signal transduction from receptor kinases to transcription factors. Annu Rev Plant Biol 61:681–704
Kim T-H, Bohmer M, Hu H, Nishimura N, Schroeder JI (2010) Guard cell signal transduction network: advances in understanding abscisic acid, CO_2, and Ca^{2+} signaling. Annu Rev Plant Biol 61:561–591
Koornneef A, Pieterse CMJ (2008) Cross talk in defense signaling. Plant Physiol 146:839–844
Krouk G, Ruffel S, Gutiérrez RA, Gojon A, Crawford NM, Coruzzi GM, Lacombe B (2011) A framework integrating plant growth with hormones and nutrients. Trends Plant Sci 16:178–182
Kudo T, Kiba T, Sakakibara H (2010) Metabolism and long-distance translocation of cytokinins. J Integr Plant Biol 52:53–60
Kuppusamy KT, Walcyer CL, Nemhauser JL (2008) Cross-regulatory mechanisms in hormone signaling. Plant Mol Biol 69:375–381
Kurepin LV, Qaderi MM, Back TG, Reid DM, Pharis RP (2008) A rapid effect of applied brassinolide on abscisic acid concentrations in *Brassica napus* leaf tissue subjected to short-term heat stress. Plant Growth Regul 55:165–167
Lee B-R, Jung W-J, Lee B-H, Avice J-C, Ourry A, Kim T-H (2008) Kinetics of drought-induced pathogenesis-related proteins and its physiological significance in white clover leaves. Physiol Plant 132:329–337
Li J (2010) Regulation of the nuclear activities of brassinosteroid signaling. Curr Opin Plant Biol 13:540–547
Liu J-J, Ekramoddoullah AKM (2006) The family 10 of plant pathogenesis-related proteins: their structure, regulation, and function in response to biotic and abiotic stresses. Physiol Mol Plant Pathol 68:3–13
Liu H-T, Liu Y-Y, Pan Q-H, Yang H-R, Zhan J-C, Huang W-D (2006) Novel interrelationship between salicylic acid, abscisic acid, and PIP-specific phospholipase C in heat acclimation-induced thermotolerance in pea leaves. J Exp Bot 57:3337–3347
Liu Y, Jiang H, Zhao Z, An L (2011) Abscisic acid is involved in brassinosteroids-induced chilling tolerance in the suspension cultured cells from *Chorispora bungeana*. J Plant Physiol 168:853–862
Maksymiec W, Krupa Z (2006) The effects of short-term exposition to Cd, excess Cu ions and jasmonate on oxidative stress appearing in *Arabidopsis thaliana*. Environ Exp Bot 57:187–194
Mantyla E, Lang V, Palva ET (1995) Role of abscisic acid in drought-induced freezing tolerance, cold acclimation, and accumulation of LT178 and RABI8 proteins in *Arabidopsis thaliana*. Plant Physiol 107:141–148
Metraux JP (2002) Recent breakthroughs in study of salicylic acid biosynthesis. Trends Plant Sci 7:331–334
Miller G, Suzuki N, Ciftci-Yilmaz S, Mittler R (2010) Reactive oxygen species homeostasis and signaling during drought and salinity stresses. Plant Cell Environ 33:453–467
Moura JCMS, Bonine CAV, Viana JOF, Dornelas MC, Mazzafera P (2010) Abiotic and biotic stresses and changes in the lignin content and composition in plants. J Integr Plant Biol 52:360–376
Munns R, Tester M (2008) Mechanisms of salinity tolerance. Annu Rev Plant Biol 59:651–681
Müssig C, Altmann T (2003) Genomic brassinosteroid effects. J Plant Growth Regul 22:313–324
Nazar R, Iqbal N, Syeed S, Khan NA (2011) Salicylic acid alleviates decreases in photosynthesis under salt stress by enhancing nitrogen and sulfur assimilation and antioxidant metabolism differentially in two mungbean cultivars. J Plant Physiol 168:807–815

Nemhauser JL, Hong F, Chory J (2006) Different plant hormones regulate similar processes through largely nonoverlapping transcriptional responses. Cell 126:467–475
Pauwels L, Inze D, Goossens A (2009) Jasmonate-inducible gene: what does it mean? Trends Plant Sci 14:87–91
Peleg Z, Blumwald E (2011) Hormone balance and abiotic stress tolerance in crop plants. Curr Opin Plant Biol 14:290–295
Peng Z-Y, Zhou X, Li L, Yu X, Li H, Cao G, Bai M, Wang X, Jiang C, Jiang Z, Lu H, Hou X, Qu L, Wang Z, Zuo J, Fu X, Su Z, Li S, Guo H (2009) Arabidopsis hormone database: a comprehensive genetic and phenotypic information database for plant hormone research in Arabidopsis. Nucleic Acids Res 37(suppl 1):D975–D982. doi:10.1093/nar/gkn873
Peng Z, Han C, Yuan L, Zhang K, Huang H, Ren C (2011) Brassinosteroid enhances jasmonate-induced anthocyanin accumulation in Arabidopsis seedlings. J Integr Plant Biol. doi:10.1111/j.17447909.2011.01042.x
Piotrowska A, Bajguz A, Godlewska-Zylkiewicz B, Czerpak R, Kaminska M (2009) Jasmonic acid as modulator of lead toxicity in aquatic plant *Wolffia arrhiza* (Lemnaceae). Environ Exp Bot 66:507–513
Potters G, Pasternak T, Guisez Y, Jansen MAK (2009) Different stresses, similar morphogenic responses: integrating a plethora of pathways. Plant Cell Environ 32:158–169
Raskin I (1992) Role of salicylic acid in plants. Annu Rev Plant Physiol Plant Mol Biol 43:439–463
Reski R (2006) Small molecules on the move: homeostasis, crosstalk, and molecular action of phytohormones. Plant Biol 8:277–280
Rivero RM, Shulaev V, Blumwald E (2009) Cytokinin-dependent photorespiration and the protection of photosynthesis during water deficit. Plant Physiol 150:1530–1540
Rock CD, Sakata Y, Quatrano RS (2010) Stress signaling I: the role of abscisic acid (ABA). In: Pareek A, Sopory SA, Bohner HJ, Govindjee (eds) Abiotic stress adaptation in plants: physiological, molecular and genomic foundation. Springer, Dordrecht
Rubio V, Bustos R, Irigoyen ML, Ximena C-L, Rojas-Triana M, Paz-Ares J (2009) Plant hormones and nutrient signaling. Plant Mol Biol 69:361–373
Rubio-Wilhelmi MM, Sanchez-Rodriguez E, Rosales MA, Blasco B, Rios JJ, Romero L, Blumwald E, Ruiz JM (2011) Effect of cytokinins on oxidative stress in tobacco plants under nitrogen deficiency. Environ Exp Bot 72:167–173
Sasaki-Sekimoto Y, Taki N, Obayashi T, Aono M, Matsumoto F, Sakurai N, Suzuki H, Hirai MY, Noji M, Saito K, Masuda T, Takamiya K, Shibata D, Ohta H (2005) Coordinated activation of metabolic pathways for antioxidants and defence compounds by jasmonates and their roles in stress tolerance in Arabidopsis. Plant J 44:653–668
Saygideger S, Deniz F (2008) Effect of 24-epibrassinolide on biomass, growth and free proline concentration in *Spirulina platensis* (*Cyanophyta*) under NaCl stress. Plant Growth Regul 56:219–223
Seo PJ, Lee A-K, Xiang F, Park C-M (2008) Molecular and functional profiling of Arabidopsis pathogenesis-related genes: insights into their roles in salt response of seed germination. Plant Cell Physiol 49:334–344
Shakirova FM (2001) Nonspecific resistance of plants to stress factors and its regulation. Gilem, Ufa
Shakirova FM (2007) Role of hormonal system in manifestation of growth promoting and antistress action of salicylic acid. In: Hayat S, Ahmad A (eds) Salicylic acid – a plant hormone. Springer, The Netherlands
Shakirova FM, Bezrukova MV (1997) Induction of wheat resistance against environmental salinization by salicylic acid. Biol Bull 24:109–112
Shakirova FM, Bezrukova MV (1998) Effect of 24-epibrassinolide and salinity on the levels of ABA and lectin. Russ J Plant Physiol 45:388–391
Shakirova FM, Konrad K, Klyachko NL, Kulaeva ON (1982) Relationship between the effect of cytokinin on the growth of pumpkin isolated cotyledons and RNA and protein synthesis in them. Russ J Plant Physiol 29:52–61

Shakirova FM, Bezrukova MV, Avalbaev AM, Gimalov FR (2002) Stimulation of wheat germ agglutinin gene expression in root seedlings by 24-epibrassinolide. Russ J Plant Physiol 49:225–228

Shakirova FM, Sakhabutdinova AR, Bezrukova MV, Fatkhutdinova RA, Fatkhutdinova DR (2003) Changes in the hormonal status of wheat seedlings induced by salicylic acid and salinity. Plant Sci 164:317–322

Shakirova FM, Allagulova ChR, Bezrukova MV, Gimalov FR (2005) Induction of expression of the dehydrin gene *TADHN* and accumulation of abscisic acid in wheat plants in hypothermia. Dokl Biochem Biophys 400:69–71

Shakirova FM, Allagulova ChR, Bezrukova MV, Aval'baev AM, Gimalov FR (2009) The role of endogenous ABA in cold-induced expression of the *TADHN* dehydrin gene in wheat seedlings. Russ J Plant Physiol 56:720–723

Shakirova FM, Avalbaev AM, Bezrukova MV, Kudoyarova GR (2010a) Role of endogenous hormonal system in the realization of the antistress action of plant growth regulators on plants. Plant Stress 4:32–38

Shakirova FM, Sakhabutdinova AR, Ishdavletova LOV (2010b) Influence of pretreatment with methyl jasmonate on wheat resistance to salt stress. Agrochimiya 7:26–32

Shan C, Liang Z (2010) Jasmonic acid regulates ascorbate and glutathione metabolism in *Agropyron cristatum* leaves under water stress. Plant Sci 178:130–139

Shimizu Y, Maeda K, Kato M, Shimomura K (2011) Co-expression of GbMYB1 and GbMYC1 induces anthocyanin accumulation in roots of cultured *Gynura bicolor* DC. plantlet on methyl jasmonate treatment. Plant Physiol Biochem 49:159–167

Shinozaki K, Yamaguchi-Shinozaki K (2007) Gene networks involved in drought stress response and tolerance. J Exp Bot 58:221–227

Si Y, Zhang C, Meng S (2009) Gene expression changes in response to drought stress in *Citrullus colocynthis*. Plant Cell Rep 28:997–1009

Silva-Ortega CO, Ochoa-Alfaro AE, Reyes-Agüero JA, Aguado-Santacruz GA, Jiménez-Bremont JF (2008) Salt stress increases the expression of *p5cs* gene and induces proline accumulation in cactus pear. Plant Physiol Biochem 46:82–92

Srivastava S, Emery RJN, Rahman MH, Kav NNV (2007) A crucial role for cytokinins in pea ABR17-mediated enhanced germination and early seedling growth of *Arabidopsis thaliana* under saline and low-temperature stresses. J Plant Growth Regul 26:26–37

Stamm P, Kumar PP (2010) The phytohormone signal network regulating elongation growth during shade avoidance. J Exp Bot 61:2889–2903

Stoparih G, Maksimovih I (2008) The effect of cytokinins on the concentration of hydroxyl radicals and the intensity of lipid peroxidation in nitrogen deficient wheat. Cereal Res Commun 36:601–609

Sun J, Xu Y, Ye S, Jiang H, Chen Q, Liu F, Zhou W, Chen R, Li X, Tietz O, Wu X, Cohen JD, Palme K, Li C (2009) Arabidopsis ASA1 is importance for jasmonate-mediated regulation of auxin biosynthesis and transport during lateral root formation. Plant Cell 21:1495–1511

Syeed S, Anjum NA, Nazar R, Iqbal N, Masood A, Khan NA (2011) Salicylic acid-mediated changes in photosynthesis, nutrients content and antioxidant metabolism in two mustard (*Brassica juncea* L.) cultivars differing in salt tolerance. Acta Physiol Plant 33:877–886

Szabados L, Savoure A (2009) Proline: a multifunctional amino acid. Trends Plant Sci 15:89–97

Tamaoki M, Freeman JL, Pilon-Smits EAH (2008) Cooperative ethylene and jasmonic acid signaling regulates selenite resistance in Arabidopsis. Plant Physiol 146:1219–1230

Thomas JC, McElwain EF, Bohnert HJ (1992) Convergent induction of osmotic stress-responses. Abscisic acid, cytokinin, and the effects of NaCI. Plant Physiol 100:416–423

Thulke O, Conrath U (1998) Salicylic acid has dual role in activation of defence-related genes in parsley. Plant J 14:35–42

Tognetti VB, Muhlenbock P, Van Breusegem F (2011) Stress homeostasis – the redox and auxin perspective. Plant Cell Environ. doi:10.1111/j.1365-3040.2011.02324.x

Traw MB, Bergelson J (2003) Interactive effects of jasmonic acid, salicylic acid, and gibberellin on induction of trichomes in Arabidopsis. Plant Physiol 133:1367–1375
Tuteja N, Sopory SK (2008) Chemical signaling under abiotic stress environment in plants. Plant Signal Behav 3:525–536
Upreti KK, Murti GSR (2004) Effects of brassinosteroids on growth, nodulation, phytohormone content and nitrogenase activity in French bean under water stress. Biol Plant 48:407–411
Van der Ent S, Van Wees SCM, Pieterse CMJ (2009) Jasmonate signaling in plant interactions with resistance-inducing beneficial microbes. Phytochemistry 70:1581–1588
Verslues PE, Bray EA (2006) Role of abscisic acid (ABA) and *Arabidopsis thaliana* ABA-insensitive loci in low water potential-induced ABA and proline accumulation. J Exp Bot 57:201–212
Vert G, Walcher CL, Chory J, Namhauser JL (2008) Integration of auxin and brassinosteroid pathways by Auxin Response Factor 2. Proc Natl Acad Sci USA 105:9829–9834
Veselov DS, Veselov SYu, Vysotskaya LB, Kudoyarova GR, Farhutdinov RG (2007) Plant hormones. Regulation of the concentration and relationship with growth and water metabolism. Nauka, Moscow
Vlasankova E, Kohout L, Klems M, Eder J, Reinohl V, Hradilik J (2009) Evaluation of biological activity of new synthetic brassinolide analogs. Acta Physiol Plant 31:987–993
Vlot AC, Dempsey DA, Klessig DF (2009) Salicylic acid, a multifaceted hormone to combat disease. Annu Rev Phytopathol 47:177–206
Vysotskaya LB, Avalbaev AM, Yuldashev RA, Shakirova FM, Veselov SYu, Kudoyarova GR (2010) Regulation of cytokinin oxidase activity as a factor affecting the content of cytokinins. Russ J Plant Physiol 57:494–500
Walia H, Wilson C, Condamine P, Liu X, Ismail AM, Close TJ (2007) Large-scale expression profiling and physiological characterization of jasmonic acid-mediated adaptation of barley to salinity stress. Plant Cell Environ 30:410–421
Wang C, Yang A, Yin H, Zhang J (2008) Influence of water stress on endogenous hormone contents and cell damage of maize seedlings. J Integr Plant Biol 50:427–434
Wang L, Wang Z, Xu Y, Joo S-H, Kim S-K, Xue Z, Xu Z, Wang Z, Chong K (2009) OsGSR1 is involved in crosstalk between gibberellins and brassinosteroids in rice. Plant J 57:498–510
Wasternack C (2007) Jasmonates: an update on biosynthesis, signal transduction and action in plant stress response, growth and development. Ann Bot 100:681–697
Widodo PJH, Newbigin E, Tester M, Bacic A, Roessner U (2009) Metabolic responses to salt stress of barley (*Hordeum vulgare* L.) cultivars, Sahara and Clipper, which differ in salinity tolerance. J Exp Bot 60:4089–4103
Wilkinson S, Davies WJ (2010) Drought, ozone, ABA and ethylene: new insights from cell to plant to community. Plant Cell Environ 33:510–525
Wong CE, Singh MB, Bhalla PL (2009) Floral initiation process at the soybean shoot apical meristem may involve multiple hormonal pathways. Plant Signal Behav 4:648–651
Wu G, Zhang C, Chu L-Y, Shao H-B (2007) Responses of higher plants to abiotic stresses and agricultural sustainable development. J Plant Interact 2:135–147
Xia X-J, Wang Y-J, Zhou Y-H, Tao Y, Mao W-H, Shi K, Asami T, Chen Z, Yu J-Q (2009) Reactive oxygen species are involved in brassinosteroid-induced stress tolerance in cucumber. Plant Physiol 150:801–814
Xu Y, Gianfagna T, Huang B (2010) Proteomic changes associated with expression of a gene (*ipt*) controlling cytokinin synthesis for improving heat tolerance in a perennial grass species. J Exp Bot 61:3273–3289
Xue YJ, Tao L, Yang ZM (2008) Aluminum-induced cell wall peroxidase activity and lignin synthesis are differentially regulated by jasmonate and nitric oxide. J Agric Food Chem 56:9676–9684
Yang Y, Qi M, Mei C (2004) Endogenous salicylic acid protects rice plants from oxidative damage caused by aging as well as biotic and abiotic stress. Plant J 40:909–919

Yang D-H, Hettenhausen C, Baldwin IT, Wu J (2011) BAK1 regulates the accumulation of jasmonic acid and the levels of trypsin proteinase inhibitors in *Nicotiana attenuata*'s responses to herbivory. J Exp Bot 62:641–652
Ye N, Zhu G, Liu Y, Li Y, Zhang J (2011) ABA controls H_2O_2 accumulation through the induction of *OsCATB* in rice leaves under water stress. Plant Cell Physiol 52:689–698
Yu X, Li L, Zola J, Aluru M, Ye H, Foudree A, Guo H, Anderson S, Aluru S, Liu P, Rodermel S, Yin Y (2011) A brassinosteroid transcriptional network revealed by genome-wide identification of BESI target genes in *Arabidopsis thaliana*. Plant J 65:634–646
Yuan G-F, Jia C-G, Li Z, Sun B, Zhang L-P, Liu N, Wang Q-M (2010) Effect of brassinosteroids on drought resistance and abscisic acid concentration in tomato under water stress. Sci Hortic 126:103–108
Zalewski W, Galuszka P, Gasparis S, Orczyk W, Nadolska-Orczyk A (2010) Silencing of the HvCKX1 gene decreases the cytokinin oxidase/dehydrogenase level in barley and leads to higher plant productivity. J Exp Bot 61:1839–1851
Zavaleta-Mancera HA, López-Delgado H, Loza-Tavera H, Mora-Herrera M, Trevilla-García C, Vargas-Suárez M, Ougham H (2007) Cytokinin promotes catalase and ascorbate peroxidase activities and preserves the chloroplast integrity during dark-senescence. J Plant Physiol 164:1572–1582
Zhang M, Zhai Z, Tian X, Duan L, Li Z (2008) Brassinolide alleviated the adverse effect of water deficits on photosynthesis and the antioxidant of soybean (*Glycine max* L.). Plant Growth Regul 56:257–264
Zhang S, Wei Y, Lu Y, Wang X (2009) Mechanisms of brassinosteroids interacting with multiple hormones. Plant Signal Behav 4:1117–1120
Zhang X, Chen S, Mou Z (2010) Nuclear localization of NPR1 is required for regulation of salicylate tolerance, isochorismate synthase 1 expression and salicylate accumulation in Arabidopsis. J Plant Physiol 167:144–148
Zhang A, Zhang J, Zhang J, Ye N, Zhang H, Tan M, Jiang M (2011) Nitric oxide mediates brassinosteroid-induced ABA biosynthesis involved in oxidative stress tolerance in maize leaves. Plant Cell Physiol 52:181–192 doi:10.1093/pcp/pcq187
Ziosi V, Bonghi C, Bregoli AM, Trainotti L, Biondi S, Sutthiwal S, Kondo S, Costa G, Torrigiani P (2008) Jasmonate-induced transcriptional changes suggest a negative interference with the ripening syndrome in peach fruit. J Exp Bot 59:563–573
Zubo YO, Yamburenko MV, Selivankina SY, Shakirova FM, Avalbaev AM, Kudryakova NV, Zubkova NK, Liere K, Kulaeva ON, Kusnetsov VV, Borner T (2008) Cytokinin stimulates chloroplast transcription in detached barley leaves. Plant Physiol 148:1082–1093

Chapter 10
The Role of Phytohormones in the Control of Plant Adaptation to Oxygen Depletion

Vladislav V. Yemelyanov and Maria F. Shishova

Abstract Capacity to survive the oxygen deprivation depends on a number of developmental, morphological, and metabolic adaptations in plants. Imposition of hypoxia (deficiency of oxygen) accelerates growth of shoot axial organs and stimulates formation of adventitious roots and aerenchyma in tolerant plant species. As a result, the shoot actively transports oxygen to a flooded root. Simultaneous shifts occur in the metabolism, which are particularly severe under anoxia (total absence of oxygen). A majority of these morphological and metabolic adaptations are strictly regulated by plant hormonal system. Ethylene and gibberellins control enhanced growth, leading to the emergence of shoots of tolerant plants under flooding conditions. Recent findings show *Sub1* gene which is important to submergence tolerance in rice to be linked with ethylene and gibberellin signaling. Ethylene is also involved in formation of aerenchyma in oxygen-depleted environment. Auxin regulates adventitious rooting and petiole elongation. Abscisic acid inhibits growth but stimulates metabolic adaptations by induction of anaerobic stress protein gene expression. Complete flooding and particularly total anoxia block ethylene production. Application of exogenous ABA, auxin, and some other growth regulators improves plant survival during oxygen deficiency. Complicated crosstalk between phytohormones under oxygen depletion is discussed as a milestone of plant adaptation.

V.V. Yemelyanov
Department of Genetics and Breeding, Faculty of Biology and Soil Science, Saint-Petersburg State University, 199034 Saint-Petersburg, Russia
e-mail: bootika@mail.ru

M.F. Shishova (✉)
Department of Plant Physiology and Biochemistry, Faculty of Biology and Soil Science, Saint-Petersburg State University, 199034 Saint-Petersburg, Russia
e-mail: shishova@mail.ru

N.A. Khan et al. (eds.), *Phytohormones and Abiotic Stress Tolerance in Plants*,
DOI 10.1007/978-3-642-25829-9_10, © Springer-Verlag Berlin Heidelberg 2012

Keywords Anoxia • Hypoxia • Auxin • Ethylene • ABA • Gibberellin • Cytokinin • Signal transduction • Submergence

Abbreviations

ACC	1-Aminocyclopropane-1-carboxylate
ACO	ACC oxidase
ACS	ACC synthase
ADH	Alcohol dehydrogenase
LOES	Low-oxygen escape syndrome

10.1 Introduction

Higher plants are strict aerobic organisms, which directly depend on molecular oxygen for their respiration, other metabolic processes, and survival. Nevertheless, very often, they suffer oxygen shortage that develops constantly in aquatic and semiaquatic habitats and frequently in dry lands, including agricultural, horticultural, and industrial areas. Availability of oxygen affects distribution of plant species in natural and agricultural ecosystems and has a severe economical impact. The lack of oxygen usually results from excess of water during seasonal or perennial flooding and after heavy rain falls. It also follows ice crust formation during wintertime and begins in compact soil due to the use of heavy agricultural machinery or asphalt covering. Waterlogging of rhizosphere and partial flooding of aboveground parts of plant lead to a gradual hypoxia (deficiency of oxygen), and complete submergence brings about anoxia (total absence of the gas).

Capacity to survive the oxygen deprivation depends on a number of developmental, morphological, and metabolic adaptations in plants. Imposition of hypoxia accelerates shoot growth and hyponasty of petioles and leaves (so-called LOES, low-oxygen escape syndrome) and stimulates formation of adventitious roots and aerenchyma in wetland plant species. As a result, the shoot actively transports oxygen to a flooded root (for review, see Crawford and Braendle 1996; Drew 1997; Vartapetian and Jackson 1997; Kende et al. 1998; Sauter 2000; Gibbs and Greenway 2003; Bailey-Serres and Voesenek 2008; Jackson 2008). Simultaneous shifts occur in the metabolism, which are particularly severe under strict oxygen lack and in tolerant plants that respond to oxygen deficiency by non-LOES quiescent syndrome. These plants respond to flooding mostly by metabolic switch with little or no stimulation of shoot growth.

Metabolic adaptations include mainly avoidance of energy starvation, prevention of toxicity of anaerobic intermediate and end products, and postanoxic injury, disposal of cytosol acidification, and use of alternative electron acceptors (like nitrate, nitrite, unsaturated lipids, etc.; Chirkova 1988; Kennedy et al. 1992; Crawford and Braendle 1996; Drew 1997; Vartapetian and Jackson 1997;

Sauter 2000; Gibbs and Greenway 2003; Greenway and Gibbs 2003; Bailey-Serres and Voesenek 2008). Hypoxia-tolerant plants are notable for maintenance of their cell ultra structure (Vartapetian et al. 1976; Vartapetian and Jackson 1997), membrane stability (Chirkova 1988; Crawford and Braendle 1996), and synthesis of anaerobic stress proteins (Kennedy et al. 1992; Vartapetian and Jackson 1997; Greenway and Gibbs 2003).

Most of the anaerobic stress proteins belong to enzymes of the glycolytic or fermentative pathways, carbohydrate mobilization, and nitrogen metabolism (Dolferus et al. 1997; Vartapetian and Jackson 1997; Greenway and Gibbs 2003). All together, these metabolic adaptations allow tolerant plants to generate sufficient amount of energy, maintain mineral nutrition, and even to grow in total absence of oxygen.

The majority of these morphological and metabolic adaptations are effectively regulated by plant hormonal system. Ethylene, gibberellins (GA), and abscisic acid (ABA) along with some other growth regulators are deeply involved in control of plant adaptation to oxygen deficiency. Their effects are highlighted in this chapter.

10.2 Ethylene

Ethylene is the first phytohormone which has been studied under conditions of oxygen deficiency. Lack of oxygen triggers ethylene production in the most tested plant species, irrespective of their habitat (aquatic or terrestrial) and tolerance to oxygen lack (Jackson et al. 1985a; Grichko and Glick 2001). An increase in ethylene production has been reported under the lack of oxygen in a wide variety of cultivated plants, including beans, radish, tomato, sunflower, chrysanthemum, corn, and wheat (Kawase 1972; Jackson and Campbell 1976; Bradford et al. 1982; Jackson et al. 1985a). Hypoxia activates ethylene production also in tolerant plants, such as rice (Metraux and Kende 1983; Satler and Kende 1985), willow (Chirkova and Gutman 1972), *Ranunculus sceleratus* (Samarakoon and Horton 1984), and sorrel (Voesenek et al. 1990a, b). The majority of investigations estimate elevation of ethylene concentration in shoots at root hypoxia or throughout the plant under total submergence.

Higher plants synthesize ethylene from L-methionine via *S*-adenosyl-L-methionine and 1-aminocyclopropane-1-carboxylate (ACC) (Kende 1993). Hypoxia-induced accumulation of ACC and activation of ACC synthase (ACS) are shown in deepwater rice (Metraux and Kende 1983; Cohen and Kende 1987), various sorrel species (Voesenek et al. 1990a, 1993; Banga et al. 1996), maize (He et al. 1994, 1996a, b), and *Arabidopsis* (Muhlenbock et al. 2007). Following reaction of ACC oxidation by ACC oxidase (ACO) requires O_2 and is blocked by oxygen depletion. Nonetheless, significant ACC oxidase activity was detected in maize roots subjected to hypoxia and/or mechanical impedance (He et al. 1996a, b) and in submerged rice internodes (Mekhedov and Kende 1996). Expression analysis revealed downregulation of *OsACS2* and upregulation of *OsACS1*

(Zarembinski and Theologis 1997) and *OsACS5* genes (Van der Straeten et al. 2001) in flooded rice. Hypoxia-induced stimulation was also demonstrated for expression of *OsACO1* in rice (Mekhedov and Kende 1996); *RpACS1* and *RpACO1* in *Rumex palustris* (Rieu et al. 2005); *ACS2*, *ACS6*, *ACS7*, and *ACS9* in *Arabidopsis* (Peng et al. 2005); and *LeACS7*, *LeACS3*, and *LeACS2* in tomato (Olson et al. 1995).

The paradoxical effect of oxygen deficiency on the ethylene production looks in the following way: both the key enzymes of ethylene biosynthesis, ACS and ACO, are induced by oxygen lack at transcriptional level, but their posttranscriptional regulation is different: ACS is activated and ACO is inhibited, particularly by total absence of molecular oxygen. This leads to spatial differentiation of ethylene synthesis in partially flooded plant: ACC is synthesized in hypoxic root, then it is transported with xylem sap to aerated shoots, where it turns into the ethylene (Jackson and Campbell 1976; Bradford et al. 1982; Grichko and Glick 2001). This system works under soil waterlogging and root anaerobiosis. ACC synthesized in stressed roots and transported to nonstressed shoots may be assumed itself as a hypoxic root-to-shoot signal (Jackson 1997, 2002).

In addition to ethylene synthesis from ACC, the hormone might be produced from other precursors (Chernys and Kende 1996), including metabolites accumulated under anoxia (Gurevich 1979). Besides that, plants are capable of absorbing ethylene from soil where its concentration raises under flooding due to microbial activity (Chirkova 1988; Crawford and Braendle 1996; Vartapetian and Jackson 1997).

Complete flooding inhibits the conversion of ACC into ethylene due to depletion of O_2 (Voesenek et al. 1993; Banga et al. 1996). This leads to ACC accumulation and its further conjugation. The concentrations of free and conjugated forms of ACC increase dramatically in less tolerant *Rumex acetosella* species (Banga et al. 1996). Accumulation of ACC has to stimulate ACO, but absence or limitation of O_2 leads to inhibition of the enzyme. Nevertheless, hormone concentration in plant tissues increases because of physical entrapment of ethylene inside plant tissues. Speed of ethylene diffusion in the air is $0.16\ cm^3 \times s^{-1}$, and in the water, it is only $0.17 \times 10^{-4}\ cm^3 \times s^{-1}$ (Voesenek et al. 1990b). Therefore, in nonsubmerged plant, 90% of synthesized hormone would escape within 1 min (Voesenek et al. 1993). In hypoxia-tolerant *R. palustris* species, ethylene accumulation occurred only in the first 8 h of flooding. Further on, the high level of phytohormone negatively regulates activity of ACS that leads to decrease in ACC concentration and ethylene production (Banga et al. 1996). Moreover, depletion of oxygen concentration strengthens sensitivity of *R. palustris* plants to ethylene due to the stimulation of expression of *RpERS1* gene, encoding ethylene receptor (Voesenek et al. 1997; Vriezen et al. 1997). In the sensitive *R. acetosella*, ethylene does not inhibit activity of ACS, which leads to continuous hormone accumulation and causes senescence, epinasty, leaf abscission, and finally the death of plant (Banga et al. 1996; Visser et al. 1996a). In *R. palustris*, ethylene adjusts vertical orientation of leaves (hyponasty) and considerably accelerates leaf petiole growth.

Ethylene plays a crucial role in regulation of plant response to flooding. Nonetheless, there are some aquatic plants incapable of synthesizing this hormone.

Potamogeton pectinatus lacks ACO and accumulates high level of ACC (Summers et al. 1996). Nevertheless, this plant may grow in hypoxic or even anoxic environments.

The growth-promoting effect of ethylene has been demonstrated in a wide variety of aquatic and amphibious species of different taxonomical origin displaying LOES (Ku et al. 1970; Musgrave et al. 1972; Cookson and Osborne 1978; Metraux and Kende 1983; Raskin and Kende 1983; Samarakoon et al. 1985; Satler and Kende 1985; Voesenek et al. 1990a, b; Banga et al. 1997). The submergence-induced hormone production not only accelerates growth rate of shoot axial organs or leaves but also enlarges their final length (Jackson 2008). It either stimulates cell elongation with little or no change in their number, e.g., in *Callitriche platycarpa* (Musgrave et al. 1972), *Hydrocharis morsus-ranae*, and *Ranunculus sceleratus* (Cookson and Osborne 1978), or increases production of new cells and their subsequent elongation, like in deepwater rice (Metraux and Kende 1984), *Nymphoides peltata*, *Ranunculus repens* (Ridge 1987), and *R. pygmaeus* (Horton 1992).

Promotion of shoot extension by ethylene is linked with cell wall loosening and interaction with other phytohormones. Loosening of cell wall is provided by ethylene-dependent stimulation of pectinase, xylanase (Bragina et al. 2001, 2003), and xyloglucan endotransglucosylase/hydrolase (Saab and Sachs 1996). These enzymes are also involved in aerenchyma formation. Ethylene mediates hypoxic induction of expansin A in *Marsilea quadrifolia* and *Regnellidium diphyllum* (Kim et al. 2000), *Rumex palustris* (Vreeburg et al. 2005), *Sagittaria pygmaea* (Ookawara et al. 2005), and expansins OsEXPA11, OsEXPA15, and OsEXPB4 in deepwater rice (Lee et al. 2001). Moreover, Vreeburg et al. (2005) reported ethylene to promote fast apoplastic acidification in flooded *Rumex palustris*, which is important for growth promotion.

Furthermore, ethylene is supposed to be the master signal for rearrangement of hormone crosstalk during flooding stress. LOES is regulated by ethylene-dependent increase in the ratio of growth-promoting hormone (GA) to growth-inhibiting one (ABA) (Kende et al. 1998). Since early studies, submergence-induced fast underwater shoot elongation in LOES species was shown to be mediated by an interaction of ethylene and GAs. Flooding- and ethylene-induced shoot growth is partially or totally blocked by treatment with GA-biosynthesis inhibitors in *Callitriche platycarpa* (Musgrave et al. 1972), *Ranunculus sceleratus* (Samarakoon and Horton 1983), rice (Raskin and Kende 1984; Suge 1985), and *Rumex palustris* (Rijnders et al. 1997). Ethylene stimulates synthesis of GA and increases sensitivity to GAs in deepwater rice stems (Hoffmann-Benning and Kende 1992; Kende et al. 1998) and petioles of *Rumex palustris* (Rijnders et al. 1997). This stimulation at least partly may be mediated via ethylene effect on ABA level. Elevation of ethylene production upon submergence leads to reduction in endogenous ABA levels in LOES plants like deepwater rice (Hoffmann-Benning and Kende 1992; Lee et al. 1994; Kende et al. 1998) and *Rumex palustris* (Benschop et al. 2005; Bailey-Serres and Voesenek 2008) by inhibition of its synthesis and stimulation of degradation to phaseic acid (Benschop et al. 2005; Saika et al. 2007; Chen et al. 2010). The decline of the endogenous ABA concentration in deepwater rice occurs within the first hour of submergence or ethylene treatment, while accumulation of GA occurs after 3 h (Hoffmann-Benning and Kende 1992; Sauter 2000). Moreover, the decrease in ABA in *Rumex palustris* is required to stimulate the expression of *RpGA3ox1* gene encoding gibberellin 3-oxidase, an enzyme converting GA_{20} to bioactive gibberellin GA_1 (Benschop et al. 2006). Notably, in nontolerant *R. acetosa,* contrary to *R. palustris*, submergence and ethylene do not depress levels of ABA (Benschop et al. 2005) nor they do alter GA, while responsiveness to GA even lowers (Rijnders et al. 1997).

Further investigations reveal ethylene signaling to be closely connected with plant tolerance to oxygen shortage. Some *indica* rice cultivars, like FR13A and Kurkaruppan, are known for their high tolerance to flooding. These cultivars do not accelerate their elongation when submerged unlike LOES deepwater rice, and their growth is less sensitive to ethylene in comparison to other rice lines. On the other hand, ethylene does not stimulate senescence as much as in nontolerant plants (Jackson et al. 1987). They demonstrate non-LOES quiescent syndrome and survive up to 2 weeks of complete submergence. Recent studies of these cultivars discovered a major quantitative trait locus designated *Sub1* (*Submergence 1*) conferring flooding tolerance in rice (Xu et al. 2006). This locus contains two or three genes encoding ethylene response transcription factors belonging to the B-2 subgroup of the ethylene response factors (ERFs)/ethylene-responsive element binding proteins (EREBPs)/apetala 2-like proteins (AP2) (Perata and Voesenek 2007). In some submergence-intolerant *indica* and in all *japonica* cultivars, this locus encodes two ERF genes, *Sub1B* and *Sub1C*. In the tolerant *indica* cultivars, including FR13A and Kurkaruppan mentioned above, the locus encodes three ERF genes, with addition of *Sub1A* (Xu et al. 2006). *Sub1A* and *Sub1C* transcript levels are upregulated by submergence and ethylene (Fukao et al. 2006) and downregulated by ABA (Fukao and Bailey-Serres 2008). *Sub1A* confers tolerance to prolonged complete submergence in lowland rice. It dampens ethylene production and represses ethylene- and GA-mediated responses in cell elongation and carbohydrate breakdown, including expansin and amylase genes expression. On the other hand, *Sub1C* gene is upregulated by submergence, ethylene, and GA and stimulates shoot elongation in LOES rice. In quiescent-tolerant rice cultivars, *Sub1C* is repressed by Sub1A (Fukao et al. 2006). *Sub1A* was reported to increase the accumulation of the signaling proteins Slender Rice-1 (SLR1) and SLR1 Like-1 (SLRL1) which repress GA signaling and LOES (Fukao and Bailey-Serres 2008). At the same time, *Sub1A* induces metabolic adaptation by upregulation of genes, encoding pyruvate decarboxylase and alcohol dehydrogenase (ADH) involved in alcoholic fermentation (Fukao et al. 2006). Thus, *Sub1A* blocks LOES and stimulates quiescence syndrome, transiently limiting growth that allows to keep the energy and capacity for regrowth upon desubmergence. Also, *Sub1* is believed to be the major factor in the integration of ethylene, ABA, and GA signaling during submergence.

In spite of ultimate role of ethylene as primary regulator of plant adaptation to oxygen deficiency, there are some cases when fast underwater elongation cannot easily be linked with ethylene mediation (Jackson 2008). In addition to the above-mentioned ethylene-independent *Potamogeton pectinatus*, coleoptiles of rice, *Echinochloa oryzicola*, and shoots of *Sagittaria pygmaea* may elongate quickly in the total anoxic environment (Costes and Vartapetian 1978; Pearce and Jackson 1991; Ookawara et al. 2005). Moreover, even treatment with exogenous hormone does not affect the elongation of rice coleoptiles under total absence of oxygen (Pearce et al. 1992; Bertani et al. 1997).

Ethylene stimulates seed germination of aquatic plants, e.g., *Epilobium hirsutum* (Etherington 1983), *Nymphaea alba*, and *Nuphar lutea* (Smits et al. 1995).

Submergence-induced synthesis and/or physical entrapment of ethylene cause limitation of primary and lateral root growth (Visser et al. 1997). At the same time, ethylene stimulates formation of adventitious roots in maize (Drew et al. 1979), *Epilobium hirsutum* (Etherington 1983), *Rumex palustris* (Visser et al. 1996a) deepwater rice (Lorbiecke and Sauter 1999), and tomato (Vidoz et al. 2010), via ethylene-induced redistribution of auxin transport (Visser et al. 1995, 1996a; Vidoz et al. 2010), accumulation (Wample and Reid 1979; Vidoz et al. 2010), and an increase in sensitivity to auxin in root-forming tissues (Visser et al. 1996b). Exogenously supplied GA, auxin, and cytokinin fail to stimulate adventitious rooting in deepwater rice (Lorbiecke and Sauter 1999). Use of inhibitors of ethylene biosynthesis or action and of auxin polar transport results in a reduction of adventitious root formation in waterlogged plants (Visser et al. 1996b; Vidoz et al. 2010). The number of roots formed upon ethylene treatment in nontolerant plants *Chamerion angustifolium* and *R. thyrsiflorus* is much lower than in wetland species *Epilobium hirsutum* and *R. palustris* (Etherington 1983; Visser et al. 1996a). Ethylene also affects level and distribution of auxin during stem hypertrophy in flooded sunflower (Wample and Reid 1979).

Being a typical plant response to partial or complete submergence, epinasty is also determined by ethylene, particularly in species with little tolerance to oxygen lack (Jackson and Campbell 1976; Bradford and Hsiao 1982; Visser et al. 1996a).

Aerenchyma is a special pneumatic tissue providing air transport and storage facility for plants under the lack of oxygen. Spaces within the aerenchymatous organ appear either by cell separation at the middle lamella (schizogeny) or by cell death and decomposition of cell wall (lysigeny). In wetland species like rice, willow (Vartapetian and Jackson 1997), and flooding-tolerant *Rumex* species, aerenchyma is formed schizogenously (Laan et al. 1989), and mechanisms of its formation are largely unknown (Bailey-Serres and Voesenek 2008). Intolerant terrestrial species develop lysigenous aerenchyma upon imposition of submergence. Formation of lysigenous aerenchyma in maize root cortex is triggered by ethylene (Drew et al. 1979; Jackson et al. 1985a; Drew 1997). It requires ethylene synthesis via ACC (He et al. 1996a) and signal transduction with implication of Ca^{2+} ions, G-proteins, and protein kinases (He et al. 1996b). Softening of cell wall is provided by ethylene-dependent increases in activity of pectinase, xylanase (Bragina et al. 2001, 2003), cellulase (He et al. 1994, 1996a), and the upregulation of xyloglucan endotransglucosylase/hydrolase gene expression (Saab and Sachs 1996). In root cortex cells, ethylene induces programmed cell death (Drew et al. 2000). On the other hand, ethylene has little effect on the formation of schizogenous aerenchyma in rice (Jackson et al. 1985b). It is still unclear whether ethylene is involved in constitutive schizogenous aerenchyma formation in wetland species.

10.3 Abscisic Acid

ABA is well-known growth inhibitor antagonizing many plant growth substances like auxin, GAs, and cytokinins. An increase of ABA concentration particularly accompanied by reduction in levels of hormone stimulators promotes significant deceleration of growth and metabolism intensity. As already mentioned above, LOES plants respond to submergence by the fast drop in endogenous ABA level. This was reported for coleoptiles (Hoson et al. 1993; Lee et al. 1994) and internodes of rice (Hoffmann-Benning and Kende 1992), shoots of *Scirpus mucronatus* (Lee et al. 1996), and petioles of *Rumex palustris* (Benschop et al. 2005; Chen et al. 2010). Nonetheless, susceptible to hypoxia plants usually respond to oxygen shortage by drastic increase of ABA concentration. Root hypoxia results in manifold elevation of ABA level in tomato shoots (Bradford 1983), roots and leaves of pea (Zhang and Davies 1987), leaves of French bean (Neuman and Smit 1991) and a hybrid poplar *Populus trichocarpa* × *Populus deltoides* (Smit et al. 1990), and roots and needles of pine (Menyajlo and Shulgina 1978). In susceptible to hypoxia *Rumex acetosa*, ABA remains stable during complete submergence (Benschop et al. 2005). Total anoxia provokes ABA accumulation in coleoptiles and roots of aerobically germinated rice seedlings (Mapelli et al. 1986). More susceptible to oxygen deprivation, wheat accumulates higher amount of ABA during submergence or total anoxia than less susceptible oat (Yemelyanov and Chirkova 1996; Bakhtenko et al. 2008) and rice (Emel'yanov et al. 2003).

The decrease of endogenous ABA level in submergence-escaping plants occurs within the first 1–3 h of oxygen deficiency (Hoffmann-Benning and Kende 1992; Sauter 2000; Benschop et al. 2005) and is triggered by ethylene since it may be imitated by applying ethylene to nonflooded plants and prevented by pretreating submerged plants with inhibitors of ethylene synthesis or action (Jackson 2008). Flooding-induced ethylene suppresses ABA synthesis at two points: first by downregulating *OsZEP* gene encoding zeaxanthin epoxidase and second by repressing genes of 9-*cis*-epoxycarotenoid dioxygenase in rice (*OsNCED*1, *OsNCED*2, and *OsNCED3*; Saika et al. 2007) and in *Rumex palustris* (*RpNCED1-4* and *RpNCED6-10*, Benschop et al. 2005; Chen et al. 2010). Simultaneously, ethylene promotes ABA breakdown by upregulation of *OsABA8ox1* gene encoding ABA 8′-hydroxylase (CYP707A5), which converts ABA to its catabolites, phaseic and dihydrophaseic acids (Saika et al. 2007). Levels of phaseic and dihydrophaseic acids rise in rice shoots during submergence (Saika et al. 2007) and in rice coleoptiles under anoxia (Mapelli et al. 1986). ABA accumulation in shoots along with its catabolites may be prevented by its release from the plant into external medium, like it was discovered in rice (Mapelli et al. 1986). Notably, the most susceptible plants like wheat are incapable of ABA extrusion into the medium (Mapelli et al. 1995). Total anoxia completely blocks oxygen-dependent synthesis of ABA, and high concentration of a free form in wheat plant is maintained by hydrolysis of its ester-bound forms (Emel'yanov et al. 2003). On the contrary, decrease of free ABA in rice may be a result of its conjugation (Emel'yanov et al. 2003; Saika et al. 2007).

ABA is a very important regulator of plant growth during submergence. High level of this hormone in plant tissue would cease growth, and the opposite, reduced concentration would stimulate elongation. Exogenously applied ABA inhibits growth of coleoptiles (Mapelli et al. 1993; Lee et al. 1994) and stem sections of rice (Hoffmann-Benning and Kende 1992), petioles and stems of *Ranunculus sceleratus* (Smulders and Horton 1991), stems of *Potamogeton pectinatus* (Summers and Jackson 1996) and *Scirpus mucronatus* (Lee et al. 1996), and petioles of *R. palustris* (Benschop et al. 2005; Chen et al. 2010). Contrariwise, use of fluridone, an ABA biosynthesis inhibitor, promotes growth in LOES plants (Hoffmann-Benning and Kende 1992; Lee et al. 1996; Benschop et al. 2005; Chen et al. 2010). There is difference in ABA content between different accessions of LOES plants originating from habitats varying in duration and depth of submergence. The slow elongating accession of *R. palustris* accumulates a relatively high amount of ABA than fast elongating one (Chen et al. 2010). On the other hand, LOES deepwater rice Plai Ngam variety contains virtually same levels of ABA during submergence as non-LOES lowland IR36 cultivar (Van Der Straeten et al. 2001). ABA may be one of the key regulators of adaptation strategy (LOES or no) by inhibiting mRNA accumulation of *OS-ACS5* (Van Der Straeten et al. 2001) and *Sub1* (Fukao and Bailey-Serres 2008).

The adaptive effect of ABA accumulation at oxygen deficiency stress is due not only to inhibition of growth. Exogenously applied ABA induces anoxic tolerance and seedling survival in maize (Hwang and Van Toai 1991), *Arabidopsis* (Ellis et al. 1999), and lettuce (Kato-Noguchi 2000a). ABA-mediated tolerance in maize depends on protein synthesis (Hwang and Van Toai 1991). Treatment with ABA increases ADH activity (Hwang and Van Toai 1991; Kato-Noguchi 2000b) and *AtADH1* gene expression (De Bruxelles et al. 1996; Dolferus et al. 1997). At the same time, *ADH* gene can be upregulated during oxygen shortage in ABA-independent manner (De Bruxelles et al. 1996; Dolferus et al. 1997).

10.4 Gibberellins

Effects of oxygen deficiency on the gibberellin content depend on plant tolerance, adaptation strategy (LOES or no), and intensity of impact. Pines, growing at marshy habitat, contain less GA in comparison to that of growing in wood (Menyajlo and Shulgina 1978). Root hypoxia resulted in slight reduction of GA_1 and GA_3 shoot concentrations in French bean and a hybrid poplar *Populus trichocarpa* × *Populus deltoides* (Neuman et al. 1990). Shoots of wheat contain less GA in comparison with more tolerant oat plants under soil waterlogging (Bakhtenko et al. 2008). Wetland plant species displaying LOES accumulate high amount of GA while submerged, as it has been shown for rice (Suge 1985; Hoffmann-Benning and Kende 1992) and *R. palustris* (Rijnders et al. 1997). In *R. palustris*, petioles of submerged plants accumulate mostly 13-OH GAs, especially GA_1 and GA_{20}, compared with drained plants. There are no differences between levels of GAs petioles of susceptible *R. acetosa* in drained or flooded conditions (Rijnders et al. 1997). Similarly, submergence results in accumulation of GA_{20} in shoots of deepwater rice Plai Ngam cultivar and does not alter its level in non-LOES lowland IR36 cultivar (Van Der Straeten et al. 2001).

As already mentioned above (see 10.2), shift in GA concentration in submergence-escaping plants is upregulated by ethylene and downregulated by ABA and *Sub1A*. Growth stimulation in LOES plants by flooding or ethylene is somewhat arrested by use of GA-biosynthesis inhibitors, and the effect of inhibitors is reversible by treatment with exogenous GA (Jackson 2008). On the opposite, treatment with fluridone (inhibitor of ABA biosynthesis) increase GA responsiveness and growth (Hoffmann-Benning and Kende 1992; Chen et al. 2010). At the same time, externally applied ABA inhibits the increase in concentration of bioactive GA_1 in submerged *Rumex* plants (Benschop et al. 2005, 2006) by repression of *RpGA3ox1* gene encoding gibberellin 3-oxidase, an enzyme converting GA_{20} to GA_1 (Benschop et al. 2006). Submergence-induced accumulation of ethylene brings about a rapid drop in ABA level (Benschop et al. 2005), which relieves the inhibition from *RpGA3ox1* and favors the concentration of GA_1 to increase.

The majority of ethylene effects on fast underwater shoot growth are mediated by GAs. Treatment with exogenous GA promotes elongation of in stems of *Callitriche platycarpa* (Musgrave et al. 1972), rice (Raskin and Kende 1984; Suge 1985; Van Der Straeten et al. 2001), petioles of *Ranunculus sceleratus* (Samarakoon and Horton 1983), and *R. palustris* (Rijnders et al. 1997; Benschop et al. 2006; Chen et al. 2010). Growth-promoting GA action in deepwater rice stem is connected with increased cell elongation and cell production rate in the intercalary meristem (Raskin and Kende 1984). Exogenous GA_3 first promotes cell elongation in the intercalary meristem (within 2 h after treatment); after 4 h, it stimulates cells in the G_2 phase to enter mitosis, and subsequent activation of DNA synthesis ($G_1 \rightarrow S$ transition) takes place between 4 and 7 h of GA_3 treatment (Sauter and Kende 1992). Elongation is strongly supported by GA-mediated enhancement in the expression of genes encoding β-expansins (Lee and Kende 2001), α-expansins, and expansin-like proteins (Cho and Kende 1997; Lee and Kende 2002). Cell division cycle in intercalary meristem of deepwater rice is activated by GA at two key points, DNA replication and mitosis. GA_3 treatment increases $p34^{cdc2/CDC28}$-like histone H1 kinase activity and upregulates genes, encoding cyclin-dependent protein kinases and B2-type mitotic cyclins (Sauter et al. 1995; Fabian et al. 2000). GA also activates the expression of genes encoding proteins involved in signal transduction: (1) a receptor-like transmembrane protein kinase, Os-TMK (*Oryza sativa*-transmembrane kinase) (van der Knaap et al. 1999); (2) a heterotrimeric protein RPA1 (replication protein A1) implicated in DNA replication, recombination, repair, and also in regulation of transcription (van der Knaap et al. 1997); and (3) a transcription factor Os-GRF1 (*O. sativa*-growth-regulating factor1) containing a nuclear localization signal and participating in the regulation of GA-induced stem elongation (van der Knaap et al. 2000).

It is important to note that GAs are not always involved in regulation of submergence/ethylene-induced growth. For example, fast underwater ethylene-mediated elongation in *Regnellidium diphyllum* (Musgrave and Walters 1974) and *Potamogeton pectinatus* (Summers and Jackson 1996) does not depend on GA. GA fails to stimulate growth in rice coleoptiles under anoxia (Mapelli et al. 1993; Loreti et al. 2003). Moreover, flooding-induced growth of coleoptiles of deepwater rice is not reduced by inhibitors of GA biosynthesis (Hoffmann-Benning and Kende 1992).

Exogenous GA does not affect anoxia-induced expression of *Amy3D* gene, encoding α-amylase in aleurone and embryo from germinating rice grains (Loreti et al. 2003). Similarly, treatment with GA has no influence on ADH activity in lettuce seedlings (Kato-Noguchi 2000b). Thus, opposite to ethylene and ABA, GA plays minor role in regulation of metabolism during oxygen deficiency.

10.5 Auxin

Auxin stays aside from the mainstream of plant hormone research under oxygen deprivation, and its role in regulation of plant adaptation is generally underestimated. This can be due to the fact that submergence reduces total level of indoleacetic acid (IAA) in pine needles (Menyajlo and Shulgina 1978), rice coleoptiles (Hoson et al. 1992), and petioles of *R. palustris* (Cox et al. 2004). On the other hand, waterlogging affects redistribution of IAA via ethylene mediation, leading to its accumulation in the stem on the border between aerated and flooded zone, like in hypocotyls of sunflower (Wample and Reid 1979) and tomato (Vidoz et al. 2010). Similarly, submergence induces raise of auxin level in the base of petioles of *R. palustris* petioles of *R. palustris* (Cox et al. 2004). There is a difference in auxin concentrations in plants during oxygen deficiency depending on their tolerance. More susceptible wheat contents less IAA than more tolerant oat when submerged (Bakhtenko et al. 2008) or at total anaerobiosis (Yemelyanov and Chirkova 1996). Rice seedlings accumulate even higher amounts of IAA in totally anoxic environment (Mapelli et al. 1986; Emel'yanov et al. 2003).

Auxin is not required for the ethylene-mediated submergence-induced growth of coleoptiles (Katsura and Suge 1979) and internodes of the most studied LOES plant, deepwater rice (Jackson 2008). However, studies with different wetland species demonstrate its necessity for the submergence/ethylene response. IAA stimulates growth in leaves of fern *Regnellidium diphyllum* (Walters and Osborne 1979), stems of *Potamogeton pectinatus* (Summers and Jackson 1996), petioles of *Nymphoides peltata* (Ridge and Osborne 1989), *Ranunculus sceleratus* (Horton and Samarakoon 1982; Smulders and Horton 1991; Rijnders et al. 1996), and

R. palustris (Cox et al. 2006). In *R. palustris*, exogenously applied auxin increases upregulation of ethylene-inducible expansin gene *RpEXPA1* (Vreeburg et al. 2005). Auxin can stimulate elongation even after treatment with inhibitors of ethylene biosynthesis or action, suggesting a role downstream from ethylene as with GAs (Jackson 2008). Although early studies suggest that independent modes of action are also possible (Cookson and Osborne 1978). Under total anoxia, IAA has no effect on the growth of excised coleoptiles (Mapelli et al. 1993), but there is slight stimulation in intact rice seedlings (Bertani et al. 1997).

In *Rumex palustris*, auxin and ethylene regulate hyponastic growth of petioles (Cox et al. 2004). Surprisingly, this regulation is independent of their role in petiole elongation (Cox et al. 2006). On the other hand, epinasty in susceptible plants like tomato also depends on IAA–ethylene interactions (Lee et al. 2008). These interactions are also involved in submergence-induced stem hypertrophy and adventitious rooting (Wample and Reid 1979; Visser et al. 1995, 1996a, b; Vidoz et al. 2010), as it was discussed earlier (see 10.2).

Similar to ABA, IAA participates in metabolic regulation during oxygen deficiency. Treatment with IAA increases ADH activity in lettuce (Kato-Noguchi 2000b) and protein synthesis in rice coleoptiles (Mapelli and Locatelli 1995).

10.6 Cytokinin

There are few scrappy evidences about cytokinins in plants during oxygen deficiency. Cytokinin content is not significantly altered by root hypoxia in leaves of *Phaseolus vulgaris* and hybrid poplar (Neuman et al. 1990; Smit et al. 1990) and by total anoxia in seedlings of wheat, oast, and rice (Yemelyanov and Chirkova 1996; Emel'yanov et al. 2003). In the roots of *Amaranthus paniculatus*, hypoxia leads to accumulation of cytokinins (Rudolf et al. 1987). This can be explained by reduction of cytokinin delivery from roots observed in tomato (Bradford 1983) bean and poplar (Neuman et al. 1990). The role of cytokinin in regulation of plant adaptation to the lack of oxygen is also unobvious. Exogenously applied cytokinins do not affect the growth of rice coleoptiles (Mapelli et al. 1993) and shoots of bean and poplar (Neuman et al. 1990) under the lack of oxygen. But kinetin inhibits anaerobic germination of rice seeds (Miyoshi and Sato 1997). The decline in photosynthetic capacity and stomatal conductance in both flooded plants may be alleviated by cytokinins in tomato (Bradford 1983), but not in French bean and poplar (Neuman et al. 1990). Applied to lettuce, cytokinin increases ADH activity (Kato-Noguchi 2000b).

10.7 Conclusions

Plant classical hormones take active part in regulation of plant adaptation to oxygen deficiency. Common among wetland species, fast shoot elongation, which is the main component of LOES, is ultimately driven by hormones. Submergence induces

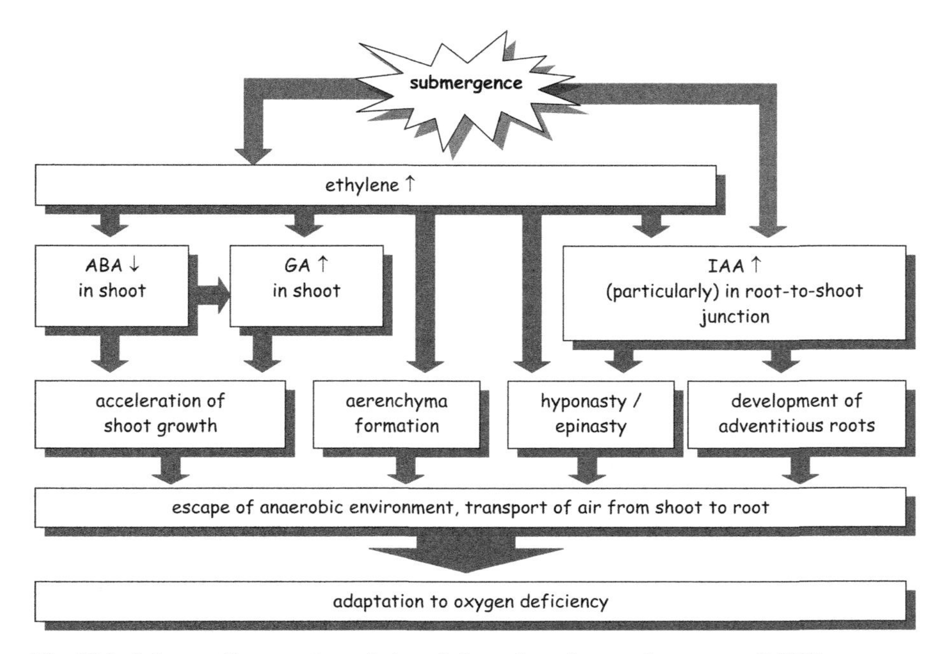

Fig. 10.1 Scheme of hormonal regulation of plant adaptation to submergence (LOES)

ethylene synthesis (Fig. 10.1), which, in turn, leads to decrease in ABA and increase in GA level. This interaction between ethylene, ABA, and GA accelerates the growth of stems and leaf petioles by stimulation of cell division and elongation. Differences between various accessions of LOES plants in their capacity to elongate are possibly determined by diversity in endogenous ABA and GA content. In some plants, growth is also supported by accumulation of auxin, caused by submergence either in ethylene-dependent or -independent manner. Ethylene–auxin interaction regulates epi- or hyponastic response and adventitious rooting depending on plant tolerance. Ethylene itself induces formation of lysigenous aerenchyma. Ethylene, ABA, auxin, and cytokinin regulate metabolic switch by activating a number of enzymes of glycolysis and ethanol fermentation. Taken together, these responses allow plant to reach water surface, escaping from oxygen-deficient environment, and provide oxygen transport from shoot to flooded root, which finally results in plant adaptation and survival. Contrarily intolerant species accumulate high amount of ethylene and ABA and decline the levels of growth stimulators (GA and auxin) under the lack of oxygen, leading to stimulation of senescence, epinasty, leaf abscission, and finally the death of plant. Nonetheless, our knowledge on involvement of hormones in the regulation of metabolism during oxygen deprivation and on the role of hormones other than ethylene, ABA, and GA is insufficient and still to be elucidated.

References

Bailey-Serres J, Voesenek LACJ (2008) Flooding stress: acclimations and genetic diversity. Annu Rev Plant Biol 59:313–339

Bakhtenko EYu, Skorobogatova IV, Karsunkina NP, Platonov AV (2008) Hormonal balance of wheat (*Triticum aestivum* L.) and oat (*Avena sativa* L.) under submergence and reparation. Agrochemistry 8:33–41 [In Russian]

Banga M, Slaa EJ, Bloom CWPM, Voesenek LACJ (1996) Ethylene biosynthesis and accumulation under drained and submerged conditions. A comparative study of two Rumex species. Plant Physiol 112:229–237

Banga M, Bogemann GM, Blom CWPM, Voesenek LACJ (1997) Flooding resistance of *Rumex* species strongly depends on their response to ethylene: Rapid shoot elongation or foliar senescence. Physiol Plant 99:415–422

Benschop JJ, Jackson MB, Gühl K, Vreeburg RA, Croker SJ, Peeters AJM, Voesenek LACJ (2005) Contrasting interactions between ethylene and abscisic acid in *Rumex* species differing in submergence tolerance. Plant J 44:756–768

Benschop JJ, Bou J, Peeters AJM, Wagemaker N, Gühl K, Ward D, Hedden P, Moritz T, Voesenek LACJ (2006) Long-term submergence-induced elongation in *Rumex palustris* requires abscisic acid-dependent biosynthesis of gibberellin 1. Plant Physiol 141:1644–1652

Bertani A, Brambilla I, Mapelli S, Reggiani R (1997) Elongation growth in the absence of oxygen - the rice coleoptile. Russ J Plant Physiol 44:543–547

Bradford KJ (1983) Involvement of plant growth substances in the alteration of leaf gas exchange of flooded tomato plants. Plant Physiol 73:480–483

Bradford KJ, Hsiao TC (1982) Stomatal behavior and water relations of waterlogged tomato plants. Plant Physiol 70:1508–1513

Bradford KJ, Hsiao TC, Yang SF (1982) Inhibition of ethylene synthesis in tomato plants subjected to anaerobic root stress. Plant Physiol 70:1503–1507

Bragina TV, Martinovich LI, Rodionova NA, Bezborodov AM, Grineva GM (2001) Ethylene-induced activation of xylanase in adventitious roots of maize as a response to the stress effect of root submersion. Prikl Biokhim Mikrobiol 37:722–725 [In Russian]

Bragina TV, Rodionova NA, Grinieva GM (2003) Ethylene production and activation of hydrolytic enzymes during acclimation of maize seedlings to partial flooding. Russ J Plant Physiol 50:794–798

Chen X, Pierik R, Peeters AJM, PoorterH VEJW, Huber H, de Kroon H, Voesenek LACJ (2010) Endogenous abscisic acid as a key switch for natural variation in flooding-induced shoot elongation. Plant Physiol 154:969–977

Chernys J, Kende H (1996) Ethylene biosynthesis in *Regnellidium diphyllum* and *Marsilea quadrifolia*. Planta 200:113–118

Chirkova TV (1988) Plant adaptation to hypoxia and anoxia. Leningrad State University Press, Leningrad

Chirkova TV, Gutman TS (1972) On the physiological role of lenticels on the branches of willow and poplar under root anaerobiosis. Sov Plant Physiol 19:352–359

Cho H-T, Kende H (1997) Expression of expansin genes is correlated with growth in deepwater rice. Plant Cell 9:1661–1671

Cohen E, Kende H (1987) *In vivo* 1-aminocyclopropane-1-carboxilic acid synthase activity in internodes of deepwater rice. Enhancement by submergence and low oxygen levels. Plant Physiol 84:282–286

Cookson C, Osborne DJ (1978) The stimulation of cell extension by ethylene and auxin in aquatic plants. Planta 144:39–47

Costes C, Vartapetian BB (1978) Plant grown in a vacuum: ultrastructure and functions of mitochondria. Plant Sci Lett 11:115–119

Cox MCH, Benschop JJ, Vreeburg RAM, Wagemaker CAM, Moritz T, Peeters AJM, Voesenek LACJ (2004) The roles of ethylene, auxin, abscisic acid, and gibberellin in the hyponastic growth of submerged *Rumex palustris* petioles. Plant Physiol 136:2948–2960
Cox MCH, Peeters AJM, Voesenek LACJ (2006) The stimulating effects of ethylene and auxin on petiole elongation and on hyponastic curvature are independent processes in submerged *Rumex palustris*. Plant Cell Environ 29:282–290
Crawford RMM, Braendle R (1996) Oxygen deprivation stress in a changing environment. J Exp Bot 47:145–159
De Bruxelles GL, Peacock WJ, Dennis ES, Dolferus R (1996) Abscisic acid induces the alcohol dehydrogenase gene in *Arabidopsis*. Pant Physiol 111:381–391
Dolferus R, Ellis M, Bruxelles D, Trevaskis B, Hoeren F, Dennis ES, Peacock WJ (1997) Strategies of gene action in *Arabidopsis* during hypoxia. Ann Bot 79:21–31
Drew MC (1997) Oxygen deficiency and root metabolism: injury and acclimation under hypoxia and anoxia. Annu Rev Plant Physiol Plant Mol Biol 48:223–250
Drew MC, Jackson MB, Giffard S (1979) Ethylene-promoted adventitious rooting and development of cortical air spaces (aerenchyma) in roots may be adaptive responses to flooding in *Zea mays* L. Planta 147:83–88
Drew MC, He CJ, Morgan PW (2000) Programmed cell death and aerenchyma formation in roots. Trends Plant Sci 5:123–127
Ellis MH, Dennis ES, Peacock WJ (1999) *Arabidopsis* roots and shoots have different mechanisms for hypoxic stress tolerance. Plant Physiol 119:57–64
Emel'yanov VV, Kirchikhina NA, Lastochkin VV, Chirkova TV (2003) Hormonal status in wheat and rice seedlings under anoxia. Russ J Plant Physiol 50:827–834
Etherington JR (1983) Control of germination and seedling morphology by ethylene: differential responses, related to habitat of *Epilobium hirsutum* L. and *Chamerion angustifolium* (L.) (J. Holub). Ann Bot 52:653–658
Fabian T, Lorbiecke R, Umeda M, Sauter M (2000) The cell cycle genes *cycA1;1* and *cdc2Os-3* are coordinately regulated by gibberellin *in planta*. Planta 211:376–383
Fukao T, Bailey-Serres J (2008) Submergence tolerance conferred by *Sub1A* is mediated by SLR1 and SLRL1 restriction of gibberellin responses in rice. Proc Natl Acad Sci USA 105:16814–16819
Fukao T, Xu K, Ronald PC, Bailey-Serres J (2006) A variable cluster of ethylene response factor-like genes regulates metabolic and developmental acclimation responses to submergence in rice. Plant Cell 18:2021–2034
Gibbs J, Greenway H (2003) Mechanisms of anoxia tolerance in plants. I. Growth, survival and anaerobic catabolism. Funct Plant Biol 30:1–47
Greenway H, Gibbs J (2003) Mechanisms of anoxia tolerance in plants. II. Energy requirements for maintenance and energy distribution to essential processes. Funct Plant Biol 30:999–1036
Grichko VP, Glick BR (2001) Ethylene and flooding stress in plants. Plant Physiol Biochem 39:1–9
Gurevich LS (1979) The role of hormonal balance of auxin and ethylene in adaptive reactions of higher plants. Botanical J 64:1600–1614 [In Russian]
He CJ, Drew MC, Morgan PW (1994) Induction of enzymes associated with lysigenous aerenchyma formation in roots of *Zea mays* during hypoxia or nitrogen starvation. Plant Physiol 105:861–865
He CJ, Finlayson SA, Drew MC, Jordan WR, Morgan PW (1996a) Ethylene biosynthesis during aerenchyma formation in roots of maize subjected to mechanical impedance and hypoxia. Plant Physiol 112:1679–1685
He CJ, Morgan PW, Drew MC (1996b) Transduction of an ethylene signal is required for cell death and lysis in the root cortex of maize during aerenchyma formation induced by hypoxia. Plant Physiol 112:463–472
Hoffmann-Benning S, Kende H (1992) On the role of abscisic acid and gibberellin in the regulation of growth in rice. Plant Physiol 99:1156–1161

Horton RF (1992) Submergence-promoted growth of petioles of *Ranunculus pygmaeus* Wahl. Aquat Bot 44:23–30
Horton RF, Samarakoon AB (1982) Petiole growth in the celery-leaved crowfoot (*Ranunculus sceleratus* L.): effects of auxin-transport inhibitors. Aquat Bot 13:97–104
Hoson T, Masuda Y, Pilet PE (1992) Auxin content in air and water grown rice coleoptiles. J Plant Physiol 139:685–689
Hoson T, Masuda Y, Pilet PE (1993) Abscisic acid content in air- and water-grown rice coleoptiles. J Plant Physiol 142:593–596
Hwang S-Y, Van Toai TT (1991) Abscisic acid induces anaerobiosis tolerance in corn. Plant Physiol 97:593–597
Jackson M (1997) Hormones from roots as signals for the shoots of stressed plants. Trends Plant Sci 2:22–28
Jackson MB (2002) Long-distance signalling from roots to shoots assessed: the flooding story. J Exp Bot 53:175–181
Jackson MB (2008) Ethylene-promoted elongation: an adaptation to submergence stress. Ann Bot 101:229–248
Jackson MB, Campbell DJ (1976) Waterlogging and petiole epinasty in tomato: the role of ethylene and low oxygen. New Phytol 76:21–29
Jackson MB, Fenning TM, Drew MC, Saker LR (1985a) Stimulation of ethylene production and gas-space (aerenchyma) formation in adventitious roots of *Zea mays*. Planta 165:486–492
Jackson MB, Fenning TM, Jenkins W (1985b) Aerenchyma (gas-space) formation in adventitious roots of rice (*Oryza sativa* L.) is not controlled by ethylene or small partial pressures of oxygen. J Exp Bot 36:1566–1572
Jackson MB, Waters I, Setter T, Greenway H (1987) Injury to rice plants caused by complete submergence; a contribution by ethylerie (ethene). J Exp Bot 38:1826–1838
Kato-Noguchi H (2000a) Abscisic acid and hypoxic induction of anoxia tolerance in roots of lettuce seedlings. J Exp Bot 51:1939–1944
Kato-Noguchi H (2000b) Effects of plant hormones on the activity of alcohol dehydrogenase in lettuce seedlings. J Plant Physiol 157:223–225
Katsura N, Suge H (1979) Does ethylene induce elongation of the rice coleoptile through auxin? Plant Cell Physiol 20:1147–1150
Kawase M (1972) Effects of flooding on ethylene concentration in horticultural plants. J Am Soc Hortic Sci 97:584–588
Kende H (1993) Ethylene biosynthesis. Annu Rev Plant Physiol Plant Mol Biol 44:283–307
Kende H, van der Knaap E, Cho H-T (1998) Deepwater rice: A model plant to study stem elongation. Plant Physiol 118:1105–1110
Kennedy RA, Rumpho ME, Fox TC (1992) Anaerobic metabolism in plants. Plant Physiol 100:1–6
Kim JH, Cho HT, Kende H (2000) α-Expansins in the semiaquatic ferns *Marsilea quadrifolia* and *Regnellidium diphyllum*: evolutionary aspects and physiological role in rachis elongation. Planta 212:85–92
Ku HS, Suge H, Rappaport L, Pratt HK (1970) Stimulation of rice coleoptile growth by ethylene. Planta 90:333–339
Laan P, Berrefoets MJ, Lythe S, Armstrong W, Blom CWPM (1989) Root morphology and aerenchyma formation as indicators of the flood-tolerance of *Rumex* species. J Ecol 11:693–703
Lee Y, Kende H (2001) Expression of beta-expansins is correlated with internodal elongation in deepwater rice. Plant Physiol 127:645–654
Lee Y, Kende H (2002) Expression of α-expansin and expansin-like genes in deepwater rice. Plant Physiol 130:1396–1405
Lee TM, Lur H, Shieng Y, Chu C (1994) Levels of abscisic acid in anoxia- or ethylene-treated rice (*Oryza sativa* L.) seedlings. Plant Sci 95:125–131

Lee TM, Shieng Y, Chou CH (1996) Abscisic acid inhibits shoot elongation of *Scripus mucronatus*. Physiol Pant 97:1–4
Lee Y, Choi D, Kende H (2001) Expansins: ever-expanding numbers and functions. Curr Opin Plant Biol 4:527–532
Lee Y, Jung JW, Kim SK, Hwang YS, Lee JS, Kim SH (2008) Ethylene-induced opposite redistributions of calcium and auxin are essential components in the development of tomato petiolar epinastic curvature. Plant Physiol Biochem 46:685–693
Lorbiecke R, Sauter M (1999) Adventitious root growth and cell-cycle induction in deepwater rice. Plant Physiol 119:21–29
Loreti E, Yamaguchi J, Alpi A, Perata P (2003) Gibberellins are not required for rice germination under anoxia. Plant Soil 253:137–143
Mapelli S, Locatelli F (1995) The relation of rice coleoptiles, auxin-binding protein, and protein synthesis to anoxia and indoleacetic acid. Russ J Plant Physiol 42:624–629
Mapelli S, Rocchi P, Bertani A (1986) ABA and IAA in rice seedlings under anaerobic conditions. Biol Plant 28:57–61
Mapelli S, Lombardi L, Bertani A (1993) The effect of growth substances on rice germination and growth under anoxia. Plant Physiol (Life Sci Adv) 12:9–15
Mapelli S, Locatelli F, Bertani A (1995) Effect of anaerobic environment on germination and growth of rice and wheat: endogenous levels of ABA and IAA. Bulg J Plant Physiol 21:33–41
Mekhedov SL, Kende H (1996) Submergence enhances expression of a gene encoding 1-aminocyclopropane-1-carboxylate oxidase in deepwater rice. Plant Cell Physiol 37:531–537
Menyajlo LN, Shulgina GG (1978) The effect of bog reclamation on the hormone metabolism of pine. Sov Plant Physiol 25:123–127
Metraux JP, Kende H (1983) The role of ethylene in the growth response of submerged deep water rice. Plant Physiol 72:444–446
Metraux JP, Kende H (1984) The cellular basis of elongation response in submerged water rice. Planta 160:73–77
Miyoshi K, Sato T (1997) The effect of kinetin and gibberellin on the germination of dehusked seeds of *Indica* and *Japonica* rice (*Oryza sativa* L.) under anaerobic and aerobic conditions. Ann Bot 80:479–483
Mühlenbock P, Plaszczyca M, Plaszczyca M, Mellerowicz E, Karpinski S (2007) Lysigenous aerenchyma formation in Arabidopsis is controlled by LESION SIMULATING DISEASE1. Plant Cell 19:3819–3830
Musgrave A, Walters J (1974) Ethylene and buoyancy control rachis elongation of semi-aquatic fern *Regnillidium diphyllum*. Planta 121:51–56
Musgrave A, Jackson MB, Ling E (1972) *Callitriche* stem elongation is controlled by ethylene and gibberellin. Nat New Biol 238:93–96
Neuman DS, Smit BA (1991) The influence of leaf water status and ABA on leaf growth and stomata of *Phaseolus* seedling with hypoxic root. J Exp Bot 42:1499–1506
Neuman DS, Rood SB, Smit BA (1990) Does cytokinin transport from root-to-shoot in the xylem sap regulate leaf responses to root hypoxia? J Exp Bot 41:1325–1333
Olson DC, Oetiker JH, Yang SF (1995) Analysis of *Le-ACS*, a 1-aminocyclopropane-1-carboxylic acid synthase gene expressed during flooding in the roots of tomato plants. J Biol Chem 270:14056–14061
Ookawara R, Satoh S, Yoshioka T, Ishizawa K (2005) Expression of alpha-expansin and xyloglucan endotransglucosylase/hydrolase genes associated with shoot elongation enhanced by anoxia, ethylene and carbon dioxide in arrowhead (*Sagittaria pygmaea* Miq.) tubers. Ann Bot 96:693–702
Pearce DME, Jackson MB (1991) Comparison of growth responses of barnyard grass (*Echinochloa oryzoides*) and rice (*Oryza sativa*) to submergence, ethylene, carbon dioxide and oxygen shortage. Ann Bot 68:201–209

Pearce DME, Hall KC, Jackson MB (1992) The effects of oxygen, carbon dioxide and ethylene on ethylene biosynthesis in relation to shoot extension in seedlings of rice (*Oryza sativa*) and barnyard grass (*Echinochloa oryzoides*). Ann Bot 69:441–447

Peng H-P, Lin T-Y, Wang N-N, Shih M-C (2005) Differential expression of genes encoding 1-aminocyclopropane-1-carboxylate synthase in *Arabidopsis* during hypoxia. Plant Mol Biol 58:15–25

Perata P, Voesenek LACJ (2007) Submergence tolerance in rice requires *Sub1A*, an ethylene-response-factor-like gene. Trends Plant Sci 12:43–46

Raskin I, Kende H (1983) Regulation of growth in rice seedlings. J Plant Growth Regul 2:193–203

Raskin I, Kende H (1984) Role of gibberellin in the growth response of submerged deep water rice. Plant Physiol 76:947–950

Ridge I (1987) Ethylene and growth control in amphibious plants. In: Crawford RMM (ed) Plant life in aquatic and amphibious habitats. Blackwell Scientific Publications, Oxford, pp 53–76

Ridge I, Osborne DJ (1989) Wall extensibility, wall pH and tissue osmolality—significance for auxin and ethylene-enhanced petiole growth in semi-aquatic plants. Plant Cell Environ 12:383–393

Rieu I, Cristescu SM, Harren FJM, Huibers W, Voesenek LACJ, Mariani C, Vriezen WH (2005) *Rp-ACS1*, a flooding-induced 1-aminocyclopropane-1-carboxylate synthase gene of *Rumex palustris*, is involved in rhythmic ethylene production. J Exp Bot 56:841–849

Rijnders JGHM, Barendse GWM, Blom CWPM, Voesenek LACJ (1996) The contrasting role of auxin in submergence-induced petiole elongation in two species from frequently flooded wetlands. Physiol Plant 96:467–473

Rijnders JGHM, Yang Y-Y, Kamiya Y, Takahashi N, Barendse GWM, Blom CWPM, Voesenek LACJ (1997) Ethylene enhances gibberellin levels and petiole sensitivity in flooding-tolerant *Rumex palustris* but not in flooding-intolerant *R. acetosa*. Planta 203:20–25

Rudolf W, Knacker T, Schaub H (1987) The effect of low oxygen concentration on the cytokinin content of the C_4 plant *Amaranthus paniculatus* L. Biochem Physiol Pflanz 182:203–211

Saab IN, Sachs MM (1996) A flooding-induced xyloglucan *endo*-transglycosylase homolog in maize is responsive to ethylene and associated with aerenchyma. Plant Physiol 112:385–391

Saika H, Okamoto M, Miyoshi K, Kushiro T, Shinoda S, Jikumaru Y, Fujimoto M, Arikawa T, Takahashi H, Ando M, Arimura S, Miyao A, Hirochika H, Kamiya Y, Tsutsumi N, Nambara E, Nakazono M (2007) Ethylene promotes submergence-induced expression of *OsABA8ox1*, a gene that encodes ABA 8′-hydroxylase in rice. Plant Cell Physiol 48:287–298

Samarakoon AB, Horton RF (1983) Petiole growth in *Ranunculus sceleratus*: the role of growth regulators and the leaf blade. Can J Bot 61:3326–3331

Samarakoon AB, Horton RF (1984) Petiole growth in *Ranunculus sceleratus* L.: ethylene synthesis and submergence. Ann Bot 54:263–270

Samarakoon AB, Woodrow L, Horton RF (1985) Ethylene- and submergence-promoted growth in *Ranunculus sceleratus* L. petioles: the effect of cobalt ions. Aquat Bot 21:33–41

Satler SO, Kende H (1985) Ethylene and the growth of rice seedlings. Plant Physiol 79:194–198

Sauter M (2000) Rice in deep water: "How to take heed against a sea of troubles". Naturwissenschaften 87:289–303

Sauter M, Kende H (1992) Gibberellin-induced growth and regulation of the cell division cycle in deepwater rice. Planta 188:362–368

Sauter M, Mekhedov SL, Kende H (1995) Gibberellin promotes histone H1 kinase activity and the expression of *cdc2* and cyclin genes during the induction of rapid growth in deepwater rice internodes. Plant J 7:623–632

Smit BA, Neuman DS, Stachowiak ML (1990) Root hypoxia reduces leaf growth. Role of factors in the transpiration stream. Plant Physiol 92:1021–1028

Smits AJM, Schmitz GHW, Vandervelde G, Voesenek LACJ (1995) Influence of ethanol and ethylene on the seed germination on three nymphaeoid water plants. Freshwater Biol 34:39–46

Smulders MJM, Horton RF (1991) Ethylene promotes elongation growth and auxin promotes radial growth in *Ranunculus sceleratus* petioles. Plant Physiol 96:806–811
Suge H (1985) Ethylene and gibberellin: regulation of internodal elongation and nodal root development in floating rice. Plant Cell Physiol 26:607–614
Summers JE, Jackson MB (1996) Anaerobic promotion of stem extension in *Potamogeton pectinatus*. Roles for carbone dioxide, acidification and hormones. Physiol Plant 96:615–622
Summers JE, Voesenek LACJ, Blom CWPM, Lewis MJ, Jackson MB (1996) *Potamogeton pectinatus* is constitutively incapable of synthesizing ethylene and lacks 1-aminocyclopropane-1-carboxylic acid oxidase. Plant Physiol 111:901–908
van der Knaap E, Jagoueix S, Kende H (1997) Expression of an ortholog of replication protein A1 (RPA1) is induced by gibberellin in deepwater rice. Proc Natl Acad Sci USA 94:9979–9983
van der Knaap E, Song WY, Ruan DL, Sauter M, Ronald PC, Kende H (1999) Expression of a gibberellin-induced leucine-rich repeat receptor-like protein kinase in deepwater rice and its interaction with kinase-associated protein phosphatase. Plant Physiol 120:559–569
van der Knaap E, Kim JH, Kende H (2000) A novel gibberellin-induced gene from rice and its potential regulatory role in stem growth. Plant Physiol 122:695–704
Van Der Straeten D, Zhou Z, Prinsen E, Van Onckelen HA, Van Montagu MC (2001) A comparative molecular-physiological study of submergence response in lowland and deepwater rice. Plant Physiol 125:955–968
Vartapetian BB, Jackson BM (1997) Plant adaptations to anaerobic stress. Ann Bot 79:3–20
Vartapetian BB, Andreeva IN, Kozlova GI (1976) The resistance to anoxia and the mitochondrial fine structure of rice seedlings. Protoplasma 88:215–224
Vidoz ML, Loreti E, Mensuali A, Alpi A, Perata P (2010) Hormonal interplay during adventitious root formation in flooded tomato plants. Plant J 63:551–562
Visser EJW, Heijink CJ, van Hout KJGM, Voesenek LACJ, Barendse GWM, Blom CWPM (1995) Regulatory role of auxin in adventitious root formation in two spices of *Rumex*, differing in their sensitivity to waterlogging. Physiol Plant 93:116–122
Visser EJW, Bogemann GM, Blom CWPM, Voesenek LACJ (1996a) Ethylene accumulation in waterlogged *Rumex* plants promotes formation of adventitious roots. J Exp Bot 47:403–410
Visser EJW, Cohen JD, Barendse GWM, Blom CWPM, Voesenek LACJ (1996b) An ethylene-mediated increase in sensitivity to auxin induces adventitious root formation in flooded *Rumex palustris* Sm. Plant Physiol 112:1687–1692
Visser EJW, Nabben RHM, Blom CWPM, Voesenek LACJ (1997) Elongation by primary lateral roots and adventitious roots during conditions of hypoxia and high ethylene concentrations. Plant Cell Environ 20:647–653
Voesenek LACJ, Harren FJM, Bogemann GM, Blom CWPM, Reuss J (1990a) Ethylene production and petiole growth in *Rumex* plants induced by soil waterlogging: the application of a continuous flow system and a laser driven intracavity photoacoustic detection system. Plant Physiol 94:1071–1077
Voesenek LACJ, Perik PJM, Blom CWPM, Sassen MMA (1990b) Petiole elongation in *Rumex* species during submergence and ethylene exposure: The elongative contributions of cell division and cell expansion. J Plant Growth Regul 9:13–17
Voesenek LACJ, Banga M, Thier RH, Mudde CM, Harren FJM, Barendse GWM, Blom CWPM (1993) Submergence-induced ethylene synthesis entrapment and growth in two plant species with contrasting flooding resistances. Plant Physiol 103:783–789
Voesenek LACJ, Vriezen WH, Smekens MJE, Huitink FHM, Bogemann GM, Blom CWPM (1997) Ethylene sensitivity and response sensor expression in petioles of *Rumex* species at low O_2 and high CO_2 concentrations. Physiol Plant 114:1501–1509
Vreeburg RA, Benschop JJ, Peeters AJM, Colmer TD, Ammerlaan AH, Staal M, Elzenga TM, Staals RH, Darley CP, McQueen-Mason SJ, Voesenek LACJ (2005) Ethylene regulates fast apoplastic acidification and expansin A transcription during submergence-induced petiole elongation in *Rumex palustris*. Plant J 43:597–610

Vriezen WH, van Rijn CPE, Voesenek LACJ, Mariani C (1997) A homolog of the *Arabidopsis thaliana ERS* gene is actively regulated in *Rumex palustris* upon flooding. Plant J 11:1265–1271

Walters J, Osborne DJ (1979) Ethylene and auxin-induced cell growth in relation to auxin transport and metabolism and ethylene production in the semi-aquatic plant, *Regnellidium diphyllum*. Planta 146:309–317

Wample RL, Reid DM (1979) The role of endogenous auxins and ethylene in the formation of adventitious roots and hypocotyl hypertrophy in flooded sunflower plants (*Helianthus annus* L.). Physiol Plant 45:219–226

Xu K, Xu X, Fukao T, Canlas P, Maghirang-Rodriguez R, Heuer S, Ismail AM, Bailey-Serres J, Ronald PC, Mackill DJ (2006) Sub1A is an ethylene-response-factor-like gene that confers submergence tolerance to rice. Nature 442:705–708

Yemelyanov VV, Chirkova TV (1996) Free forms of phytohormones in plants with different tolerance to the lack of oxygen under aeration and anaerobiosis. Bulletin of St. Petersburg University. Ser.3. (Biology) Iss. 2:73-81 [in Russian]

Zarembinski TI, Theologis A (1997) Expression characteristics of *Os-ACS1* and *Os-ACS2*, two members of the 1-aminocyclopropane-1-carboxylate synthase gene family in rice (*Oryza sativa* L. cv. Habiganj Aman II) during partial submergence. Plant Mol Biol 33:71–77

Zhang J, Davies WJ (1987) ABA in roots and leaves of flooded pea plants. J Exp Bot 38:649–659

Chapter 11
Stress Hormone Levels Associated with Drought Tolerance vs. Sensitivity in Sunflower (*Helianthus annuus* L.)

Cristian Fernández, Sergio Alemano, Ana Vigliocco, Andrea Andrade, and Guillermina Abdala

Abstract Six inbred lines (B59, B67, B71, R432, R419, and HAR4) of sunflower (*Helianthus annuus* L.) were evaluated in field and laboratory experiments under drought vs. irrigation. In field studies, relative seed yield per hectare and oil yield per hectare were reduced under drought in B59, B67, and R419, but not in R432, B71, or HAR4. In lab studies, germination percentage was reduced under 200 and 400 mM mannitol treatment (which simulates drought) for B59 and under 400 mM mannitol for R432, B71, and HAR4. B59 and B71 were used as typical drought-sensitive and drought-tolerant lines, respectively, for subsequent experiments. Levels of the phytohormones jasmonates (JAs), abscisic acid (ABA), and ABA catabolites were evaluated in dry and germinated seeds from B59 and B71 parent plants grown under drought and irrigation. For dry seeds from plants grown under drought, ABA was the major compound accumulated in B71, whereas 12-OH-JA was the major compound in B59. Germinated seeds of both lines, compared to dry seeds, showed increased 12-oxophytodienoic (OPDA) and decreased ABA. Our results indicate that soil moisture conditions under which parent plants grow affect hormonal content of seeds produced, and JAs and ABA levels during germination are variable. F_3 seedling families obtained by crossing R432 (drought-tolerant) and A59 (drought-sensitive) lines were assayed for germination percentage and endogenous levels of salicylic acid (SA), JA, and ABA following drought treatment (400 mM mannitol). Germination percentage showed a typical segregation pattern of quantitative inheritance to drought tolerance in the phenological stage of seedling. Levels of SA and ABA under drought compared to control condition increased in F_3 tolerant families but decreased in F_3 sensitive families. JA levels changed under drought condition, but the direction of change was not consistent within

C. Fernández • S. Alemano • A. Vigliocco • A. Andrade • G. Abdala (✉)
Departamento de Ciencias Naturales, Facultad de Ciencias Exactas, Físico-Químicas y Naturales, Universidad Nacional de Río Cuarto, Río Cuarto, Argentina
e-mail: gabdala@exa.unrc.edu.ar

N.A. Khan et al. (eds.), *Phytohormones and Abiotic Stress Tolerance in Plants*,
DOI 10.1007/978-3-642-25829-9_11,

tolerant or sensitive families. Our results provide important information for strategies to maintain or increase yield of sunflower crops under drought conditions.

11.1 Introduction

Environmental stress factors are the primary cause of crop failure, causing average yield losses of over 60% for major crops worldwide (Bray et al. 2000). Among these factors, water deficit (drought) probably has the greatest limiting effect on crop quality and productivity (Roche et al. 2009). A drop in water potential induces a variety of metabolic, morphological, and/or physiological responses, including generation of reactive oxygen species (Papadakis and Roubelakis-Angelakis 2005), accumulation of osmotically active solutes (Sánchez-Díaz et al. 2008), changes in endogenous levels of plant hormones (phytohormones) (Perales et al. 2005; Seki et al. 2007; Dobra et al. 2010), altered expression of stress-responsive genes (Xiong et al. 2002; Yamaguchi-Shinozaki and Shinozaki 2005; Huang et al. 2008), and reduced vegetative growth (Mahajan and Tuteja 2005). Some of these responses are triggered directly by altered water status in plant tissues, while others are mediated by phytohormones (Chaves et al. 2003). In particular, the hormones abscisic acid (ABA), jasmonates (JAs), and salicylic acid (SA) are key components of a complex signal-transduction network that coordinates growth and development with plant responses to environmental factors (Agrawal et al. 2002; Jiang and Zhang 2002; Fujita et al. 2006; Szalai et al. 2010).

Plant responses to drought vary depending on intensity and duration of stress, plant species, and development stage (Chaves et al. 2003). Seed germination is the earliest and most sensitive stage in the plant life cycle (Ashraf and Mehmood 1990), and establishment of seedlings is highly susceptible to drought and other environmental stresses (Albuquerque and de Carvalho 2003). Sunflower (*Helianthus annuus* L.) is better able to tolerate drought than many agricultural crop species and is often cultivated in arid regions. However, like other oil crops, sunflower is sensitive to water deficit at the germination stage. Poor weather and soil can cause unsynchronized crop establishment (Mwale et al. 2003), and drought stress can reduce yield and seed quality (Roche et al. 2009).

Development of new genotypes with enhanced tolerance to drought stress is an important strategy for expanding the agricultural area for sunflower crop planting. Selection on the basis of seedling traits is a useful technique for evaluating a large number of genotypes for drought tolerance (Tomar and Kumar 2004; Basal et al. 2005; Longenberger et al. 2006). Certain traits selected for improvement of general yield also give yield increases in dry environments. Recently developed new technologies in plant cell biology and molecular biology provide powerful tools to complement traditional methods of crop improvement (Rauf 2008). In particular, Marker-assisted selection (MAS) is useful for segregation and association mapping studies of germplasm to identify useful alleles in cultivated strains and their wild-type relatives. Association mapping is intrinsically a more powerful tool than "classical" genetic linkage mapping because it scrutinizes the results of thousands of generations of recombination and selection (Syvänen 2005). However, most data accumulated so far on drought tolerance are based on segregation mapping and QTL (quantitative trait loci) analysis (Cattivelli et al. 2008).

11.2 Role of Phytohormones in Response to Drought

Plants are subject to a variety of abiotic and biotic stress factors, including attack by pathogens and herbivores, and levels of light, water, temperature, nutrients, or salts that are too low or too high. Plants, because they are not able to move, depend on a repertoire of rapid, complex, and highly adapted responses to survive and grow in spite of these environmental stress factors. Perception of stress signals often triggers biosynthesis of signaling molecules, including a variety of phytohormones, which not only regulate developmental processes but are also involved in signaling networks that mediate plant stress responses. The importance of SA, JAs, and ABA as primary signals in regulation of plant responses is well established (Bari and Jones 2009; Pieterse et al. 2009). These hormones are part of a signal-transduction network triggering a cascade of events responsible for physiological adaptation of plants to stress. The outcome of defensive responses depends on the composition and kinetics of the hormones produced (De Vos et al. 2005; Mur et al. 2006; Koornneef et al. 2008; Leon-Reyes et al. 2010).

11.3 Salicylic Acid

Salicylic acid (SA) is well known as a key signaling molecule in induction of plant defense mechanisms (Klessing et al. 2000; Shah 2003), including responses to a variety of pathogenic infections (Singh et al. 2004; Pasquer et al. 2005; Makandar et al. 2006) and systemic acquired resistance (SAR) (Vlot et al. 2008). Increasing evidence in recent years shows that SA elicits defense mechanisms in response to abiotic stresses such as excessive levels of heavy metals (Krantev et al. 2008; Zhou et al. 2009), salts (El-Tayeb 2005; Szepesi et al. 2009), heat (Larkindale and Huang 2004), cold (Wang et al. 2006), and UV radiation or ozone (Ervin et al. 2004) and salt (Nazar et al. 2011; Syeed et al. 2011). SA also influences physiological processes such as seed germination, seedling growth, fruit ripening, flowering, ion uptake and transport, photosynthesis rate, stomatal conductance, and chloroplast biogenesis (Fariduddin et al. 2003; Khodary 2004; Shakirova 2007).

There are several routes for SA biosynthesis (Shah 2003). (1) SA can be synthesized from phenylalanine via cinnamic acid. Decarboxylation of the side chain of cinnamic acid generates benzoic acid, which then undergoes hydroxylation at the C-2 position to form SA (Yalpani et al. 1993; Ribnicky et al. 1998). (2) Cinnamic acid can undergo 2-hydroxylation to o-coumaric acid, which is then decarboxylated to give SA (Alibert and Ranjeva 1971, 1972). (3) In chloroplasts of *Arabidopsis* (Brassicaceae), SA can be synthesized from chorismate via isochorismate (Wildermuth 2006; Mustafa et al. 2009). (4) SA can be conjugated with a variety of molecules by either glycosylation or esterification (Popova et al. 1997) and is metabolized to 2,3-dihydrobenzoic acid or 2,5-dihydrobenzoic acid in leaves of *Astilbe chinensis* and *Lycopersicon esculentum* (Billek and Schmook 1967).

Because it enhances tolerance of several abiotic stresses, SA is a promising compound for reducing susceptibility of crops to stress (Horváth et al. 2007). Several methods of SA application (soaking the seed before sowing, adding to hydroponic solution, irrigating, spraying with SA solution) have been reported to mitigate damaging effects of abiotic stresses (Szalai et al. 2010). Such mitigation depends on concentration applied, mode of application, plant developmental stage, oxidative balance of cells, and other factors.

SA helps promote tolerance of plants to drought conditions. Plants raised from seeds soaked in acetyl-SA solution displayed enhanced tolerance to drought and dry matter accumulation (Hamada 1998; Hamada and Al-Hakimi 2001). In tomato and bean plants, drought tolerance was enhanced by exogenous addition of SA at low concentrations, but not at high concentrations (Senaratna et al. 2000).

SA was shown to be involved in promotion of drought-induced leaf senescence in *Salvia officinalis* plants under field conditions (Abreu and Munne-Bosch 2008). SA-treated wheat seedlings under drought stress, in comparison to nontreated controls, displayed higher moisture content, dry matter accumulation, ribulose-1,5-bisphosphate carboxylase oxygenase (RuBisCO) activity, superoxide dismutase (SOD) activity, and total chlorophyll content. SA treatment also protected nitrate reductase activity and thereby helped maintain normal levels of various proteins in leaves (Singh and Usha 2003). Exogenously applied SA mitigated the deleterious effects of water deficit on cell membranes of barley plants (Bandurska and Stroiński 2005) and enhanced drought tolerance in cell suspensions from turgid leaves of *Sporobolus stapfianus* (Ghasempour et al. 2001).

Endogenous SA content in plants is altered by the majority of abiotic stresses, indicating its involvement in stress signaling (Horváth et al. 2007).

11.4 Jasmonates

Jasmonic acid (JA) and its cyclic precursors and derivatives, collectively referred to as jasmonates (JAs), constitute a family of bioactive oxylipins that regulate plant responses to environmental and developmental cues (Turner et al. 2002; Devoto and Turner 2003). These signaling molecules affect numerous plant processes, including responses to wounding and abiotic stresses, and defenses against insects (McConn et al. 1997) and necrotrophic pathogens (Thomma et al. 1999). Developmental processes shown to be influenced by JAs in studies of various *Arabidopsis* mutants include root growth, pollen maturation and dehiscence, carbon portioning, ovule development, germination, seedling development, and senescence (Turner et al. 2002; Wasternack and Hause 2002; Andrade et al. 2005; Wasternack 2006; Browse 2009). Plant responses mediated by JAs are associated with extensive reprogramming of gene expression (Reymond et al. 2004; Yan et al. 2007).

The biosynthetic pathway for JA was initially described by Vick and Zimmerman (1983) and reviewed recently by Wasternack and Kombrinck (2010). The substrates are α-linolenic acid (α-LeA; C18:3) or hexadecatrienoic acid (C16:3) released from plastidial galactolipids by phospholipases. Following oxidation by lipoxygenase (LOX) of α-LeA to 13(S)-hydroperoxyoctadecatrienoic acid (13(S)-HPOT), the first committed step of JA biosynthesis is conversion of the LOX product to the allene oxide 12,13(S)-epoxyoctadecatrienoic acid (12,13(S)-EOT) by allene oxide synthase (AOS). This unstable allylic epoxide can decompose into various products, such as α- and γ-ketols, or spontaneously rearrange to racemic 12-oxophytodienoic acid or undergo enzymatic cyclization by allene oxide cyclase (AOC) to optically pure *cis*-(+)-12-oxophytodienoic acid ((9*S*,13*S*)-OPDA), which is the end product of the plastid-localized part of the JA biosynthetic pathway and has the same stereochemical configuration as naturally occurring (+)-7-*iso*-JA.

Translocation of OPDA into peroxisomes, where the subsequent part of the JA biosynthetic pathway occurs, is mediated by the ABC transporter COMATOSE and/or an ion-trapping mechanism (Theodoulou et al. 2005). Reduction of OPDA is catalyzed by a peroxisomal OPDA reductase (OPR) to produce 3-oxo-2(2[Z]-pentenyl)cyclopentane-1-octanoic acid (OPC-8:0). Next, three rounds of β-oxidation catalyzed by acyl-CoA oxidase (ACX), multifunctional protein (MFP), and L-3-ketoacyl-CoA thiolase (KAT) lead to production of jasmonoyl-CoA. From this, a yet-unknown thioesterase releases (+)-7-iso-JA ((3R,7S)-JA), which equilibrates to the more stable (-)-JA ((3R,7R)-JA). Homeostasis among various JA metabolites is a common mechanism in plants to sustain the content of active hormones. The metabolic routes for conversion of JA have been elucidated by identification of the corresponding products in plant tissues: (1) methylation by JA-specific methyltransferase (JAME) (Seo et al. 2001); (2) hydroxylation at C-11 or C-12 of the pentenyl side chain (Sembdner and Parthier 1993), followed by sulfation (Gidda et al. 2003); (3) glucosylation at the carboxylic acid group, leading to jasmonoyl-β-glucose ester; (4) glucosylation of 12-OH-JA to 12-*O*-glucoside (12-*O*-Glc-JA) (Swiatek et al. 2004); (5) conjugation with the ethylene precursor 1-aminocyclopropane-1-carboxylic acid (ACC) (Staswick and Tiryaki 2004); (6) reduction of the keto group of pentenone ring, leading to cucurbic acid (Dathe et al. 1991); (7) conjugation with amino acids such as valine, leucine, tyrosine, and isoleucine (Staswick and Tiryaki 2004; Wasternack 2007; Fonseca et al. 2009; Wasternack and Kombrinck 2010); and (8) decarboxylation to *cis*-jasmone (Koch et al. 1997). Most of these metabolic products exhibit biological activities that are similar but not identical, raising the question whether metabolic conversion of JA plays a regulatory role in JA signaling (Miersch et al. 2008).

In addition to its well-established role in biotic stress responses, JA is involved in mediation of numerous abiotic stress responses. In regard to drought response, treatment of barley leaves with sorbitol or mannitol was found to increase level of endogenous JAs, followed by synthesis of jasmonate-induced proteins (JIPs) (Lehmann et al. 1995). Sorbitol treatment caused enhancement of various octadecanoids and JAs, and this threshold was necessary and sufficient to initiate JA-responsive gene expression (Kramell et al. 2000). Endogenous JA content increased in maize root cells under drought stress (Xin et al. 1997), and JAs elicited betaine accumulation in pear leaves (Gao et al. 2004).

We recently observed differences in basal JAs content, and patterns of response to drought stress, in two populations of *Pinus pinaster* Ait. at different sites and suggested that such differences reflect adaptations to diverse ecological conditions (Pedranzani et al. 2007).

Under drought stress, accumulation of JA increased transcript levels and activities of antioxidant enzymes (Shan and Liang 2010), and exogenous JA or MeJA increased antioxidative ability of plants (Bandurska et al. 2003; Wang 1999). Ascorbate and glutathione metabolism are important components of antioxidant metabolism in plants, and JA was shown to be involved in regulation of such metabolism under ozone stress (Sasaki-Sekimoto et al. 2005). JAs play important signaling roles in drought-induced antioxidant responses, including ascorbate metabolism (Li et al. 1998; Ai et al. 2008).

11.5 Abscisic Acid

Abscisic acid (ABA) influences the ability of plants to withstand abiotic stresses, including drought, cold, and salt stress (Zhu 2002; Finkelstein et al. 2002). Plants are therefore able to control their response to such stresses by modulating endogenous ABA content (Xiong et al. 2002; Zhang et al. 2006; Thameur et al. 2011). This hormone is also involved in regulating expression of many stress-induced genes, providing adaptability to environmental stresses (Bray 2003; Shinozaki and Yamaguchi-Shinozaki 2007; Huang et al. 2008). ABA is also involved in regulating many aspects of plant growth and development, including embryo and seed development, desiccation tolerance and dormancy, vegetative development, leaf senescence, stomatal aperture, and pathogen defense response (Wasilewska et al. 2008; Ton et al. 2009; Fan et al. 2009; Cutler et al. 2010).

The biosynthetic pathway of ABA in higher plants has been established using forward-genetic approaches (Seo and Koshiba 2002; Nambara and Marion-Poll 2005). ABA is synthesized from zeaxanthin, a C_{40} carotenoid, whose conversion to xanthoxin, a C_{15} intermediate, is catalyzed in plastids by zeaxanthin epoxidase (Agrawal et al. 2001; Xiong et al. 2002) and 9-*cis*-epoxycarotenoid dioxygenase (NCED) (Schwartz et al. 1997; Qin and Zeevaart 1999; Iuchi et al. 2001). Xanthoxin is then converted to ABA via abscisic aldehyde in the cytosol. The oxidation of xanthoxin produces abscisic aldehyde, which can be converted to ABA by aldehyde oxidase 3 (AAO3) (Seo et al. 2000). There are two pathways for ABA catabolism: an oxidative pathway and conjugation (Kushiro et al. 2004; Nambara and Marion-Poll 2005). The most common oxidative pathway is initiated by oxidation of 8′-hydroxy ABA (8′-OH ABA), which can reversibly cyclize to phaseic acid (PA) (Zaharia et al. 2005). This compound is then reduced to the dihydrophaseic acid (DPA) as major product, with minor amounts of *epi*-dihydrophaseic acid (*epi*-DPA).

Minor oxidation pathways include formation of 7′-hydroxy ABA (7′-OH ABA) and of 9′-hydroxy ABA (9′-OH ABA). The latter can cyclize reversibly to *neo*phaseic acid (*neo*PA) (Zhou et al. 2004). ABA and hydroxy ABA can also be conjugated with glucose, to form corresponding glucose esters at C-1 (ABA-GE) or glycosides at C-1′ or C-4′ (Zeevaart 1999; Oritani and Kiyota 2003). Isolation and characterization of several new ABA catabolites were reported by Kikuzaki et al. (2004).

ABA action is the one of the most extensively studied topics in abiotic stress response research (Hirayama and Shinozaki 2007; Wasilewska et al. 2008). The earliest reports on effects of ABA in drought stress involved the morphological and physiological aspects. Many investigators have observed increases in endogenous ABA content in woody and herbaceous species under various drought-induced environmental conditions, including high salinity (Hassine and Lutts 2010), freezing (Janowiak et al. 2002), dehydration (Nayyar and Walia 2004), and osmoticity (Hsu et al. 2003). In *Arabidopsis thaliana* seedlings, drought stress promoted both ABA biosynthesis and catabolism, leading to increased levels of ABA and its catabolites PA, DPA, and ABA-GE (Huang et al. 2008). Leaf ABA concentrations were higher in drought-stressed *Laurus azorica* (Seub.) Franco plants than in controls (Sánchez-Díaz et al. 2008). Exogenous application of ABA enhanced drought tolerance of triploid bermuda grass (*Cynodon dactylon*) (Lu et al. 2009).

Although not all studies show direct correlation between stress tolerance and high ABA content, there are several reports that drought-tolerant strains have higher ABA level than susceptible strains (Perales et al. 2005; Veselov et al. 2008; Thameur et al. 2011).

Under drought conditions, ABA plays a role in closing of stomata, which reduces water loss through transpiration (Assmann 2003; Hartung et al. 2005). ABA produced by roots in drying soil is transported to the xylem and regulates stomatal opening (Zhang et al. 1987; Zhang and Davies 1989, 1990a, b). Changes in pH of xylem sap are frequently observed under drought stress; this change may be an important component of root-to-shoot signaling and act synergistically with ABA. In *Helianthus annuus*, *Phaseolus coccineus*, and *Commelina communis*, xylem sap pH becomes more alkaline under drought stress, leading to enhanced stomatal closure and reduced growth (Schachtman and Goodger 2008).

The function of ABA in stomatal closure under drought has been proposed to involve localization of ABA receptors. Two ABA receptors are located inside the cell, but a third is found on the cell surface (Liu et al. 2007), suggesting that plant cells can sense both extracellular and intracellular ABA concentration. The increased stomatal closure that occurs because of increased sap pH under drought suggests that extracellular ABA is sensed by guard cells via receptors on the plasma membrane (Schachtman and Goodger 2008). Hydrogen peroxide (H_2O_2) and nitric oxide (NO) appear to be involved in ABA-induced stomatal closure (Assmann

2003; Desikan et al. 2004; Bright et al. 2006). Increased ABA concentration in response to drought was correlated with increased proline concentration (Ober and Sharp 1994), and ABA was required for enhanced proline accumulation at low water potential (Verslues and Bray 2006; Sánchez-Díaz et al. 2008).

ABA has a dual role in regulation of plant growth (Cheng et al. 2002; Finkelstein et al. 2002). It displays an inhibitory function when accumulated under stress conditions, to promote plant survival through inhibition of stomatal opening and size increase (Zhang et al. 2006). Its promoting function, displayed at low concentration under "normal" conditions, is essential for primary root growth (Sharp et al. 2000; Spollen et al. 2000) and postgermination seedling development (Cheng et al. 2002).

At a molecular level, ABA is involved in numerous changes in gene expression in response to drought (Zhang et al. 2006). Signaling under drought stress occurs through an ABA-dependent and also an ABA-independent pathway (Shinozaki and Yamaguchi-Shinozaki 2000; Riera et al. 2005; Yamaguchi-Shinozaki and Shinozaki 2005). ABA-responsive element (ABRE) is a major *cis*-element in the ABA-dependent gene expression pathway, e.g., two ABRE motifs are important *cis*-acting elements controlling ABA-dependent expression of the *Arabidopsis RD29B* gene. Two basic leucine "zipper" (bZIP) transcription factors, AREB/ABF, can bind to ABRE, thereby activating ABA-dependent gene expression (Uno et al. 2000). The drought-inducible gene *RD22* is mediated by ABA and requires protein biosynthesis for its ABA-dependent expression. Another drought-inducible gene, *RD26*, was identified, whose expression is induced by drought, high salinity, and ABA or JA treatment (Shinozaki and Yamaguchi-Shinozaki 2007). Some drought-inducible genes may be regulated through both ABA-dependent and ABA-independent pathways. For example, transcriptional activation of some stress responsive genes is well understood, owing to studies on the RD29A/COR78/LTI78 (responsive to dehydration/cold-regulated/low-temperature-induced) gene. The promoter of this gene contains both an ABRE (abscisic acid responsive element) and a DRE/CRT (dehydrationresponsive element/ C-repeat). ABRE and DRE/CRT are cis-acting elements that function in abscisic acid (ABA)-dependent and ABA-independent gene expression in response to stress, respectively (Yamaguchi-Shinozaki and Shinozaki 1994).

11.6 Characterization of Drought Tolerance in Sunflower Inbred Lines

11.6.1 Evaluation of Agronomic Parameters

Assessment of comparative degree of drought tolerance of different genotypes is crucial. Identification of tolerant vs. susceptible cultivars is typically based on a few physiological measures related to drought response (Cattivelli et al. 2008). In studies evaluating a large number of sunflower genotypes with diverse origins fact, high variability was observed in traits related to water status, e.g., osmotic adjustment, root characteristics, gas exchange parameters, seedling traits, and drought susceptibility index (Chimenti et al. 2002; Lambrides et al. 2004; Kiani et al. 2007; Rauf and Sadaqat 2007, 2008).

We evaluated drought tolerance at germination and seedling growth stages in five inbred sunflower (*H. annuus*) lines from the Experimental Station INTA-Manfredi (B59, B67, B71, R419, and R432) and one line from the US Department of Agriculture (Fargo, ND) (HAR4), in field and laboratory experiments.

In field experiments, plants of the six lines were grown under drought and irrigation conditions, using a split-plot experimental design with complete randomization and two replications. Drought and irrigation were applied once plants reached the V4 stage of development (fourth pair of leaves). For drought, soil was covered with polypropylene until harvest. For irrigation, plants were watered when soil moisture reached 60% of field capacity. Parameters used to evaluate differential responses to drought were plant height, weight of 1,000 seeds, number of seeds per head, oil heads per hectare (ha), and seed yield per hectare. Significant differences for plants grown under drought vs. irrigation were observed only for oil heads per hectare and seed yield per hectare. These parameters were therefore used to compare responses to drought. Lines B59, R419, and B67 showed reduction of seed yield per hectare under drought, whereas R432, B71, and HAR4 lines did not (Fig. 11.1). Oil yield per hectare showed a reduction under drought in B59, R419 and B67, but not in B71 (Fig. 11.2). No differences were observed for drought vs. irrigation in plant height, 1,000 seed weight, or number of seed per head.

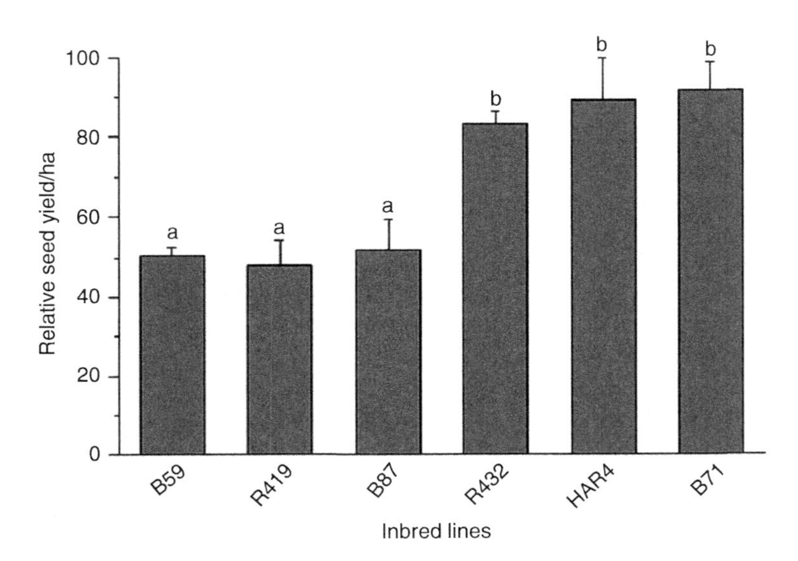

Fig. 11.1 Relative seed yield per hectare for inbred sunflower lines (see text) grown under drought vs. irrigation, in field studies, during 2003–2004 and 2004–2005. Data are means of five replicates. Values with the same letter are not significantly different ($P \leq 0.05$)

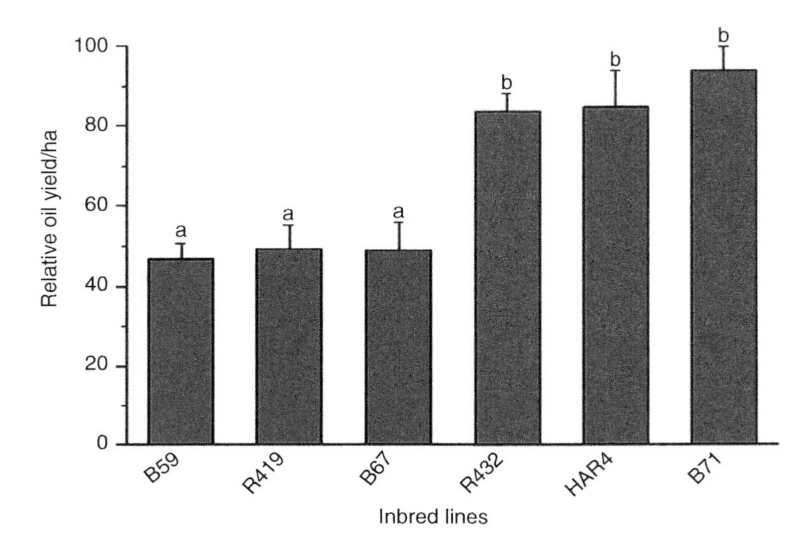

Fig. 11.2 Relative oil yield per hectare for lines grown under drought vs. irrigation, in field studies, during 2003–2004 and 2004–2005. Data are means of five replicates. Values with the same letter are not significantly different ($P \leq 0.05$)

11.6.2 Evaluation of Physiological Parameters

Laboratory studies were conducted using lines B59, R432, HAR4, and B71. For germination experiments, 50 seeds were placed in pots containing sand, in a controlled environmental chamber with a cycle of 16 h light/28°C/60% relative humidity, and 8 h dark/18°C/70% relative humidity. Moisture content at the time of sowing was 60% of field capacity. At day 5 after sowing, and intervals of 3 days thereafter, seedlings were watered by capillary ascent with half-strength Hoagland solution supplemented with 200 or 400 mM mannitol. Seedlings that received distilled water were used as controls. Germination percentage was determined at days 4 and 11, and dry weight and fresh weight were determined at day 11. The only parameter showing significant difference under drought vs. control was germination percentage. B59 showed substantial reduction in this parameter for both 200 and 400 mM mannitol treatment. R432, B71, and HAR4 showed reduction in this parameter for 400 mM but not 200 mM mannitol treatment (Fig. 11.3). Dry weight and fresh weight did not show significant differences for drought vs. control seedlings. Based on the above findings, we characterized B59 as drought sensitive and R432, HAR4, and B71 as drought tolerant, under field and laboratory conditions.

Several physiological processes, including germination, seedling development, and drought response, are regulated by ABA and JAs (Wasilewska et al. 2008; Sánchez-Díaz et al. 2008; Wasternack 2006; Browse 2009; Shan and Liang 2010). Determination of endogenous levels of these phytohormones in genotypes with different degrees of drought tolerance is therefore useful for selective breeding of drought-tolerant crop varieties.

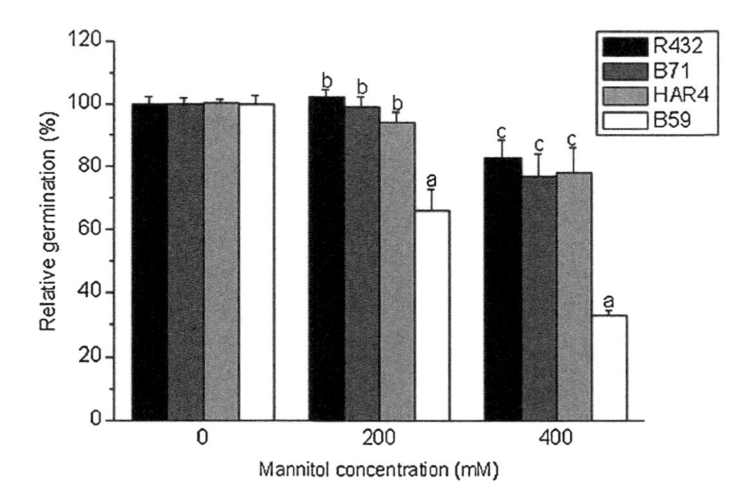

Fig. 11.3 Relative germination percentage of lines under drought condition (mannitol treatment), in laboratory studies. Data are means of five replicates. Values with the same letter are not significantly different ($P \leq 0.05$)

We determined levels of ABA, ABA catabolites (PA; DPA; ABA-GE), JA precursor (OPDA), JA, and jasmonic acid derivatives (11-OH-JA, 12-OH-JA) in dry and germinated seeds of B59 and B71 plants grown at different soil moisture conditions.

ABA content was higher in dry seeds from plants grown under drought than under irrigation (Figs. 11.4 and 11.5), particularly for B71 (Fig. 11.5a). Most of this accumulation was in the pericarp (data not shown). 12-OH-JA was the major phytohormone found in dry seeds from B59 plants grown under drought (Fig. 11.4b), and may interact with ABA as part of drought response in this line. These findings are consistent with previous reports that soil moisture level during plant growth affects hormonal content of seeds (Benech Arnold et al. 1991; Amzallag et al. 1998).

For both B59 and B71, there was a substantial increase of OPDA during the period of water uptake in germinated seeds, compared to dry seeds, from plants grown under both drought and irrigation (Figs. 11.4 and 11.5). OPDA has been shown to function as a signaling molecule in various JA-independent processes, including alkaloid biosynthesis (Memelink et al. 2001), tendril coiling (Stelmach et al. 1998; Blechert et al. 1999), inhibition of apoptosis (programmed cell death) (Reinbothe et al. 2009), and gene expression (Böttcher and Pollmann 2009).

In contrast, ABA content was lower in germinated seed than in dry seeds of both B59 and B71 grown under either soil moisture condition (Figs. 11.4 and 11.5). This is consistent with the well-established role of ABA in inhibition of Raz et al. (2001). In conclusion, the differences observed in ABA profile between dry vs. germinated seeds were independent of genotype or soil moisture conditions under which parent plants were grown.

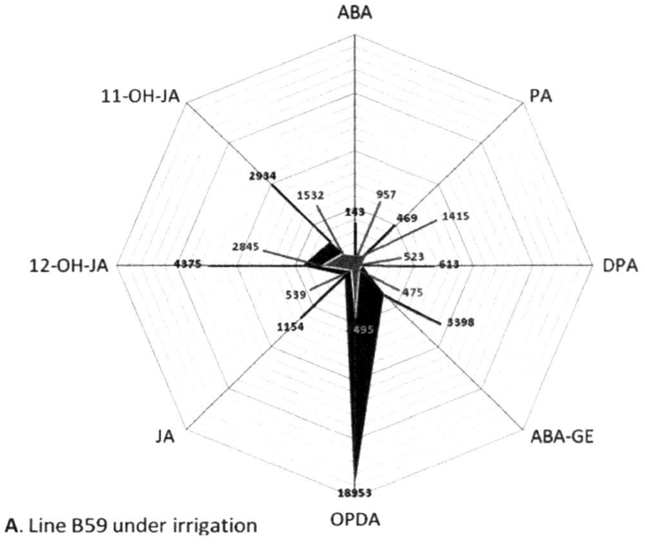

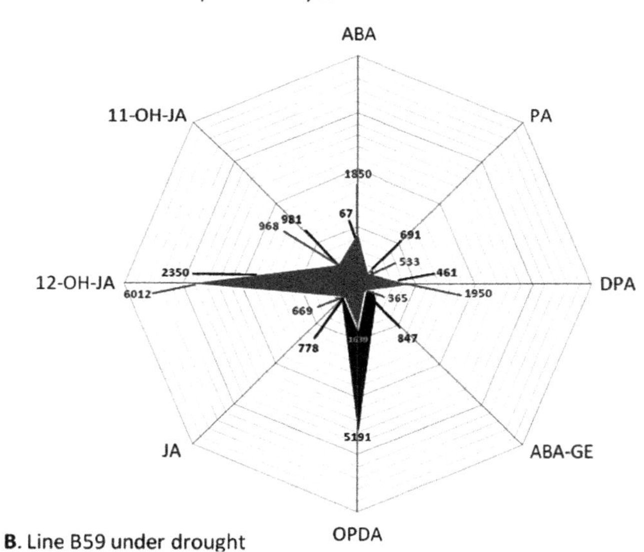

Fig. 11.4 Ratios of JAs, ABA, and their catabolites. (**a**) Seeds from parent plants of drought-sensitive line B59 grown under irrigation. (**b**) Seeds from B59 plants grown under drought. Values are expressed in pmol g^{-1} dry weight. Different scales are shown for convenience

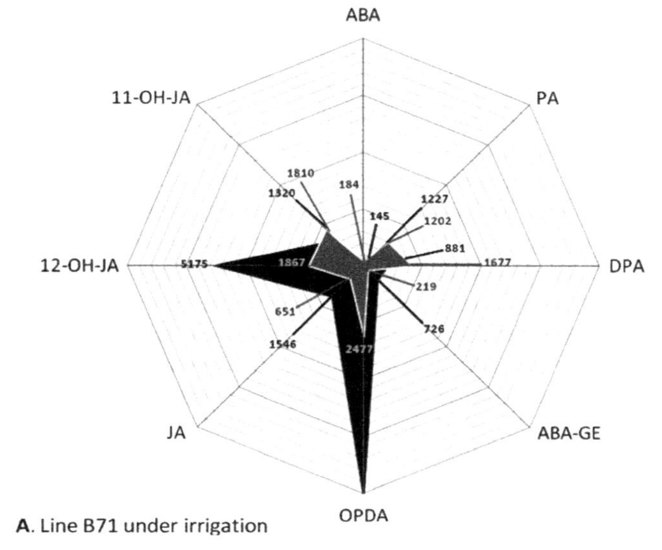

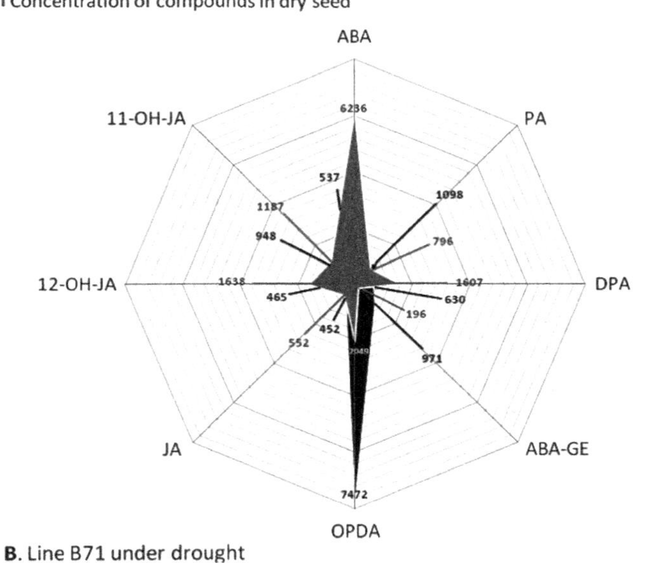

Fig. 11.5 Ratios of JAs, ABA, and their catabolites. (**a**) Seeds from parent plants of drought-tolerant line B71 grown under irrigation. (**b**) Seeds from B71 plants grown under drought. Values are expressed in pmol g^{-1} dry weight

11.7 Evaluation of Segregating Populations

Recently developed new technologies in plant cell biology and molecular biology provide powerful tools to complement traditional methods of crop improvement. Studies using molecular markers have facilitated genetic selection and improvement of agricultural strains through their ability to elucidate genetic variability at the DNA level (Paran et al. 1991). Markers useful for such selection must be tightly linked to the gene of interest (Kumar 1999).

Near-isogenic lines (NILs) have been developed to identify markers which differ in the presence vs. absence of the target gene and a small flanking region (Muehlbauer et al. 1988). Markers can reveal polymorphisms between NILs and the recurrent parents, if the sources of the genes are sufficiently divergent. Application of molecular markers associated with traits of interest leads to better characterization of gene pool diversity and genetic distance between populations. Thus, molecular markers are useful to identify all genotypes in an F_2 segregating population and to select those of interest, resulting in a population highly enriched in desired alleles (Sorrells 1998). Reduction of the number of genotypes moving in each selection cycle, permits smaller-size field experiments and a more efficient selection process.

Seed germination and emergence characteristics are complex traits, typically under the control of multiple genes. If the loci determining genetic determinants of germination are identified, plant breeders can select specifically for alleles contributing to improved germination. Availability of locus-specific molecular markers for germination allows rapid screening of beneficial combination of alleles in breeding programs (Al-Chaarani et al. 2005).

We evaluated sunflower F_3 seedling families for germination percentage and endogenous content of SA, JA, and ABA. These F_3 families were obtained by crossing two inbred lines: R432 (drought tolerant) and A59 (drought sensitive) in the National Agricultural Technology Institute by Dr. Jorge Gieco. The F_1 seedlings obtained from this crossing were selfed to produce F_2 seedlings, and the F_2 families were selfed to obtain F_3 families.

Relative germination percentages of F_3 families were evaluated under drought (400 mM mannitol) compared to control conditions. Differences in this parameter were observed among F_3 families. The 33 F_3 families showed a typical 1:2:1 segregation pattern (Fig.11.6). Families 255, 148, 333, 336, 95, and 300 were characterized as drought sensitive. Families 290, 314, 310, 174, 199, and 224 were characterized as drought tolerant (Fig. 11.7). To avoid loss of variability in the drought tolerance characteristic, families 95 and 300 were included as drought sensitive, and family 314 as drought tolerant.

Some of the tolerant F_3 families showed performance superior to that of parental tolerant (transgressive segregation), a phenomenon which may result from intra-allelic interactions (e.g., additivity, overdominance) or interallelic interactions (e.g., epistasis). The observed variability in germination percentage indicates the importance of testing effects of drought in the germination stage. Drought has been

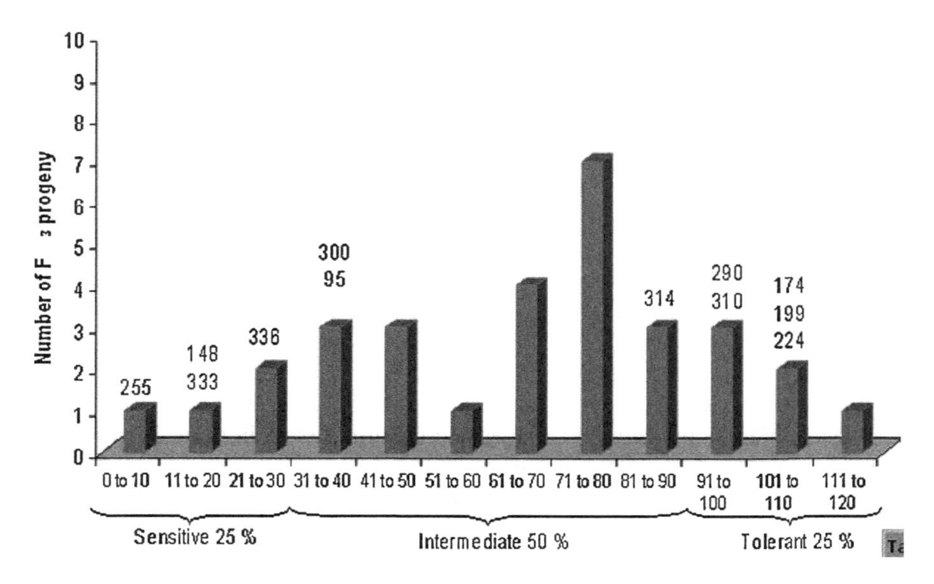

Fig. 11.6 Thirty-three F_3 families (see text) grouped by intervals of relative germination percentage (0–120%)

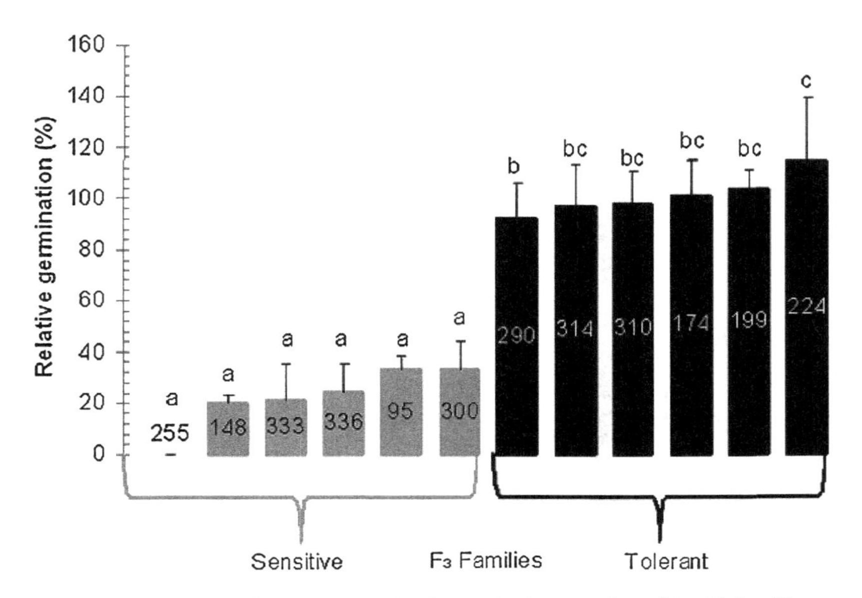

Fig. 11.7 Relative germination percentage in plants of tolerant and sensitive F_3 families grown under drought. Data are means of four replicates. Values with the same letter are not significantly different ($P \leq 0.05$)

shown to produce poor and erratic germination and lack of synchronization in seedling establishment (Mwale et al. 2003; Mustapha et al. 2009).

SA, JA, and ABA levels in the F_3 families were assessed using liquid chromatography–electrospray ionization–tandem mass spectrometry (LC–ESI–MS/MS), which allows quantification of multiple compounds with high sensitivity and selectivity (Chiwocha et al. 2003; Durgbanshi et al. 2005; Feurtado et al. 2007).

Under drought condition, seedlings of most of the drought-tolerant families showed endogenous SA level higher than that under control condition. The difference was significant for families 174 and 310, and 174 showed the highest level (150,161 pmol g^{-1} DW) (Fig. 11.8).

The drought-sensitive families showed SA level under drought condition lower than that under control condition—usually below 40,000 pmol g^{-1} DW. The difference was significant for families 336 and 95 (Fig. 11.9).

These findings indicate that SA plays a significant role in drought response of sunflower, consistent with previous studies. For example, Borsani et al. (2001) found that SA is the main phytohormone involved in osmotic stress response in germination of *Arabidopsis* seedlings. Mikolajczyk et al. (2000) showed that osmotic stress induces high synthesis of SA, which activates protein kinases for appropriate response to these adverse conditions. Oxidative stress generated in *Arabidopsis* plants under drought stress (Trejo and Davies 1991) was alleviated by SA (Sharma et al. 1996; Rao and Davis 1999).

JA levels in tolerant and sensitive F_3 families under drought condition, compared to control condition, did not show a clear response pattern. Among the tolerant

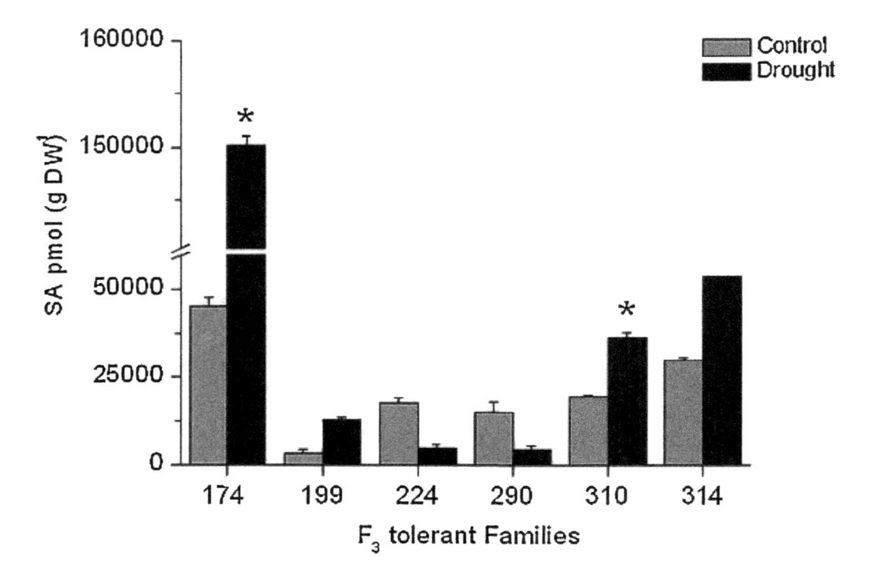

Fig. 11.8 Salicylic acid (SA) levels in plants of tolerant F_3 families grown under control and drought conditions. Data are means of four replicates. *Difference from control is statistically significant ($P \leq 0.05$)

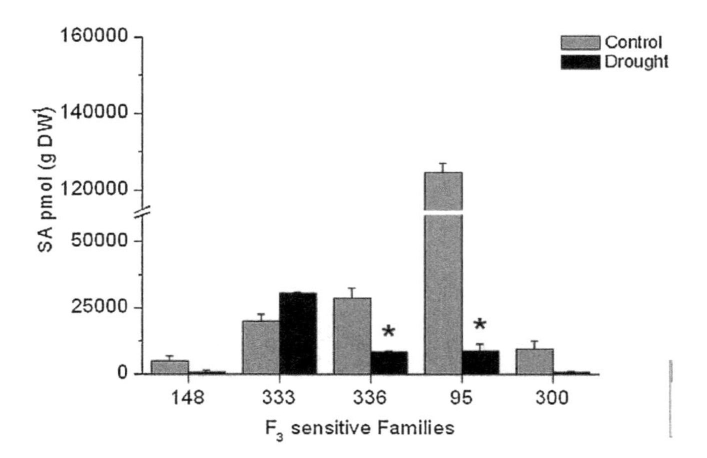

Fig. 11.9 Salicylic acid (SA) levels in sensitive F_3 families grown under control vs. drought condition. Data are means of four replicates. *Difference from control is statistically significant ($P \leq 0.05$)

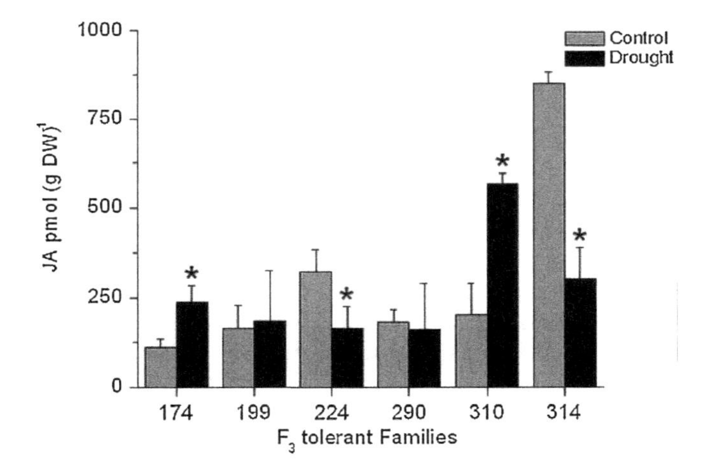

Fig. 11.10 Jasmonic acid (JA) levels in tolerant F_3 families grown under control vs. drought condition. Data are means of four replicates. *Difference from control is statistically significant ($P \leq 0.05$)

families, 174 and 310 showed a significant increase in JA under drought, whereas 314 and 224 showed a decrease (Fig. 11.10). Among the sensitive families, 336 showed an increase in JA under drought, whereas 333 and 300 showed a decrease (Fig. 11.11). Thus, drought caused changes in endogenous JA content of seedlings, but the changes were not consistent within tolerant and sensitive F_3 families. Many previous studies have shown differential patterns of JAs in response to abiotic stress (Xin et al. 1997; Kramell et al. 2000; Pedranzani et al. 2003). For example, two wild

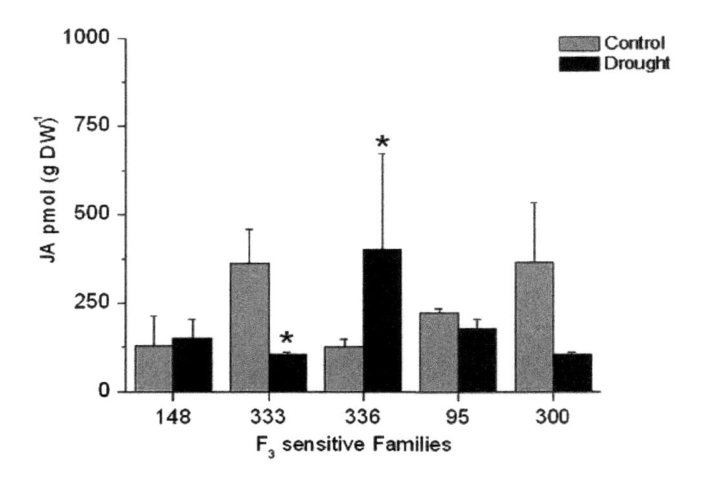

Fig. 11.11 Jasmonic acid (JA) level in sensitive F_3 families grown under control vs. drought condition. Data are means of four replicates. *Difference from control is statistically significant ($P \leq 0.05$)

populations of *Pinus pinaster* Ait. with differing climatic conditions in their place of origin showed differences in JAs basal content and in pattern of response to drought stress (Pedranzani et al. 2007). JA has been suggested to play a role in plant response to drought stress, which induces expression of several genes that also respond to JA (Turner et al. 2002; Hashimoto et al. 2004). JA is also associated with cold stress response. Levels of JA and its derivative 12-OH-JA were higher in leaves of cold-tolerant *Digitaria eriantha* cv. Mejorada INTA than cold-sensitive cv. Sudafricana (Garbero et al. 2010).

ABA levels of most tolerant families were increased under drought. The differences for 199, 314, 310, and 290 in comparison to controls were significant. ABA levels under drought were highest for 199 (706 pmol g^{-1} DW) and 310 (656 pmol g^{-1} DW) (Fig. 11.12). In contrast, ABA levels in sensitive families under drought were lower (usually <500 pmol g^{-1} DW) than in controls; this difference was significant for 333, 336, and 300 (Fig. 11.13).

The above findings suggest that increase of ABA in tolerant F_3 families confers an advantage for coping with adverse conditions (Jacobsen et al. 2002). ABA accumulation may result from increased ABA biosynthesis (Nambara and Marion-Poll 2005; Marion-Poll and Leung 2006) or from decreased catabolism (Lee et al. 2006; Ren et al. 2007). Plants can make dynamic adjustments of endogenous ABA level in response to changes in environmental conditions. The differences we observed between sensitive and tolerant F_3 families are consistent with results from comparisons of stress-sensitive vs. stress-tolerant cultivars in other species (Zheng and Li 2000; Zhu 2002; Perales et al. 2005).

In conclusion, we found that physiological efficiency of drought-tolerant vs. drought-sensitive sunflower lines can be quantified in terms of relative oil yield,

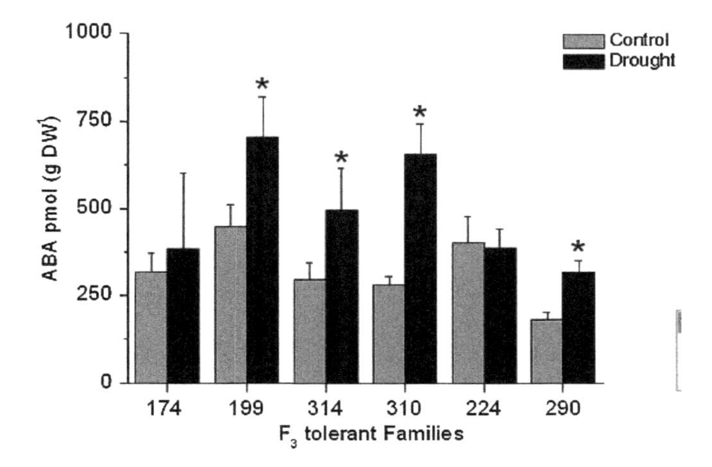

Fig. 11.12 Abscisic acid (ABA) level in tolerant F_3 families grown under control vs. drought condition. Data are means of four replicates. *Difference from control is statistically significant ($P \leq 0.05$)

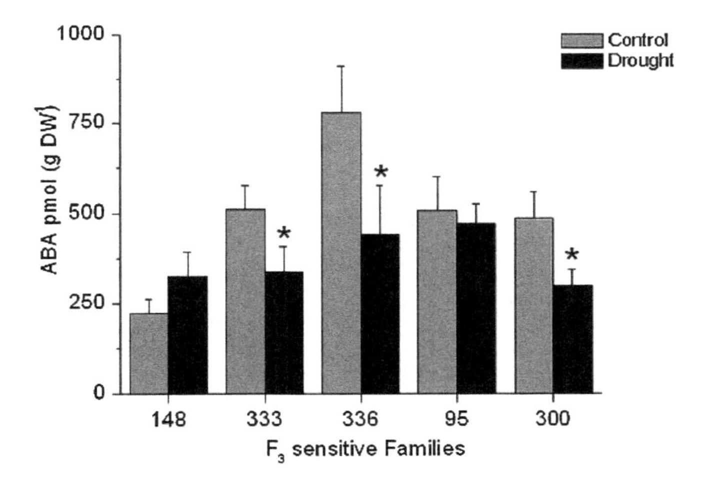

Fig. 11.13 Abscisic acid (ABA) level in sensitive F_3 families grown under control vs. drought condition. Data are means of four replicates. *Difference from control is statistically significant ($P \leq 0.05$)

seed yield, and germination percentage. These parameters were used to distinguish drought-tolerant line R432 vs. drought-sensitive B59. Levels of JAs, ABA, and their catabolites differed in seeds from sunflower plants grown under drought vs. irrigation. These phytohormones show variable content during germination, and environmental conditions encountered by the parent plant affect hormonal content of seeds. Relative germination percentage was used to characterize F_3 families as drought tolerant vs. drought sensitive. Under drought, compared to control condition, tolerant families showed increased levels of SA and ABA, whereas sensitive families did not. JA level was changed in seedlings of plants grown under drought, but the direction of change was not consistent within tolerant or sensitive families. The present results provide important information for strategies to maintain or even increase yield under drought of sunflower, which has been studied less extensively than many other agricultural crop species.

References

Abreu ME, Munne-Bosch S (2008) Salicylic acid may be involved in the regulation of drought-induced leaf senescence in perennials: a case study in field-grown *Salvia officinalis* L. plants. Environ Exp Bot 64:105–112

Agrawal GK, Yamazaki M, Kobayashi M, Hirochika R, Miyao A, Hirochika H (2001) Screening of the rice viviparous mutants generated by endogenous retrotransposon Tos17 insertion. Tagging of a zeaxanthin epoxidase gene and a novel ostatc gene. Plant Physiol 125:1248–1257

Agrawal GK, Rakwal R, Jwa NS, Han KS, Agrawal VP (2002) Molecular cloning and mRNA expression analysis of the first rice jasmonate biosynthetic pathway gene allene oxide synthase. Plant Physiol Biochem 40:771–782

Ai L, Li ZH, Xie ZX, Tian XL, Eneji AE, Duan LS (2008) Coronatine alleviates polyethylene glycol-induced water stress in two rice (*Oryza sativa* L.) cultivars. J Agron Crop Sci 194:360–368

Albuquerque FMC, de Carvalho NM (2003) Effect of type of environmental stress on the emergence of sunflower (*Helianthus annuus* L.), soybean (*Glycine max* (L.) Merril) and maize (*Zea mays* L.) seeds with different levels of vigor. Seed Sci Technol 31:465–467

Al-Chaarani GR, Gentzbittel L, Wedzony M, Sarrafi A (2005) Identification of QTLs for germination and seedling development in sunflower (*Helianthus annuus* L.). Plant Sci 169:221–227

Alibert G, Ranjeva R (1971) Recharches sur les enzymes catalysant la biosyntheses des acid phenoliques chez *Quarcus pedunculata* (Ehrn): I—formation des series cinnamique et benzoique. FEBS Lett 19:11–14

Alibert G, Ranjeva R (1972) Recharches sur les enzymes catalysant la biosyntheses des acid phenoliques chez *Quarcus pedunculata* (Ehrn): II—localization intercelulaire de la phenyalanin mmonique-lyase, de la cinnamate 4-hydroxylase, et de la "benzoote synthase". Biochem Biophys Acta 279:282–289

Amzallag GN, Nachmias A, Lerner HR (1998) Influence of the mode of salinization on reproductive traits of field-grown progeny in *Sorghum bicolor*. Isr J Plant Sci 46:9–16

Andrade A, Vigliocco A, Alemano S, Miersch O, Abdala G (2005) Endogenous jasmonates and octadecanoids during germination and seedling development: their relation with hypersensitive tomato mutants to abiotic stress. Seed Sci Res 15:309–318

Ashraf M, Mehmood S (1990) Response of four *Brassica* species to drought stress. Environ Exp Bot 30:93–100

Assmann SM (2003) OPEN STOMATA1 opens the door to ABA signalling in *Arabidopsis* guard cells. Trends Plant Sci 5:151–153

Bandurska H, Stroiński A (2005) The effect of salicylic acid on barley response to water deficit. Acta Physiol Plant 27:379–386

Bandurska H, Stroiński A, Kubiś J (2003) The effect of jasmonic acid on the accumulation of ABA, proline and spermidine and its influence on membrane injury under water deficit in two barley genotypes. Acta Physiol Plant 25:279–285

Bari R, Jones JDG (2009) Role of hormones in plant defense responses. Plant Mol Biol 69:473–488

Basal H, Smith CW, Thaxton PS, Hemphill JK (2005) Seedling drought tolerance in upland cotton. Crop Sci 45:766–771

Benech Arnold RL, Fenner M, Edwards PJ (1991) Changes in germinability, ABA content and ABA embryonic sensitivity in developing seeds of *Sorghum bicolor* (Moench) induced by water stress during grain filling. New Phytol 118:339–347

Billek G, Schmook FP (1967) Zur biosynthese der gentisinaure. Monatsh Chem 98:1651–1664

Blechert S, Brodschelm W, Holder S, Kammerer L, Kutchan TM, Muller MJ, Xia Z, Zen K (1999) The octadecanoic pathway: signal molecules for the regulation of secondary pathways. Proc Natl Acad Sci USA 92:4099–4105

Borsani O, Valpuesta V, Botella MA (2001) Evidence for the role of salicylic acid in the oxidative damage generated by ClNa and osmotic stress in *Arabidopsis* seedlings. Plant Physiol 126:1024–1030

Böttcher C, Pollmann S (2009) Plant oxylipins: plant responses to 12-oxo-phytodienoic acid are governed by its specific structural and functional properties. FEBS J 276:4693–4704

Bray EA (2003) Abscisic acid regulation of gene expression during water deficit stress in the era of the *Arabidopsis* genome. Plant Cell Environ 25:153–161

Bray EA, Bailey-Serres J, Weretilnyk E (2000) Responses to abiotic stresses. In: Gruissem W, Buchannan B, Jones R (eds) Biochemistry and molecular biology of plants. American Society of Plant Physiologist, Rockville, MD, pp 1158–1249

Bright J, Desikan R, Tancock JT, Weir IS, Neill SJ (2006) ABA-induced NO generation and stomatal closure in *Arabidopsis* are dependent on H2O2 synthesis. Plant J 45:113–122

Browse J (2009) The power of mutants for investigating jasmonate biosynthesis and signaling. Phytochemistry 70:1539–1546

Cattivelli L, Rizza F, Badeck F-W, Mazzucotelli E, Mastrangelo AM, Francia E, Marè C, Tondelli A, Stanca AM (2008) Drought tolerance improvement in crop plants: an integrated view from breeding to genomics. Field Crop Res 105:1–14

Chaves MM, Maroco JP, Pereira JS (2003) Understanding plant responses to drought-from genes to the whole plants. Funct Plant Biol 30:239–264

Cheng W-H, Endo A, Zhou L, Penney J, Chen H-C, Arroyo A, Leon P, Nambara E, Asami T, Seo M, Koshiba T, Sheen J (2002) A unique short-chain dehydrogenase/reductase in *Arabidopsis* glucose signaling and abscisic acid biosynthesis and functions. Plant Cell 14:2723–2743

Chimenti CA, Pearson J, Hall AJ (2002) Osmotic adjustment and yield maintenance under drought in sunflower. Field Crop Res 75:235–246

Chiwocha S, Abrams S, Ambrose S, Cutler A, Loewen A, Ross A, Kermode A (2003) A method for profiling classes of plant hormones and their metabolites using liquid chromatography-electrospray ionization tandem mass spectrometry: an analysis of hormone regulation of thermodormancy of lettuce (*Lactuca sativa* L.) seeds. Plant J 35:405–417

Cutler SR, Rodriguez PL, Finkelstein RR, Abrams SR (2010) Abscisic acid: emergence of a core signaling network. Annu Rev Plant Biol 61:651–679

Dathe W, Schindler C, Schneider G, Schmidt J, Porzel A, Jensen E, Yamaguchi I (1991) Cucurbic acid and its 6,7-stereoisomers. Phytochemistry 30:1990–1914

De Vos M, Van Oosten VR, Van Poecke RM, Van Pelt JA, Pozo MJ, Mueller MJ, Buchala AJ, Métraux JP, Van Loon LC, Dicke M, Pieterse CM (2005) Signal signature and transcriptome changes of *Arabidopsis* during pathogen and insect attack. Mol Plant Microbe Interact 18:923–937

Desikan R, Cheung MK, Bright J, Henson D, Hancock JT, Neill SJ (2004) ABA hydrogen peroxide and nitric oxide signaling in stomatal guard cells. J Exp Bot 55:205–212

Devoto A, Turner JG (2003) Regulation of jasmonate mediated plant responses in *Arabidopsis*. Ann Bot 92:329–337

Dobra J, Motyka V, Dobrev P, Malbeck J, Prasil IT, Haisel D, Gaudinova A, Havlova M, Gubis J, Vankova R (2010) Comparison of hormonal response to heat, drought and combined stress in tobacco plants with elevated proline content. J Plant Physiol 167:1360–1370

Durgbanshi A, Arbona V, Pozo O, Miersch O, Sancho JV, Gómez-Cadenas A (2005) Simultaneous determination of multiple phytohormones in plant extracts by liquid chromatography-electrospray tandem mass spectrometry. J Agric Food Chem 53:8437–8442

El-Tayeb MA (2005) Response of barley grains to the interactive effect of salinity and salicylic acid. Plant Growth Regul 45:215–224

Ervin EH, Zhang XZ, Fike JH (2004) Ultraviolet-B radiation damage on Kentucky Bluegrass II: hormone supplement effects. Hort Sci 113:120–128

Fan J, Hill L, Crooks C, Doerner P, Lamb C (2009) Abscisic acid has a key role in modulating diverse plant-pathogen interactions. Plant Physiol 150:1750–1761

Fariduddin Q, Hayat S, Ahmad A (2003) Salicylic acid influences net photosynthetic rate, carboxylation efficiency, nitrate reductase activity and seed yield in *Brassica juncea*. Photosynthetica 41:281–284

Feurtado JA, Yang J, Ambrose SJ, Cutler A, Abrams S, Kermode AR (2007) Disrupting abscisic acid homeostasis in western white pine (*Pinus monticola*) seeds induces dormancy termination and changes in abscisic acid catabolites. J Plant Growth Regul 26:46–54

Finkelstein R, Gampala S, Rock C (2002) Abscisic acid signaling in seeds and seedlings. Plant Cell 14:S15–S45

Fonseca S, Chini A, Hamberg M, Adie B, Porzel A, Kramell R, Miersch O, Wasternack C, Solano R (2009) (+)-7-Iso-Jasmonoyl-L-isoleucine is the endogenous bioactive jasmonate. Nat Chem Biol 5:344–350

Fujita M, Fujita Y, Noutoshi Y, Takahashi F, Narusaka Y, Yamaguchi-Shinozaki K, Shinozaki K (2006) Crosstalk between abiotic and biotic stress responses: a current view from the points of convergence in the stress signaling networks. Curr Opin Plant Biol 9:436–442

Gao XP, Wang XF, Lu YF, Zhang LY, Shen YY, Liang Z, Zhang DP (2004) Jasmonic acid is involved in the water-stress-induced betaine accumulation in pear leaves plant. Plant Cell Environ 27:497–507

Garbero M, Pedranzani H, Zirulnik F, Molina A, Pérez-Chaca MV, Vigliocco A, Abdala G (2010) Short-term cold stress in two cultivars of *Digitaria eriantha*: effects on stress-related hormones and antioxidant defense system. Acta Physiol Plant 33:497–507

Ghasempour HR, Anderson EM, Gaff DF (2001) Effects of growth substances on the protoplasmic drought tolerance of leave cells of the resurrection grass, *Sporobolus stapfianus*. Aust J Plant Physiol 28:1115–1120

Gidda SK, Miersch O, Levitin A, Schmidt J, Wasternack C, Varin L (2003) Biochemical and molecular characterization of a hydroxy jasmonate sulfotransferase from *Arabidopsis thaliana*. J Biol Chem 278:17895–17900

Hamada AM (1998) Effects of exogenously added ascorbic acid, thiamin or aspirin on photosynthesis and some related activities of drought-stressed wheat plants. In: Garab G (ed) Photosynthesis: mechanisms and effects. Kluwer, Dordrecht, Netherlands, pp 2581–2584

Hamada AM, Al-Hakimi AMA (2001) Salicylic acid versus salinity-drought induced stress on wheat seedlings. Rostl Výr 47:444–450

Hartung W, Scharaut D, Jiang F (2005) Physiology of abscisic acid (ABA) in roots under stress—a review of the relationship between root ABA and radial water and ABA flows. Aust J Agric Res 56:1253–1259

Hashimoto M, Larisa K, Shinichiro S, Toshiko F, Setsuko K, Tomokazu K (2004) A novel rice PR10 protein, RSOsPR10, specifically induced in roots by biotic and abiotic stresses, possibly via the jasmonic acid signaling pathway. Plant Cell Physiol 45:550–559

Hassine AB, Lutts S (2010) Differential responses of saltbush *Atriplex halimus* L. exposed to salinity and water stress in relation to senescing hormones abscisic acid and ethylene. J Plant Physiol 167:1448–1456

Hirayama T, Shinozaki K (2007) Perception and transduction of abscisic acid signals: keys to the function of the versatile plant hormone ABA. Trends Plant Sci 12:343–351

Horváth E, Szalai G, Janda T (2007) Induction of abiotic stress tolerance by salicylic acid signaling. J Plant Growth Regul 26:290–300

Hsu SY, Hsu YT, Kao CH (2003) Ammonium ion, ethylene, and abscisic acid in polyethylene glycol-treated rice leaves. Biol Plant 46:239–242

Huang D, Wu W, Abrams SR, Cutler AJ (2008) The relationship of drought-related gene expression in *Arabidopsis thaliana* to hormonal and environmental factors. J Exp Bot 11:2991–3007

Iuchi S, Kobayshi M, Taji T, Naramoto M, Seki M, Kato T, Tabata S, Kakubari Y, Yamaguchi-Shinozaki K, Shinozaki K (2001) Regulation of drought tolerance by gene manipulation of 9-cis-epoxycarotenoid, a key in abscisic acid biosynthesis in *Arabidopsis*. Plant J 27:325–333

Jacobsen JV, Pearce DW, Poole AT, Pharis R, Mander LN (2002) Abscisic acid, phaseic acid and gibberellin contents associated with dormancy and germination in barley. Physiol Plant 115:428–441

Janowiak F, Maas B, Dörffling K (2002) Importance of abscisic acid for chilling tolerance of maize seedlings. J Plant Physiol 159:635–643

Jiang M, Zhang J (2002) Water stress-induced abscisic acid accumulation triggers the increased generation of reactive species and up-regulates the activities of antioxidant enzymes in maize leaves. J Exp Bot 53:2401–2410

Khodary SFA (2004) Effect of salicylic acid on the growth, photosynthesis and carbohydrate metabolism in salt stressed maize plants. Int J Agric Biol 6:5–8

Kiani SP, Talia P, Maury P, Grieu P, Heinz R, Perrault A, Nishinakamasu V, Hopp E, Gentzbittel L, Paniego N, Sarrafi A (2007) Genetic-analysis of plant water status and osmotic adjustment in recombinant inbred lines in sunflower under 2 water treatments. Plant Sci 172:773–778

Kikuzaki H, Kayano S, Fukutsuka N, Aoki A, Kasamatsu K, Yamasaki Y, Mitani T, Nakatani N (2004) Abscisic acid related compounds and lignans in prunes (*Prunus domestica* L.) and their oxygen radical absorbance capacity (ORAC). J Agric Food Chem 52:344–349

Klessing DF, Durner J, Noad R, Navarre DA, Wendehenne D, Kumar D, Zhou JM, Shah J, Zhang S, Kachroo P, Trifa Y, Pontier D, Lam E, Silva H (2000) Nitric oxide and salicylic acid signaling in plant defense. Proc Natl Acad Sci 97:8849–8855

Koch T, Bandemer K, Boland W (1997) Biosynthesis of cis-jasmone: a pathway for the inactivation and the disposal of the plant stress hormone jasmonic acid to the gas phase? Helv Chim Acta 80:838–850

Koornneef A, Leon-Reyes A, Ritsema T, Verhage A, Den Otter FC, Van Loon LC, Pietersen CM (2008) Kinetics of salicylate-mediated suppression of jasmonate signaling reveals a role for redox modulation. Plant Physiol 147:1358–1363

Kramell R, Miersch O, Atzorn R, Parthier B, Wasternack C (2000) Octadecanoid-derived alteration of gene expression and the “oxylipin signature” in stressed barley leaves. Implications for different signaling pathways. Plant Physiol 123:177–187

Krantev A, Yordanova R, Janda T, Szalai G, Popova L (2008) Treatment with salicylic acid decreases the effect of cadmium on photosynthesis in maize plants. J Plant Physiol 165:929–931

Kumar L (1999) DNA markers in plant improvement: an overview. Biotechnol Adv 17:143–182

Kushiro T, Okamoto M, Nakabayashi K, Yamagishi K, Kimatura S, Asami T, Hirai N, Koshiba T, Kamiya Y, Nambara E (2004) The *Arabidopsis* cytochrome P450 CYP707A encodes ABA 8′-hydroxylases: key enzymes in ABA catabolism. EMBO J 23:1647–1656

Lambrides CJ, Chapman SC, Shorter R (2004) Genetic variation for carbon isotope discrimination in sunflower: association with transpiration efficiency and evidence for cytoplasmic inheritance. Crop Sci 44:1642–1653

Larkindale J, Huang B (2004) Thermotolerance and antioxidant system in *Agrostis stolonifera*: involvement of salicylic acid, calcium, hydrogen peroxide, and ethylene. J Plant Physiol 161:405–413

Lee KH, Piao HL, Kin H-Y, Choi SM, Jiang E, Hartung W, Hwang I, Kwak JM, Lee I-J, Hwang I (2006) Activation of glucosidase via stress-induced polymerization rapidly increases active pools of abscisic acid. Cell 126:1109–1120

Lehmann J, Atzorn R, Brückner C, Reinbothe S, Leopold J, Wasternack C, Parthier B (1995) Accumulation of Jasmonate, abscisic acid, specific transcripts and proteins in osmotically stressed barley leaf segments. Planta 197:156–162

Leon-Reyes A, Du Y, Koornneef A, Proietti S, Körbes AP, Memelink J, Pieterse CMJ, Ritsema T (2010) Ethylene signaling renders the jasmonate response of *Arabidopsis* insensitive to future suppression by salicylic acid. Mol Plant Microbe Interact 23:187–197

Li L, Staden JV, Jager AK (1998) Effect of plant growth regulators on the antioxidant system in seedlings of two maize cultivars subjected to water stress. J Plant Growth Regul 25:81–87

Liu XG, Yue YL, Li B, Nie YL, Li W, Wu WH, Ma LG (2007) A G protein-coupled receptor is a plasma membrane receptor for the plant hormone abscisic acid. Science 315:1712–1716

Longenberger PS, Smith CW, Thaxton PS, McMichael BL (2006) Development of a screening method for drought tolerance in cotton seedlings. Crop Sci 46:2104–2110

Lu S, Su W, Li H, Guo Z (2009) Abscisic acid improves drought tolerance of triploid bermudagrass and involves H_2O_2- and NO-induced antioxidant enzyme activities. Plant Physiol Biochem 47:132–138

Mahajan S, Tuteja N (2005) Cold, salinity and drought stresses: an overview. Arch Biochem Biophys 444:139–158

Makandar R, Essig JS, Schapaugh MA, Trick HN, Shah J (2006) Genetically engineered resistance to *Fusarium* head blight in wheat by expression of *Arabidopsis* NPR1. Mol Plant-Microbe Interact 19:123–129

Marion-Poll A, Leung J (2006) Abscisic acid synthesis, metabolism and signal transduction. In: Hedden P, Thomas SG (eds) Annual plant reviews: plant hormone signaling. Blackwell, Oxford, UK, pp 1–35

McConn M, Creelman RA, Bell E, Mullet JE, Browse J (1997) Jasmonate is essential for insect defense in *Arabidopsis*. Proc Natl Acad Sci USA 94:5473–5477

Memelink J, Verpoorte R, Kijn JW (2001) ORCAnization of jasmonate responsive gene expression in alkaloid metabolism. Trends Plant Sci 6:212–219

Miersch O, Neumerkel J, Dippe M, Stenzel I, Wasternack C (2008) Hydroxylated jasmonates are commonly occurring metabolites of jasmonic acid and contribute to a partial switch-off in jasmonate signaling. New Phytol 177:114–127

Mikolajczyk M, Awotunde OS, Muszynska G, Klessig DF, Dobrowolsja G (2000) Osmotic stress induces rapid activation of salicylic acid-induced protein kinase and a homolog of protein kinase ASK1 in tobacco cells. Plant Cell 12:165–178

Muehlbauer GJ, Specht JE, Thomas-Compton MA, Staswick PE, Bernard RL (1988) Near-isogenic lines: a potential resource in the integration of conventional and molecular marker linkage maps. Crop Sci 28:729–735

Mur LAJ, Kenton P, Atzorn R, Miersch O, Wasternack C (2006) The outcomes of concentration-specific interactions between salicylate and jasmonate signaling include synergy, antagonism and oxidative stress leading to cell death. Plant Physiol 140:249–262

Mustafa NR, Kim HK, Choi YH, Erkelens C, Lefeber AWM, Spijksma G, van der Heijden R, Verpoorte R (2009) Biosynthesis of salicylic acid in fungus elicited *Catharanthus roseus* cells. Phytochemistry 70:532–539

Mustapha G, Tahar T, Mohamed N (2009) Influence of water stress on seed germination characteristic in invasive *Diplotaxis harra* (Forssk.) Boiss (Brassicaceae) in arid zone of Tunisia. J Phytol 1:249–254

Mwale SS, Hamusimbi C, Mwansa K (2003) Germination emergence and growth of sunflower (*Helianthus annuus* L.) in response to osmotic seed priming. Seed Sci Technol 31:199–206

Nambara E, Marion-Poll A (2005) Abscisic acid biosynthesis and catabolism. Annu Rev Plant Physiol Mol Biol 56:165–185

Nayyar H, Walia DP (2004) Genotypic variation in wheat in response to water stress and abscisic acid-induced accumulation of osmolytes in developing grains. J Agron Crop Sci 190:39–45

Nazar R, Iqbal N, Syeed S, Khan NA (2011) Salicylic acid alleviates decreases in photosynthesis under salt stress by enhancing nitrogen and sulfur assimilation and antioxidant metabolism differentially in two mungbean cultivars. J Plant Physiol 168:807–815

Ober ES, Sharp RE (1994) Proline accumulation in maize (*Zea mays* L.) primary roots at low water potentials (I. Requirement for increased levels of abscisic acid). Plant Physiol 105:981–987

Oritani T, Kiyota H (2003) Biosynthesis and metabolism of abscisic acid and related compounds. Nat Prod Rep 20:414–425

Papadakis AK, Roubelakis-Angelakis KA (2005) Polyamines inhibit NADPH oxidase-mediated superoxides generation and putrescine prevents programmed cell death syndrome induced by the polyamine oxidase generated hydrogen peroxide. Planta 220:826–837

Paran I, Kesseli R, Michelmore R (1991) Identification of restriction fragment length polymorphism and random amplified polymorphic DNA markers linked to downy mildew resistance genes in lettuce using near isogenic lines. Genome 34:1021–1027

Pasquer F, Isidore E, Zarn J, Keller B (2005) Specific patterns of changes in wheat gene expression after treatment with three antifungal compounds. Plant Mol Biol 57:693–707

Pedranzani H, Racagni G, Alemano S, Miersch O, Ramírez I, Peña CH, Machado-Domenech E, Abdala G (2003) Salt tolerant tomato plants show increased levels of jasmonic acid. Plant Growth Regul 41:149–158

Pedranzani H, Sierra-de-Grado R, Vigliocco A, Miersch O, Abdala G (2007) Cold and water stresses produce changes in endogenous jasmonates in two populations of *Pinus pinaster* Ait. Plant Growth Regul 52:111–112

Perales L, Arbona B, Gómez-Cadenas A, Cornejo MJ, Sanz A (2005) A relationship between tolerance to dehydration of rice lines and ability for ABA synthesis under stress. Plant Physiol Biochem 43:786–792

Pieterse CMJ, Leon-Reyes A, Van der Ent S, Van Wees SCM (2009) Networking by small-molecule hormones in plant immunity. Nat Chem Biol 5:308–316

Popova L, Pancheva T, Uzunova A (1997) Salicylic acid: properties, biosynthesis and physiological role. Bulg J Plant Physiol 23:85–93

Qin XQ, Zeevaart JAD (1999) The 9-cis-epoxicarotenoid cleavage reaction is the key regulatory step of abscisic acid biosynthesis in water stress bean. Proc Natl Acad Sci USA 96:15354–15361

Rao MV, Davis RD (1999) Ozone-induced cell death occurs via two distinct mechanisms in *Arabidopsis*: the role of salicylic acid. Plant J 17:603–614

Rauf S (2008) Breeding sunflower (*Helianthus annuus* L.,) for drought tolerance. Communic Biom Crop Sci 3:29–44

Rauf S, Sadaqat HA (2007) Sunflower (*Helianthus annuus* L.) germplasm evaluation for drought tolerance. Communic Biom Crop Sci 2:8–16

Rauf S, Sadaqat HA (2008) Identification of physiological traits and genotypes combined to high achene yield in sunflower (*Helianthus annuus* L.) under contrasting water regimes. Aust J Crop Sci 1:23–30

Raz V, Bergervoet J, Koornneef M (2001) Sequential steps for developmental arrest in *Arabidopsis* seeds. Development 128:243–252

Reinbothe C, Springer A, Samol I, Reinbothe S (2009) Plant oxylipins: role of jasmonic acid during programmed cell death, defense and leaf senescence. FEBS J 276:4666–4681

Ren H, Gao Z, Chen L, Wei K, Liu J, Fan Y, Davies WJ, Jia W, Zhang J (2007) Dynamic analysis of ABA accumulation in relation to the rate of ABA catabolism in maize tissue under water stress. J Exp Bot 58:211–219

Reymond P, Bodenhausen N, Van Poecke RM, Krishnamurthy V, Dicke M, Farmer EE (2004) A conserved transcript pattern in response to a specialist and a generalist herbivore. Plant Cell 16:3132–3147

Ribnicky DM, Shulaev V, Raskin I (1998) Intermediates of salicylic acid biosynthesis in tobacco. Plant Physiol 118:565–572

Riera M, Valon C, Fenzi F, Giraudat J, Leung J (2005) The genetics of adaptive responses to drought stress: abscisic acid-dependent and abscisic acid-independent signaling components. Physiol Plant 123:111–119

Roche J, Hewezi T, Bouniols A, Gentzbittel L (2009) Real-time PCR monitoring of signal transduction related genes involved in water stress tolerance mechanism of sunflower. Plant Physiol Biochem 47:139–145

Sánchez-Díaz M, Tapia C, Antolín MC (2008) Abscisic acid and drought response of Canarian laurel forest tree species growing under controlled conditions. Environ Exp Bot 64:155–161

Sasaki-Sekimoto Y, Taki N, Obayashi T (2005) Coordinated activation of metabolic pathways for antioxidants and defense compounds by jasmonates and their roles in stress tolerance in *Arabidopsis*. Plant J 44:653–668

Schachtman DP, Goodger JQ (2008) Chemical root to shoot signaling under drought. Trends Plant Sci 13:281–287

Schwartz SH, Tan BC, Gage DA, Zeevaart JAD, McCarty DR (1997) Specific oxidative cleavage of carotenoids by Vp14 of maize. Science 276:1872–1874

Seki M, Umezawa T, Urano K, Shinozaki K (2007) Regulatory metabolic networks in drought stress responses. Curr Opin Plant Biol 10:296–302

Sembdner G, Parthier B (1993) Biochemistry, physiological and molecular actions of jasmonates. Annu Rev Plant Physiol Mol Biol 44:569–589

Senaratna T, Touchell D, Bunn E, Dixon K (2000) Acetyl salicylic acid (aspirin) and salicylic acid induce multiples tress tolerance in bean and tomato plants. Plant Growth Regul 30:157–161

Seo M, Koshiba T (2002) Complex regulation of ABA biosynthesis in plants. Trends Plant Sci 7:41–48

Seo M, Peeters AJM, Koiwai H, Oritani T, Marion-Poll A, Zeevart JAD, Koorneef M, Kamiya Y, Koshiba T (2000) The Arabidopsis aldehyde oxidase 3 (AAO3) gene products catalyzes the final step in abscisic acid biosynthesis in leaves. Proc Natl Acad Sci USA 97:12908–12913

Seo HS, Song JT, Cheong J-J, Lee Y-H, Lee Y-W, Hwang I, Lee JS, Choi YD (2001) Jasmonic acid carboxyl methyl transferase: a key enzyme for jasmonate-regulated plant response. Proc Natl Acad Sci USA 98:4788–4793

Shah J (2003) The salicylic acid loop in plant defense. Curr Opin Plant Biol 6:365–371

Shakirova FM (2007) Role of hormonal system in the manifestation of growth promoting and anti-stress action of salicylic acid. In: Hayat S, Ahman A (eds) Salicylic acid. A plant hormone. Springer, Dordrecht, Netherlands, pp 69–89

Shan C, Liang Z (2010) Jasmonic acid regulates ascorbate and glutathione metabolism in *Agropyron cristatum* leaves under water stress. Plant Sci 178:130–139

Sharma YK, Leon J, Raskin I, Davies KR (1996) Ozone induced responses in *Arabidopsis thaliana*: the role of salicylic acid in the accumulation of defense-related transcripts and induced resistance. Proc Natl Acad Sci USA 93:5099–5104

Sharp RE, LeNoble ME, Else MA, Thorne ET, Gherardi F (2000) Endogenous ABA maintains shoot growth in tomato independently of affects on plant water balance evidence for an interaction with ethylene. J Exp Bot 51:1575–1584

Shinozaki K, Yamaguchi-Shinozaki K (2000) Molecular response to dehydration and low temperature: differences and cross-talk between two stress signaling pathways. Curr Opin Plant Biol 3:217–223

Shinozaki K, Yamaguchi-Shinozaki K (2007) Gene networks involved in drought stress response and tolerance. J Exp Bot 58:221–227

Singh B, Usha K (2003) Salicylic acid induced physiological and biochemical changes in wheat seedlings under water stress. Plant Growth Regul 39:137–141

Singh BN, Mishra RN, Agarwal PK, Goswami M, Nair S, Sopory SK, Reddy MK (2004) A pea chloroplast translation elongation factors that is regulated by abiotic factors. Biochem Biophys Res Commun 320:523–530

Sorrells ME (1998) Marker assisted selection: is it practical? In: Kolhi MM, Francis M (eds) International workshop on the application of biotechnologies to wheat breeding. INIA La Estanzuela, Colonia, Uruguay, pp 49–56

Spollen WG, LeNoble ME, Sammuels TD, Bernstein N, Sharp RE (2000) Abscisic acid accumulation maintains maize primary roots elongation at low water potentials by restricting ethylene production. Plant Physiol 122:967–976

Staswick PE, Tiryaki I (2004) The oxylipin signal jasmonic acid is activated by an enzyme that conjugates it to isoleucine in *Arabidopsis*. Plant Cell 16:2117–2127

Stelmach BA, Müller A, Hennig P, Laudert D, Andert L, Weiler EW (1998) Quantitation of the octadecanoid 12-oxo- phytodienoic acid, a signalling compound in plant mechanotransduction. Phytochemistry 533:319–323

Swiatek A, Van Dongen W, Esmans EI, Van Onckelen H (2004) Metabolic fate of jasmonates in tobacco bright yellow-2 cells. Plant Physiol 135:161–172

Syeed S, Anjum NA, Nazar R, Iqbal N, Masood A, Khan NA (2011) Salicylic acid-mediated changes in photosynthesis, nutrients content and antioxidant metabolism in two mustard (*Brassica juncea* L.) cultivars differing in salt tolerance. Acta Physiol Plant 33:877–886

Syvänen AC (2005) Toward genome-wide SNP genotyping. Nat Genet 37:S5–S10

Szalai G, Horgosi S, Soós V, Majláth I, Balázs E, Janda T (2010) Salicylic acid treatment of pea seeds induces its de novo synthesis. J Plant Physiol 168:213–219

Szepesi Á, Csiszár J, Gémes K, Horváth E, Horváth F, Simon ML, Tari I (2009) Salicylic acid improves acclimation to salt stress by stimulating abscisic aldehyde oxidase activity and abscisic acid accumulation, and increase Na^+ content in leaves without toxicity symptoms in *Solanum lycopersicum* L. J Plant Physiol 166:914–925

Thameur A, Ferchichi A, López-Carbonell M (2011) Quantification of free and conjugated abscisic acid in five genotypes of barley (*Hordeum vulgare* L.) under water stress conditions. S Afr J Bot 77:222–228

Theodoulou FL, Job K, Slocombe SP, Footitt S, Holdsworth M, Baker A, Larson TR, Graham IA (2005) Jasmonic acid levels are reduced in COMATOSE ATP-binding cassette transporter mutants. Implications for transport of jasmonate precursors into peroxisomes. Plant Physiol 137:835–840

Thomma BPHJ, Eggermont K, Tierens KFM-J, Broekaert WF (1999) Requirement of functional ethylene-insensitive 2 genes for efficient resistance of *Arabidopsis* to infection by *Botrytis cinerea*. Plant Physiol 121:1093–1101

Tomar SMS, Kumar GT (2004) Seedling survivability as a selection criterion for drought tolerance in wheat. Plant Breed 123:392–394

Ton J, Flors V, Mauch-Mani B (2009) The multifaceted role of ABA in disease resistance. Trends Plant Sci 14:310–317

Trejo CL, Davies WJ (1991) Drought-induced closure of *Phaseolus vulgaris* L. stomata precedes leaf water deficit and any increase in xylem ABA concentration. J Exp Bot 42:1507–1515

Turner JG, Ellis Ch, Devoto A (2002) The jasmonate signal pathway. Plant Cell 14:153–164

Uno Y, Furihata T, Abe H, Yoshida R, Shinozaki K (2000) *Arabidopsis* basic leucine zipper transcription factors involved in an abscisic acid-dependent signal transduction pathway under drought and high-salinity conditions. Proc Natl Acad Sci USA 97:11632–11637

Verslues PE, Bray EA (2006) Role of abscisic acid (ABA) and *Arabidopsis thaliana* ABA-insensitive loci in low water potential-induced ABA and proline accumulation. J Exp Bot 57:201–212

Veselov DS, Sharipova GV, Veselov SU, Kudoyarova GR (2008) The effects of NaCl treatment on water relations, growth and ABA content in barley cultivars differing in drought tolerance. J Plant Growth Regul 27:380–386

Vick BA, Zimmerman DC (1983) The biosynthesis of jasmonic acid: a physiological role for plant lipoxygenase. Biochem Biophys Res Commun 111:470–77

Vlot AC, Liu P-P, Cameron RK, Park S-W, Yang Y, Kumar D, Zhou F, Padukkavidana T, Gustafsson C, Pichersky E, Klessig DF (2008) Identification of likely orthologs of tobacco salicylic acid-binding protein 2 and their role in systemic acquired resistance in *Arabidopsis thaliana*. Plant J 56:445–456

Wang SY (1999) Methyl jasmonate reduces water stress in strawberry. Plant Growth Regul 18:127–134
Wang L, Chen S, Kong W, Li S, Archbold DD (2006) Salicylic acid pretreatment alleviates chilling injury and affects the antioxidant system and heat shock proteins of peaches during cold storage. Postharvest Biol Technol 41:244–251
Wasilewska A, Vlad F, Sirichandra C, Redko Y, Jammes F, Valon C, Frei dit Frey N, Leung J (2008) An update on abscisic acid signaling in plants and more. Mol Plant 1:198–217
Wasternack C (2006) Oxylipins: biosynthesis, signal transduction and action. In: Hedden P, Thomas S (eds) Plant hormone signaling. Annual plant reviews. Blackwell, Oxford, UK, pp 185–228
Wasternack C (2007) Jasmonates: an update on biosynthesis, signal transduction and action in plant stress response, growth and development. Ann Bot 100:681–697
Wasternack C, Hause B (2002) Jasmonates and octadecanoids: signals in plant stress response and development. Prog Nucl Acid Res Mol Biol 72:165–221
Wasternack C, Kombrinck E (2010) Jasmonates: structural requirements for lipid-derived signals active in plant stress responses and development. ACS Chem Biol 5:63–77
Wildermuth MC (2006) Variations on a theme: synthesis and modifications of plant benzoic acids. Curr Opin Plant Biol 9:288–296
Xin ZY, Zhou X, Pilet PE (1997) Level changes of jasmonic, abscisic and indole-3yl-acetic acids in maize under desiccation stress. J Plant Physiol 151:120–124
Xiong L, Shumaker KS, Zhu JK (2002) Cell signaling during cold, drought and salt stress. Plant Cell 14:S165–S183
Yalpani N, Leen J, Lawthon MA, Raskin I (1993) Pathway of salicylic acid biosynthesis in healthy and virus-inoculated tobacco. Plant Physiol 103:315–321
Yamaguchi-Shinozaki K, Shinozaki K (1994) A novel cis-acting element in an Arabidopsis gene is involved in responsiveness to drought, low temperature or high-salt stress. Plant Cell 6: 251–264
Yan Y, Stolz S, Chételat A, Reymond P, Pagni M, Dubugnon L, Farmer EE (2007) A downstream mediator in the growth repression limb of the jasmonate pathway. Plant Cell 19:2470–2483
Zaharia LI, Walker-Simmon M, Rodríguez C, Abrams S (2005) Chemistry of abscisic acid, abscisic acid catabolites and analogs. Plant Growth Regul 24:274–284
Zeevaart JAD (1999) Abscisic acid metabolism and its regulation. In: Hooykaas PJJ, Hall MA, Libbenga KR (eds) Biochemistry and molecular biology of plant hormones. Elsevier Science, Amsterdam, The Netherlands, pp 189–207
Zhang J, Davies WJ (1989) Abscisic acid produced in dehydrating roots may enable the plant to measure the water status of the soil. Plant Cell Environ 12:73–81
Zhang J, Davies WJ (1990a) Changes in the concentration of ABA in xylem sap as a function of changing soil water will account for changes in leaf conductance. Plant Cell Environ 13:277–285
Zhang J, Davies WJ (1990b) Does ABA in the xylem control the rate of leaf growth in soil-dried maize and sunflower plants. J Exp Bot 41:1125–1132
Zhang J, Schurr U, Davies WJ (1987) Control of stomatal behavior by abscisic acid which apparently originates in roots. J Exp Bot 38:1174–1181
Zhang J, Jia W, Yang J, Ismail AM (2006) Role of ABA in integrating plant responses to drought and salt stress. Field Crop Res 97:111–119
Zheng Y-Z, Li T (2000) Changes in proline levels and abscisic acid content in tolerant/sensitive cultivars of soybean under osmotic conditions. Soybean Genetics NewsLetter 27. http://www.soygenetics.org/
Zhou R, Cutler A, Ambrose SJ, Galka MM, Nelson KM, Squires TM, Loewen MK, Juadhav AS, Ross AR, Taylor DC, Abrams SR (2004) A new abscisic acid catabolic pathway. Plant Physiol 134:361–369
Zhou ZS, Guo K, Elbaz AA, Yang ZM (2009) Salicylic acid alleviates mercury toxicity by preventing oxidative stress in roots of *Medicago sativa*. Environ Exp Bot 65:27–34
Zhu JK (2002) Salt and drought stress signal transduction in plants. Annu Rev Plant Physiol Plant Mol Biol 53:247–273

Chapter 12
An Insight into the Role of Salicylic Acid and Jasmonic Acid in Salt Stress Tolerance

M. Iqbal R. Khan, Shabina Syeed, Rahat Nazar, and Naser A. Anjum

Abstract Phytohormones are organic compounds that in small amount promote, inhibit, or modify physiological processes in plants. Researchers have recognized salicylic acid (SA) and jasmonic acid (JA) as a potential hormone. Application of SA and JA could provide tolerance against biotic and abiotic stresses such as salinity, temperature stress, heavy metal stress, etc. The role of SA and JA in the protection against abiotic stress is played by its ability to induce expression of genes coding proteins. A low concentration of SA and JA appears to be effective in tolerance to stress by enhancing physiological processes and improving salt tolerance by its effect on biochemical and molecular mechanisms. The present review gives an insight into the role of SA and JA in inducing various physiological responses in plants under salinity stress, and an interaction between these two phytohormones is also discussed.

12.1 Introduction

A wide range of environmental factors including biotic and abiotic stress affect plant during life cycle (Parvaiz and Satyawati 2008). Among several abiotic stresses, salinity is one of the major abiotic stresses that plants encounter. Salinity is usually of great concern and the most injurious factor in arid and semiarid regions. More than 800 million hectares of land throughout the world are salt affected, equating to more than 6% of the world's total land area (FAO 2008). Salinity causes detrimental effects (Fig. 12.1) on plant growth and productivity

M.I.R. Khan (✉) • S. Syeed • R. Nazar
Department of Botany, Aligarh Muslim University, Aligarh 202 002, India
e-mail: amu.iqbal@gmail.com

N.A. Anjum
Centre for Environmental and Marine Studies (CESAM) and Department of Chemistry, University of Aveiro, 3810-193 Aveiro, Portugal

N.A. Khan et al. (eds.), *Phytohormones and Abiotic Stress Tolerance in Plants*,
DOI 10.1007/978-3-642-25829-9_12, © Springer-Verlag Berlin Heidelberg 2012

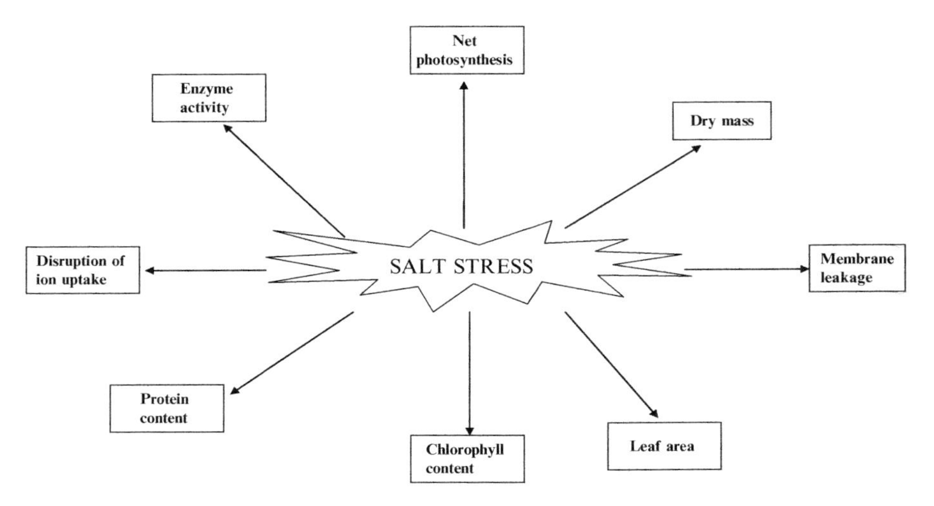

Fig. 12.1 Diagrammatic presentation of negative effects by salt stress. All given parameters are reducing by salinity

(Turkan and Demiral 2009), mainly through changes at physiological, biochemical, and molecular level (Khan et al. 2009; Syeed et al. 2011). The loss in plant productivity due to salinity is a result of imbalance in ionic and osmotic effects (Ashraf 2009). The physiological processes that are primarily adversely affected by salt stress include ion toxicity, osmotic stress, nutrient deficiency, and oxidative stress (Munns and Teste 2008; Daneshmand et al. 2009). Salinity results in the excess production of reactive oxygen species (ROS), as by-products of various metabolic pathways that are localized in different cellular compartments such as chloroplast, mitochondria, and peroxisomes (del Rio et al. 2006; Navrot et al. 2007). ROS are highly reactive and can alter normal cellular metabolism through oxidative damages to membranes, proteins, and nucleic acids and cause lipid peroxidation, protein denaturation, and DNA mutation (Ahmad et al. 2008). Plants have the ability to scavenge/detoxify ROS by producing different types of antioxidants. The use of techniques to alleviate adverse effects of salinity stress is expected to result in sustainable development.

Plant hormones play important roles in regulating developmental processes and signaling networks involved in plant responses to a wide range of biotic and abiotic stresses. Information on phytohormones of plants and salinity tolerance is scarce, particularly on JA. SA is an endogenous growth regulator of phenolic nature, which participates in the regulation of physiological processes in plants. It plays an important role in the plant response to adverse environmental conditions such as salinity (Joseph et al. 2010). JA is endogenous growth regulator identified in many plant cultivars and induces a wide variety of physiological and developmental responses (Engelberth et al. 2001). JA may act as stress modulators by suppressing or enhancing the stress responses of plants (Agrawal et al. 2002).

Significant progress has been made in indentifying the key components and understanding the role of SA and JA implicated in plant defense pathways, but their role in abiotic stress defense is less studied. Plant hormones are known to control internal metabolism; therefore, attempts need to make to explore the possibility of using these for alleviating salt stress-induced physiological effects. Among the phytohormones, SA and JA are increasingly being recognized as the major essential phytohormones after five classical group of phytohormones that not only plays an important role in growth and development of higher plants but also is associated with stress tolerance in plants (Popova et al. 2003). Adequate SA and JA improve growth and photosynthesis of plants (Kang et al. 2005; Kim et al. 2009; Syeed et al. 2011), and it has regulatory interaction with each other (Spoel et al. 2003). SA and JA have been shown to take part in the removal of excess ROS (Parra-Lobato et al. 2009; Nazar et al. 2011) and protect plants from oxidative damage.

Since salinity is considered as one of the potential threats for agricultural productivity, the present review focuses mainly to improve our understanding on the effects of salinity stress on plant physiology and metabolism and elucidates the potential mechanisms of SA and JA in modulating salinity stress response.

12.2 Biosynthesis of Salicylic Acid

SA may be synthesized via the phenylalanine or isochorismate pathways (Fig. 12.2) (Kawano et al. 2004; Mustafa et al. 2009). The phenylalanine pathway is the most common pathway in plants. After a series of reactions, SA is produced by the enzyme benzoic acid 2-hydroxylase, which catalyzes the hydroxylation of benzoic acid at the orthoposition (at C-2 position). Benzoic acid is synthesized through a series of reactions starting from cinnamic acid (*trans*-cinnamic acid) either via a β-oxidation of fatty acids or a nonoxidative pathway (Verberne et al. 1999; Mustafa et al. 2009). *Trans*-cinnamic acid is produced from phenylalanine by the action of the enzyme phenylalanine ammonia lyase (PAL). This enzyme is known to be induced by different types of biotic and abiotic stresses and is a key regulator of the phenylpropanoid pathway, which gives rise to various types of phenolics with multiple functions (Yalpani et al. 1993).

In the isochorismate pathway, chorismate is converted to isochorismate by the activity of isochorismate synthase (ICS), which is subsequently converted to SA by isochorismate pyruvate lyase (Mustafa et al. 2009). For example, the biosynthesis of the SA analog 2,3-dihydroxybenzoic in Madagascar Periwinkle takes place via isochorismate, whereas in some plants belonging to Rubiaceae family, ICS gives rise to anthraquinones (Moreno et al. 1994; Budi Muljono et al. 2002). In *Arabidopsis*, ICS has been found to be involved in the biosynthesis of SA during the plant defense process (Wildermuth et al. 2001).

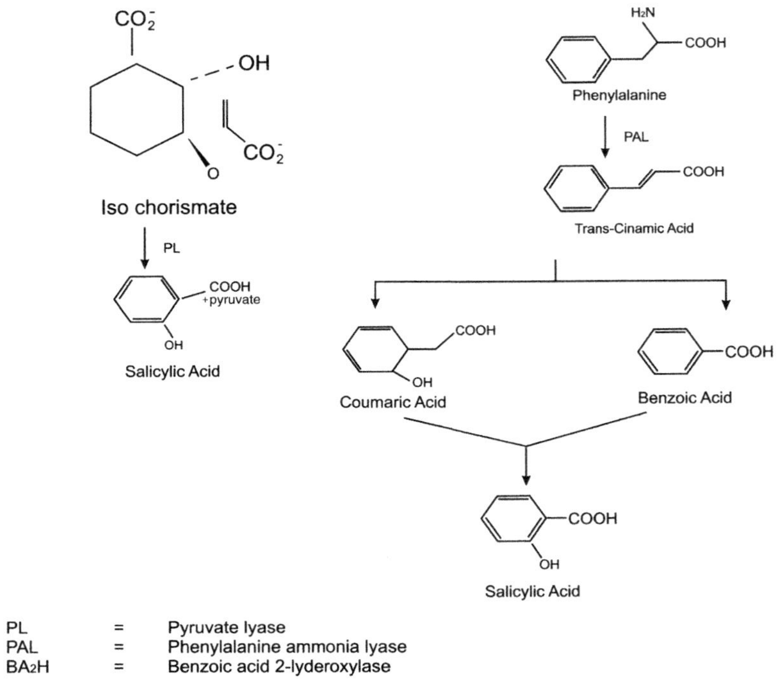

PL = Pyruvate lyase
PAL = Phenylalanine ammonia lyase
BA_2H = Benzoic acid 2-lyderoxylase

Fig. 12.2 Biosynthesis of salicylic acid

12.3 Biosynthesis of Jasmonic Acid

The pathway for JA synthesis from α-linolenic acid (18:3) was first proposed by Vick and Zimmerman (1983) (Fig. 12.3), linolenic acid, 18:3, released from plasma membrane glycerolipids (Hyun et al. 2008). The 18:3 is converted to 13-hydroperoxylinoleic acid (13-HPOT) by 13-lipoxygenase, and then allene oxide synthase (AOS) (Lee et al. 2008) produces 12,13-epoxyoctadecatrienoic acid, which is acted on by allene oxide cyclase (AOC). The AOC enzyme determines the stereoconfiguration of the product as (9S, 13S)-12-oxo-phytodienoic acid (OPDA) (Ziegler et al. 2000). The same enzymes act on 16:3 to form dinor-OPDA (Weber et al. 1997). OPDA has been identified as a substituent at sn-1 of the chloroplast lipid monogalactosyldiacylglycerol (Andersson et al. 2006), which causes the release of OPDA by lipases that contribute to JA synthesis. A specific isozyme of OPDA reductase is required to reduce (9S, 13S) OPDA to 3-oxo-2(2- pentenyl) cyclopentane-1-octanoic acid (OPC- 8:0) (Sanders et al. 2000; Stintzi and Browse 2000), which is then converted to (3R, 7S) jasmonic acid by three cycles of β-oxidation (Vick and Zimmerman 1983). Evidences indicate that the synthesis of OPDA occurs in the chloroplast (plastid) (Blee 1998), whereas the final production of JA occurs in the peroxisome, the only known site of β-oxidation in plants

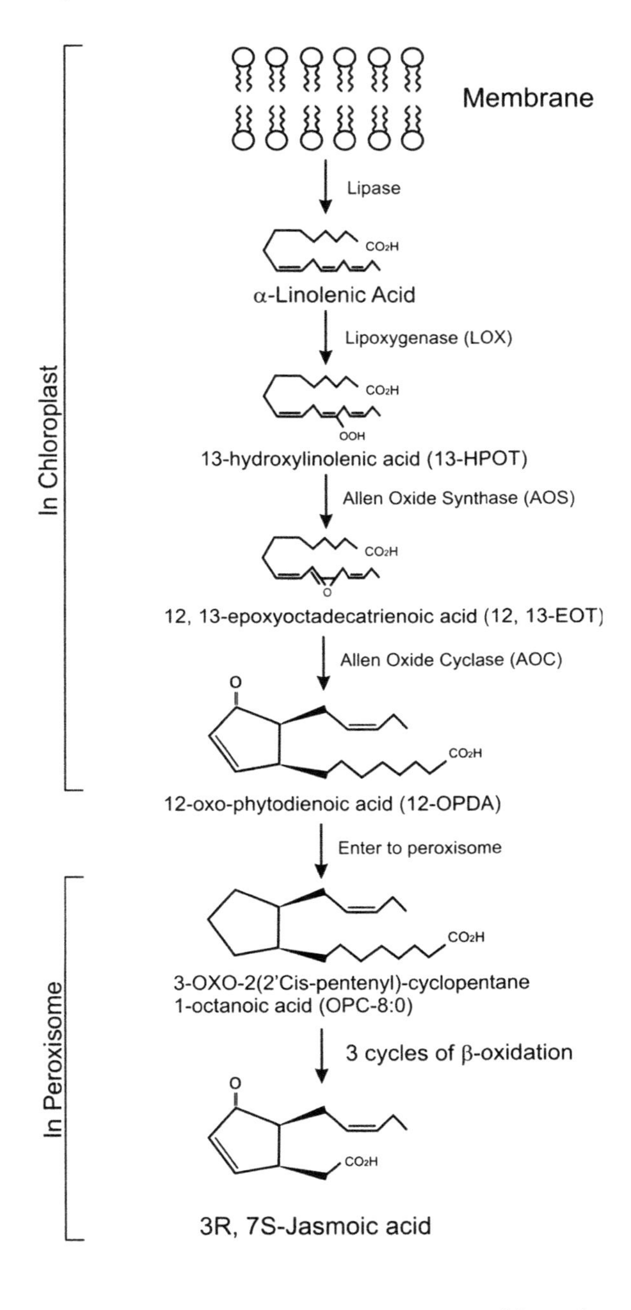

Fig. 12.3 Biosynthesis of jasmonic acid

(Ziegler et al. 2000). In plants, β-oxidation is catalyzed by acyl-CoA oxidase, the multifunctional protein (MFP) (which exhibits 2-*trans*-enoyl-CoA hydratase, l-3-hydroxyacyl-CoA dehydrogenase, D-3-hydroxyacyl-CoA epimerase, and 3-2-enoyl-CoA isomerase), and l-3-ketoacyl-CoA thiolase.

12.4 Salinity Stress and Plant Responses: A General Aspect

12.4.1 Photosynthetic Responses Under Salt Stress

Photosynthesis, together with cell growth, is among the primary processes to be affected by salinity (Munns et al. 2006). The effects of salt stress on photosynthesis are either direct (as the diffusion limitations through the stomata and the mesophyll and the alterations in photosynthetic metabolism) or secondary, such as the oxidative stress arising from the superimposition of multiple stresses (Chaves et al. 2009). Photosynthesis and its related physiological variables are invariably affected by the soil salinity in plants (Parida and Das 2005; Chaves et al. 2009). Salinity-caused reduction in photosynthetic pigments and stomatal conductance directly and/or indirectly affects photosynthesis (Flexas et al. 2007; Nazar et al. 2011). A significant reduction in leaf water potential was observed under high salinity stress in *Pueraria lobata* (Al-Hamdani 2004). It has been shown that salinity-induced reductions in photosynthesis resulted from decreased CO_2 availability (Flexas et al. 2007) or in the alterations of photosynthetic metabolism (Lawlor and Cornic 2002).

Genes or proteins associated with photosynthetic pathways were in general not among the most altered by the stress. For example, in Thellungiella (a stress-tolerant plant), photosynthesis genes correspond to 15% of all genes downregulated (Wong et al. 2006), while in rice, alterations in photosynthesis-related genes are mostly associated with stress recovery (Zhou et al. 2007). As a result of being relatively unaffected by salinity and drought, photosynthesis-related genes and proteins have not been deeply analyzed so far.

12.4.2 Reactive Oxygen Species and Salt Stress

The increased production of ROS occurs under all kinds of stresses, although their identity and compartment of origin may differ (Li et al. 2009). The toxic effect of salinity is through oxidative stress caused by enhanced production of ROS (Giraud et al. 2008). These ROS may be signals inducing ROS scavengers and other protective mechanisms, as well as damaging agents contributing to stress injury in plants (Prasad et al. 1994). Controlling ROS production and scavenging in the chloroplast are shown to be essential for tolerance to salinity in plants and in salinity-tolerant cultivars (Tseng et al. 2007). Salt stress induces water deficit and increases ionic and osmotic effects leading to the generation of ROS and oxidative stress (Parida and Das 2005). Salt stress manifested as an oxidative stress, mediated by ROS, and has deleterious effects (Lopez-Berenguera et al. 2007). It is clear that ROS contribute to stress damage, as evidenced by observations that transgenic plants overexpressing ROS scavengers or mutants with higher ROS scavenging ability show increased tolerance to environmental stresses (Kocsy et al. 2001).

The connections between ROS signal transduction and salt stress signal transduction are an exciting aspect for further studies (Miller et al. 2010).

12.4.3 Ion Homeostasis Under Salt Stress

In salt stress, the rate of increase in ambient salt concentrations can lead to ion toxicity primarily due to the accumulation of Na^+ and Cl^-, which leads to decrease in chemical activity causing cells to lose turgor and simultaneously to alterations in various physiological processes (Manchanda and Garg 2008). In fact, excess Na^+ and Cl^- causes negative impacts on the acquisition and homeostasis of essential nutrients (Greenway and Munns 1980) and causes conformational changes in protein structure and membrane depolarization (Manchanda and Garg 2008). The reduction in photosynthesis has been associated with the disturbance in homeostasis of Na + and Cl − ions and essential mineral nutrients (Gunes et al. 2007; Keutgen and Pawelzik 2009), stomatal closure (Steduto et al. 2000), reduction in leaf water potential (Silva et al. 2008), and the increased production of ROS in chloroplasts (Meneguzzo et al. 1999). It is also suggested that Cl^- ion accumulation adversely affects photosynthesis (Khayyat et al. 2009). High K^+/Na^+ selectivity in plant under salt stress has been suggested as an important selection criterion for salt tolerance (Wenxue et al. 2003). Under saline condition, due to excessive amounts of exchangeable Na^+/K^+ and Na^+/Ca^+ ratios occurring in the soil, plants subjected to such environments take up high amounts of Na^+, whereas the uptake of K^+ and Ca^+ is considerably reduced. Reasonable amounts of K^+ and Ca^+ are required to maintain the integrity and functioning of cell membranes (Wenxue et al. 2003).

Cellular ion homeostasis under salt stress can be achieved by the following strategies:

1. Utilization of Na^+ for osmotic adjustment by compartmentation of Na + into vacuole through tonoplast Na^+/H^+ antiporters
2. Exclusion of Na^+ from cell by plasma membrane-bound Na^+/H^+ antiporters or by limiting the Na^+ entry
3. Na^+ secretion

12.4.4 Salt Tolerance Mechanism by Salt Overly Sensitive Pathway

Regulation of ion transport system is fundamental to plant salt tolerance. The salt overly sensitive (SOS) salt stress signaling pathway was determined to have a pivotal regulatory function in salt tolerance, fundamental of which is the control of ion homeostasis (Hasegawa et al. 2000; Sanders 2000; Zhu 2000). The molecular identities of key ion transport systems that are fundamental to plant salt tolerance are reported by Hasegawa et al. (2000). Regulation of ion (Na^+ and K^+) homeostasis involving SOS genes has been recently suggested by the SOS pathway. The SOS signal pathway is a pivotal regulator of, at least some, key transport systems required for ion homeostasis (Sanders 2000; Zhu 2000). The input of SOS pathway is due to excessive intracellular or extracellular Na^+, which somehow triggers a cytoplasmic Ca^+ signal (Zhu 2001). SOS3 encodes a Ca^+-binding protein with sequence similarity to the regulatory B subunit of calcineurin (protein phosphatase 2B) and neuronal Ca^+ sensors (Liu and Zhu 1998; Ishitani et al. 2000). Cellular ion homeostasis of SOS pathway under salt stress regulates by SOS2 and SOS1. Molecular genetic analysis of the SOS3-SOS2 pathway was identified in the *sos1* mutant of Arabidopsis. As with *sos2* and *sos3*, *sos1* is hypersensitive to salt and all three mutants accumulate higher levels of Na^+ than is found in wild-type plants. Genetic analysis confirmed that SOS3, SOS2, and SOS1 function in a common pathway of salt tolerance (Zhu et al. 1998). Interaction of SOS3 with the SOS2 kinase (Liu et al. 2000) and SOS2 activation is Ca^+ dependent (Halfter et al. 2000). The function of SOS3 as a salt tolerance determinant is dependent on Ca^+ binding (Ishitani et al. 2000). The kinase activity of SOS2 is essential for its salt tolerance determinant function (Zhu 2000). The SOS2 C-terminal regulatory domain interacts with the kinase domain to cause autoinhibition. Regulatory domain of SOS2 is the site where SOS3 interacts with the kinase and is the autoinhibitory domain of the kinase (Guo et al. 2001). Binding of SOS3 to this motif blocks autoinhibition of SOS2 kinase activity. Deletion of the autoinhibitory domain results in constitutive SOS2 activation, independent of SOS3. A Thr/Asp mutation in the activation loop of the kinase domain constitutively activates SOS2. The plasma membrane-sited Na^+/H^+ antiporter SOS1 is controlled by the SOS pathway at the transcriptional and posttranscriptional level (Guo et al. 2001; Zhu 2001). In addition to positive control of Na^+ exclusion from cytosol, the SOS pathway may also negatively regulate Na^+ influx systems. Expression of plant high-affinity K^+ transporters, *atHKT1*, and *ecHKT1* in *Xenopus laevis* oocytes showed that they could mediate Na^+ uptake. Laurie et al. (2002) showed in transgenic wheat plants that expressing the wheat *HKT1* in antisense orientation of an ubiquitin promoter significantly less Na^+ uptake and enhanced growth under salt stress with respect to control. Functional disruption of *AtHKT1* was shown to suppress the salt-sensitive phenotype of sos3-1 mutant indicating that the SOS pathway negatively controls this Na^+ influx system (Rus et al. 2001), these results revealed that *HKT* mediates sodium uptake under salt stress, and salt tolerance can be improved by downregulation of *HKT1* expression. Similarly, the SOS pathway negatively controls expression of *AtNHX* family members that are implicated as determinants in the salt stress response (Yokoi et al. 2002) (Fig. 12.4).

SOS pathway functioning in response to Ca^+ and salt stress signaling in plants might have general implications and plays important role in plant growth and development. Plant adaptation to different stresses is dependent upon the activation of cascades of molecular networks involved in stress perception, signal transduction, and expression of specific stress-responsive genes. The low Ca^+/Na^+ ratio of a saline medium plays a significant role in growth inhibition in addition to causing significant changes in morphology and anatomy of plants (Mass and Grieve 1987;

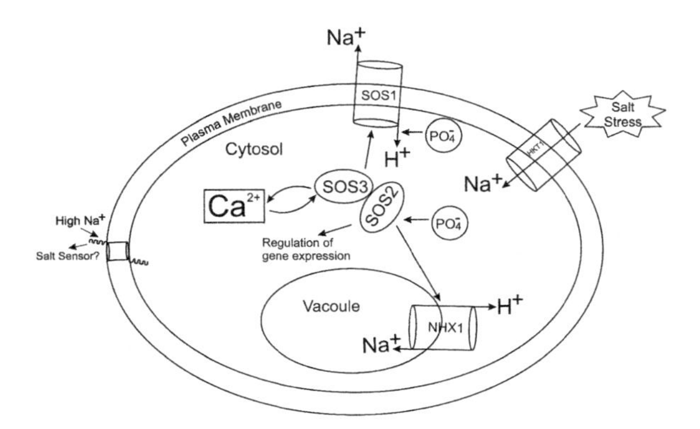

Fig. 12.4 Diagrammatic presentation of SOS pathway involved in coping up with salt stress. Under salt stress, the increase in cytosolic Na^+ level causes damage to several cellular processes. One of the major salt-detoxifying mechanisms in the cell is calcium-activated SOS3–SOS2 protein complex, which activates SOS1which is responsible for extrusion of Na^+ out of the cell. At the same time, SOS3–SOS2 complex is also involved in inhibiting HKT1, low-affinity potassium transporter, which transports Na^+ ion under high-salt condition. Recently, extrusion of Na^+ ion (by regulating SOS1) and sequestration/compartmentalization of Na^+ ion into the vacuole (activating NHX-type transporter which pump Na^+ ion into the vacuole)

Cramer 1992). The maintenance of Ca^+ acquisition and transport under salt stress is an important detriment of salinity tolerance (Unno et al. 2002).

Although, salinity in soil could be somewhat relieved through farm management practices, such as better irrigation practice, phase farming, intercropping, and precision farming (Munns 2002). It takes a long time and high cost to improve soil quality suitable for crop growth. Therefore, it is imperative that researchers develop breeding strategies and technologies to make crops more productive under stressful environments (Cushman and Bohnert 2000). Thus, in order to reduce salinity stress, plants speed up their rate of ROS production that sends signal to activate antioxidants for ROS scavenging. Plants containing high levels of antioxidants can scavenge/detoxify ROS, thereby contributing to increased salt tolerance (Demiral and Turkan 2005). These antioxidants can be speeded up by means of supplement of phytohormones.

Several reports have been published that phytohormones can enhance tolerance of plant (Afzal et al. 2005; Yoon et al. 2009; Syeed et al. 2011). Among all phytohormones, SA and JA have been established as potential enhancer of tolerance under salinity. Gemes et al. (2011) reported that SA can reduce ROS production. SA is also believed to play a role in plant responses to abiotic stresses including osmotic stress, drought, salt, and UV stress (Al-Hakimi and Hamada 2001; Horváth et al. 2007). MeJA has an important role in alleviation of salinity stress in plant. Yoon et al. (2009) have reported that MeJA counteracted the negative effects of NaCl stress on plant growth, chlorophyll content, leaf photosynthetic rate, leaf transpiration rate, and proline content. Parra-Lobato et al. (2009) concluded that exogenous MeJA may be involved in the oxidative stress processes by regulating antioxidant enzyme activities.

12.5 Phytohormones in Salinity Tolerance

Salinity is one of the major abiotic stresses which are a threat to the world climate and crop production. Salinity brings numerous hazardous changes in soil as well as in plants. Salinity adversely affects plant growth and development, hindering seed germination (Dash and Panda 2001), enzyme activity (Seckin et al. 2009), central dogma process (Anuradha and Rao 2001), and cell division (Tabur and Demir 2010). Numerous studies have shown the effects of salinity on plants (Jamil et al. 2007; Duan et al. 2008). Researcher has investigated more on the mechanisms of salt tolerance in plants (Dajic 2006; Munns and Tester 2008). Some researchers have used PGRs for reducing or eradicating the negative effects of salinity (Kabar 1987). Phytohormones suggested playing important roles in stress responses and adaptation (Sharma et al. 2005; Shaterian et al. 2005). Jung and Park (2011) showed that NTM2 (a membrane-bound NAC transcription factor) is a molecular link that incorporates auxin signal into salt stress signaling during seed germination, providing a role of auxin in modulating seed germination under high salinity. Similarly, cytokinins can also enhance resistance to salinity and high temperature in plants (Barciszewski et al. 2000). Seed enhancement (seed priming) with cytokinin is reported to increase plant salt tolerance (Iqbal et al. 2006). It is thought that the repressive effect of salinity on seed germination and plant growth could be related to a decline in endogenous levels of phytohormones (Zholkevich and Pustovoytova 1993; Jackson 1997).Wang et al. (2001) clearly defined that ABA and JA will be increased in response to salinity, whereas indole-3-acetic acid (IAA) and salicylic acid (SA) are declined. Recently, Hamayun et al. (2010) reported same observation in soya bean plant, phytohormonal analysis of soybean showed that the level of bioactive gibberellins (GA_1 and GA_4) and jasmonic acid increased in GA_3 treated plants, while the endogenous abscisic acid (ABA) and SA contents declined under the same treatment. ABA is a major internal signal enabling phytohormone of plants to survive adverse environmental conditions such as salt stress (Keskin et al. 2010). There is good evidence for the effect of enhanced accumulation of endogenous ABA on root growth under salinity (Mulholland et al. 2003). Chen et al. (2006) demonstrated the importance of applied ABA in root growth, morphology, and regulation of ion accumulation. The negative effect of NaCl salt on root nodule dry weight of common bean has also been shown to be alleviated by exogenous ABA supply (Khadri et al; 2007). Phytohormones can be applied exogenously which are benefit for plants under salt stress. For example, auxins (Khan et al. 2004), gibberellins (Afzal et al. 2005), cytokinins (Gul et al. 2000), salicylic acid (Syeed et al. 2011), and jasmonic acid (Yoon et al. 2009) produce benefit in alleviating the adverse effects of salt stress and also improve germination, growth, development, and seed yields and yield quality (Egamberdieva 2009).

12.5.1 Role of Salicylic Acid in Plants Under Salinity Stress

SA is an endogenous growth regulator of phenolic nature, which participates in the regulation of physiological processes in plants such as growth, photosynthesis, and nitrogen metabolism (Nazar et al. 2011) and also provides protection against biotic and abiotic stresses such as salinity (Khan et al. 2010; Syeed et al. 2011). SA has been known to be present in some plant tissues for quite some time but has only recently been recognized as a potential plant growth regulator (Nazar et al. 2011). Jumali et al. (2011) showed that most genes responding to acute SA treatment are related to stress and signaling pathways which eventually led to cell death. This includes genes encoding chaperone, heat shock proteins (HSPs), antioxidants, and genes involved in secondary metabolite biosynthesis, such as sinapyl alcohol dehydrogenase (SAD), cinnamyl alcohol dehydrogenase (CAD), and cytochrome P450 (CYP 450). Several methods of application (soaking the seeds prior to sowing, adding to the hydroponic solution, irrigating, or spraying with SA solution) have been shown to protect various plant cultivars against abiotic stress factors by inducing a wide range of processes involved in stress tolerance mechanisms (Horváth et al. 2007). The role of SA in defense mechanism to alleviate salt stress in plants was studied (Hussein et al. 2007). The ameliorative effects of SA have been well documented including salt tolerance in many crops such as bean (Azooz 2009), mung bean (Khan et al. 2010), and mustard (Syeed et al. 2011). Sakhabutdinova et al. (2003) investigated the effect of SA on plant resistance against environmental stress factors. Treatment of wheat plants with 0.05 mM SA increased the level of cell division within the apical meristem of seedling roots which increased the plant growth. Exogenous application of SA enhanced the photosynthetic rate and also maintained the stability of membranes, thereby improved the growth of salinity-stressed barley plants (El-Tayeb 2005) and mung bean plants (Nazar et al. 2011). SA added to the soil also had an ameliorating effect on the survival of maize and mustard plants during salt stress and decreased the Na^+ and Cl^- accumulation (Gunes et al. 2007). Lipid peroxidation and membrane permeability, which were increased by salt stress, were lower in SA-treated plants (Horváth et al. 2007). SA treatment was accompanied by a transient increase in the H_2O_2 level. As seed treatment with H_2O_2 itself had an alleviating effect on the oxidative damage caused by salt stress in wheat plants (Wahid et al. 2007), it seems possible that SA may exert its protective effect partially through the transiently increased level of H_2O_2. The endogenous level of SA increased under salt stress in rice seedlings, and the activity of the SA biosynthesis enzyme, benzoic acid 2-hydroxylase, was induced (Sawada et al. 2006). Treatment with SA essentially diminished the alteration of phytohormone levels in wheat seedlings under salinity. It was found that the SA treatment caused accumulation of both ABA and IAA in wheat seedlings under salinity. However, the SA treatment did not influence on cytokinin content (Sakhabutdinova et al. 2003). Thus, protective SA action includes the development of antistress programs and acceleration of normalization of growth processes after removal stress factors (Sakhabutdinova et al. 2003). The results obtained in the last few years strongly argue that SA could be a very promising compound for the reduction of the abiotic stress sensitivity of crops because under certain conditions, it has been found to mitigate the damaging effects of various stress factors in plants.

SA is involved in the induction of pathogen-regulated (PR) genes and also activates mitogen-activated protein kinases (MAPK) in tobacco cell cultures (Zhang and Klessig 1997). MAPKs are proteins found in all eukaryotes, which mediate responses to a variety of extracellular stimuli (Jonak et al. 2002). MAPK pathways have been implicated in signal transduction for a wide variety of stress responses (Zhang and Klessig 2001). The PR proteins are induced in pathological or related situations (Van Loon et al. 2006). The PR-1 proteins are often used as markers of an enhanced defense state conferred by pathogen-induced systemic acquired resistance (SAR), but their biological activity remains elusive (Van Loon et al. 2006). It has been reported that overexpression of OsMAPK5 results in enhanced tolerance to abiotic stresses, such as drought and salinity, in rice (Xiong and Yang 2003). However, suppression of OsMAPK5 expression reduced abiotic stress tolerance but led to constitutive PR-1 and PR-10 expression and increased resistance to fungal and bacterial pathogens (Xiong and Yang 2003). PsMAPK3 expression was not altered by salt stress however PsMAPK3 expression was higher in the absence of SA compared to the presence of SA, this suggested that SA treatment could enhance the resistance of salt-stressed plants to possible opportunistic pathogen attack, by increased PR-1b gene expression. Under the same conditions, pea showed increased sensitivity to salinity, as indicated by the higher percentage of symptomatic leaves. In parallel, the decrease in PsMAPK3 expression correlated with higher expression of the PR-1b gene in salt-stressed plants treated with SA (Barba-Espιn et al. 2011).

Several studies published indicate that the role of SA as cell stress protectant through induction of antioxidant system (Gemes et al. 2011). SA has been reported to influence the activities of antioxidative enzymes differentially. In general, authors found enhancement in the activities of antioxidant enzymes such as superoxide dismutase (SOD), glutathione reductase (GR), and ascorbate peroxidase (APX) with NaCl and SA application. The increases in the activities of antioxidant enzymes following SA application could be the indicator of buildup of a protective mechanism against oxidative damage induced by salt stress through increase in nutrient contents and antioxidative metabolism (Khan et al. 2010). Exogenously applied SA involved in the induction of at least one of cellular mechanisms that are concomitant with the accumulation of ROS (Gunes et al. 2007). It was found that inhibition of catalase (CAT), a H_2O_2 scavenging enzyme, by SA, plays a major role in the generation of ROS (Horváth et al. 2007). SA inhibited the activities of CAT and APX and increased the content of H_2O_2 (Durner and Klessig 1995; Kawano and Muto 2000; Luo et al. 2001). Moharekar et al. (2003) suggested that an increase in SA concentration might induce oxidative stress in wheat and mung bean, but the degree of oxidative stress was different in different plant species. CAT activity from cucumber (*Cucumis sativus*), tomato, *Arabidopsis*, and tobacco has been found substantially inhibited by SA, whereas those from maize and rice were found to be insensitive (Sanchez-Casas and Klessig 1994). In hypersensitive responses, SA is thought to potentiate ROS signaling (Klessig et al. 2000).

12.5.2 Role of Jasmonic Acid in Salt-Stressed Plants

Jasmonic acid (JA) and methyl jasmonate (MeJA) are collectively referred to as jasmonates and are important cellular regulators which are involved in diverse developmental processes such as seed germination, root growth, fertility, fruit ripening, and senescence (Wasternack and Hause 2002). Exposure of plants to salt stress results in changes in most physiological and biochemical processes resulting in a disturbance of normal growth and development. The response of plants to salt stress depends on multiple factors, but phytohormones are thought to be among the most important endogenous substances involved in the mechanisms of tolerance or susceptibility of various plant species (Velitcukova and Fedina 1998). It was reported that many of the proteins produced by the plant under abiotic stress are induced by phytohormones, such as abscisic acid (Jin et al. 2000), salicylic acid (Hoyos and Zhang 2000), and jasmonates (Thaler 1999). They activate plant defense mechanisms in response to insect-driven wounding, pathogens, and environmental stresses, including drought, low temperature, and salinity (Wasternack and Parthier 1997). Under salt stress, jasmonates recovered salt inhibition on dry mass production in rice (Kang et al. 2005) and diminished the inhibitory effect of NaCl on the rate of $^{14}CO_2$ fixation, protein content in *Pisum sativum* (Velitcukova and Fedina 1998). Jasmonates have been the focus of much attention because of their ability to provide protection to salinity stress (Tsonev et al. 1998) or to increase freezing tolerance in brome grass (Wilen et al. 1994), leading to the suggestion that jasmonates could mediate the defense response to various environmental stresses.

Phospholipase D (PLD) has also been shown to trigger the release of linolenic acid and to stimulate JA biosynthesis (Creelman and Mullet 1997). PLD activity has been linked with stress processes playing a main function in membrane deterioration, although there is enough evidence for its role in plant signal transduction (Wang 1999). Jasmonic acid and its derivatives also respond to salinity (Wang et al. 2001). It has been reported that jasmonate treatments (or endogenous of these compounds) are accompanied by the synthesis of abundant proteins in response to abiotic stress, called jasmonate-induced proteins (JIPs) (Sembdner and Parthier 1993). It has been reported that JA levels in tomato cultivars changed in response to salt stress, and JA increase was observed in salt-tolerant cultivar HF (Hellfrucht Fruhstamm) from the beginning of salinization, while in salt-sensitive cultivar Pera, JA level decreased after 24 h of salt treatment (Pedranzani et al. 2003). A rapid increase was seen in endogenous JA content in barley leaf segments subjected to osmotic stress with sorbitol or mannitol (Kramell et al. 2000); however, endogenous jasmonates did not increase when they were treated with a high NaCl concentration (Kramell et al. 1995). The changes of endogenous JA levels in rice plants under various salt stresses were investigated. Kang et al. (2005) reported that the concentrations of JA in salt-sensitive cultivar plants were lower than in salt-tolerant cultivar plants. In addition, MeJA levels in rice roots increased significantly in 200 mM NaCl (Moons et al. 1997). Therefore, high levels of JA in salt-tolerant plants accumulated after salt treatments can be an effective protection against high salinity. However, there seems to be little information about how salinity affects endogenous JA levels in plants. Kang et al. (2005) reported that postapplication with exogenous JA can ameliorate salt-stressed rice seedlings, especially the salt-sensitive rather than the salt-tolerant cultivar. In addition, sodium concentration dramatically decreased by exogenous JA application. On the contrary, uptake of major ions was partially increased by JA application. At the same time, pretreatment with JA reduced the inhibitory effect of high salt concentrations on growth and photosynthesis of barley (Tsonev et al. 1998). Exogenous JA application after salt treatment may change the balance of endogenous hormones, such as ABA, which provides an important clue for understanding the protection mechanisms against salt stress (Kang et al. 2005). These results clearly demonstrate that exogenous JA may be involved in the defense not only during wounding and pathogen stress but also during salt stress.

Plants respond to a variety of environmental stresses through the induction of antioxidant defense enzymes that protect against further damage (Kawano and Muto 2000). It has been observed that MeJA is not only to regulate a variety of plant developmental responses but also to be induced by pathogen attack or wounds that often lead to the generation of ROS, including hydrogen peroxide (H_2O_2), superoxide anions ($O_2.^-$), and hydroxyl free radicals (.OH) (Devoto and Turner 2003). These observations suggest that MeJA could be linked to oxidative stress. ROS have the potential to interact with many cellular components, leading to membrane damage, causing an immediate cellular response to trigger a plant-defense signal (Van Breusegem et al. 2001)

Antioxidative enzymes include catalase (CAT) and peroxidase (POD) that remove H_2O_2 and superoxide dismutase (SOD) that catalyzes the disproportion of superoxide radicals to hydrogen peroxide and dioxygen. Activity of CAT, POD, and SOD increased during biotic and abiotic stresses to protect cells from the potential hazardous effects of ROS (Menezes-Benavente et al. 2004). Popova et al. (2003) have reported the accumulation of SOD, CAT, and peroxide (POX) isoforms in barley under MeJA treatment. The peroxidase isozymic pattern showed two specific isoforms, and the activities of SOD and CAT were significantly elevated in JA-treated peanut leaves (Kumari et al. 2006).

12.6 Interaction Between Salicylic Acid and Jasmonic Acid

There are several ways in which the phytohormonal pathways are known to interact and regulate the growth and development of plants (Gfeller et al. 2010). Interaction between phytohormonal-mediated signaling pathways depends on contraction and timing of elicitation or nature of stresses. Salicylic acid (SA) and jasmonic acid (JA) pathways are two of the key biochemical response mechanisms that can be triggered by various biotic and abiotic stresses and function as necessary signaling molecules responsible for defensive responses in plants. SA and JA are phytohormones that are important regulators of biotic and abiotic stress responses in plants which are biochemically linked (Tuteja and Sopory 2008). Antagonistic interactions between SA and JA affect the expression of PR protein genes in tomato

(Thaler et al. 1999). SA induces acidic pathogen-related (PR) genes and inhibits basic PR genes, whereas JA does the opposite (Wang et al. 2001).

12.6.1 Interaction Between Salicylic Acid and Jasmonic Acid at Signaling Level

Mitogen-activated protein kinase 4 (MAPK4) has been identified as another key component involved in mediating the antagonism between salicylic acid (SA)- and jasmonic acid (JA)-mediated signaling in Arabidopsis. Results indicate that MAPK4 acts as a negative regulator of SA signaling and positive regulator of JA signaling in Arabidopsis. The Arabidopsis mpk4 mutants show elevated SA levels, constitutive expression of SA responsive PR genes, and increased resistance to Pst. In contrast, the expression of JA responsive genes and the resistance to *A. brassicicola* were found to be impaired in mpk4 mutants (Brodersen and Petersen 2006). Glutaredoxin, GRX480, is an important regulator identified to affect antagonism between SA- and JA-mediated signaling. GRX480 is disulfide reductases which catalyze thiol disulfide reductions and is involved in the redox regulation of protein activities involved in a variety of cellular processes (Meyer et al. 2008). Ndamukong et al. (2007) have shown that GRX480 interacts with TGA transcription factors involved in the regulation of SA-responsive PR genes. The expression of GRX480 is induced by SA and requires TGA transcription factors and NPR1. Furthermore, the expression of JA-responsive PDF1.2 gene was inhibited by GRX480 (Ndamukong et al. 2007). These findings suggest that SA-induced NPR1 activates GRX480, which forms a complex with TGA factors and suppresses the expression of JA-responsive genes.

Mutual antagonism between SA and JA was also evident from a microarray study of defense-related mutants infected with *P. syringae* pv. *maculicola* (Glazebrook et al. 2003). This showed that expression of a cluster of SA-related genes, including PR1, was increased in JA-insensitive mutants, while JA-related genes showed increased expression in SA pathway mutants. Inhibition of SA signaling by JA also occurs, as activation of JA signaling in bacterial nahG salicylate hydroxylase also impairs SAR induction, although nahG expression has pleiotropic effects beyond SA catabolism (Heck et al. 2003). Jasmonic acid (JA) appears to be dispensable for SAR activation (Pieterse et al. 1998).

12.6.2 Biosynthetic Interaction Between Salicylic Acid and Jasmonic Acid

SA has antagonistic effects on JA by preventing its accumulation in response to wounding. In flax, SA inhibits transcription of allene oxide synthase (AOS), which mediates the conversion of lipoxygenase-derived fatty acid hydroperoxides to unstable allene epoxides and then to JA precursors (Harms et al. 1998). The wound-induced increase of leucine aminopeptidase (LapA) RNAs, which is upregulated by JA, and salinity, was inhibited by SA (Chao et al. 1999). SA also induces acidic pathogen-related (PR) genes and inhibits basic PR genes, whereas JA does the opposite (Niki et al. 1998). SA also reduced the synthesis of tomato proteinase inhibitors (Doares et al. 1995). Antagonistic interactions between SA and JA affect the expression of PR protein genes in tomato (Thaler et al. 1999). Two drought-inducible genes in the drought-tolerant cowpea (*Vigna unguiculata*) were identified after treatment with high salinity and exogenous ABA. One of these genes was also expressed in response to heat stress, methyl jasmonate, and salicylic acid (Iuchi et al. 1996), suggesting that SA affects plant resistance to stress.

12.7 Conclusion and Future Prospects

Plant hormones regulate a number of signaling networks involving developmental processes and plant responses to environmental stresses including biotic and abiotic stresses. Significant progress has been made in identifying the key components and understanding plant hormone signaling [especially salicylic acid (SA) and jasmonic acid (JA)] and plant defense responses. Phytohormone signaling pathways are not isolated but rather interconnected with a complex regulatory network involving various defense signaling pathways and developmental processes. To understand how plants coordinate multiple hormonal components in response to various developmental and environmental cues is a major challenge for the future.

It can be concluded that changes in hormone levels in plant tissue are thought to be an initial process controlling growth reduction due to salinity. Therefore, NaCl-induced reduction in the plant growth can be mitigated by application of plant growth regulators. SA and JA are the two phytohormones having a potential role in alleviating salinity stress and can be used as a technique to improve or protect the salinity stress hazard effect on plants' productivity.

Many reports are available under biotic stress on interaction between SA and JA, but relationship under salinity is not yet known well. So it is a challenging work for today as salinity is a major threat of agriculture and productivity of plants. Therefore, it may be said that SA and JA could enhance tolerance of plants in their respective relationship.

References

Afzal I, Basra S, Iqbal A (2005) The effect of seed soaking with plant growth regulators on seedling vigor of wheat under salinity stress. J Stress Physiol Biochem 1:6–14

Agrawal GK, Rakwal R, Tamogami S, Yonekura M, Kubo A, Saji H (2002) Chitosan activates defense/stress response(s) in the leaves of *Oryza sativa* seedlings. Plant Physiol Biochem 40:1061–1069

Ahmad P, Sarwat M, Sharma S (2008) Reactive oxygen species. Antioxidants and signaling in plants. J Plant Biol 51:167–173

Al-Hakimi AMA, Hamada AM (2001) Counteraction of salinity stress on wheat plants by grain soaking in ascorbic acid, thiamin or sodium salicylate. Biol Plant 44:253–261

Al-Hamdani SA (2004) Influence of varied NaCl concentrations on selected physiological responses of kudzu. Asian J Plant Sci 3:114–119

Andersson MX, Hamberg M, Kourtchenki O, Brunnstrom A, McPhail KL (2006) Oxylipin profiling of the hypersensitive response in Arabidopsis thaliana. Formation of a novel oxo-phytodienoic acid-containing galactolipid, Arabidopside. Eur J Biol Chem 281:31528–31537

Anuradha S, Rao SSR (2001) Effect of brassinosteroids on salinity stress induced inhibition of seed germination and seedling growth of rice (*Oryza sativa* L.). Plant Growth Regul 33:151–153

Ashraf M (2009) Biotechnological approach of improving plant salt tolerance using antioxidants as markers. Biotechnol Adv 27:84–93

Azooz MM (2009) Salt stress mitigation by seed priming with salicylic acid in two faba bean genotypes differing in salt tolerance. Int J Agric Biol 11:343–350

Barba-Espɩn G, Clemente-Moreno MJ, Lvarez S, Garcɩa-Legaz MF, Hernandez JA, Dɩaz-Vivancos P (2011) Salicylic acid negatively affects the response to salt stress in pea plants. Plant Biol. doi:10.1111/j.1438-8677.2011.00461.x

Barciszewski J, Siboska G, Rattan SIS, Clark BFC (2000) Occurrence, biosynthesis and properties of kinetin (N6-furfuryladenine). Plant Growth Regul 32:257–265

Blee E (1998) Phytooxylipins and plant defense reactions. Prog Lipid Res 37:33–72

Brodersen P, Petersen M (2006) Arabidopsis MAP kinase 4 regulates salicylic acid- and jasmonic acid/ethylene-dependent responses via EDS1 and PAD4. Plant J 47:532–546

Budi Muljono RA, Scheffer JJC, Verpoorte R (2002) Isochorismate is an intermediate in 2,3-dihydroxybenzoic acid biosynthesis in *Catharanthus roseus* cell cultures. Plant Physiol Biochem 40:231–234

Cramer GR (1992) Kinetics of maize leaf elongation. II. Response of a Na-excluding cultivar and a Na including cultivar to varying Na/Ca salinities. J Exp Bot 43:857–864

Chao WS, Gu YQ, Pautot V, Bray EA, Walling LL (1999) Leucine aminopeptidase RNAs, proteins, and activities increase in response to water deficit, salinity, and the wound signals systemin, methyl jasmonate, and abscisic acid. Plant Physiol 120:979–992

Chaves MM, Flexas J, Pinheiro C (2009) Photosynthesis under drought and salt stress: regulation mechanisms from whole plant to cell. Ann Bot 103:551–560

Chen CW, Yang YW, Lur HS, Tsai YG, Chang MC (2006) A novel function of abscisic acid in the regulation of rice (*Oryza sativa* L.) root growth and development. Plant Cell Physiol 47:1–13

Creelman RA, Mullet JE (1997) Biosynthesis and action of jasmonates in plants. Annu Rev Plant Physiol Plant Mol Biol 48:355–381

Cushman JC, Bohnert HJ (2000) Genomic approaches to plant stress tolerance. Curr Opin Plant Biol 3:117–124

Dajic Z (2006) Salt stress. In: Madhava Rao KV, Raghavendra AS, Janardhan Reddy K (eds) Physiology and molecular biology of salt tolerance in plant. Springer, Netherlands

Daneshmand F, Mohammad JA, Khosrow MK (2009) Effect of acetylsalicylic acid (Aspirin) on salt and osmotic stress tolerance in *Solanum bulbocastanum* in vitro: enzymatic antioxidants. Am Eurasian J Agric Environ Sci 6:92–99

Dash M, Panda SK (2001) Salt stress induced changes in growth and enzyme activities in germinating *Phaseolus mungo* seeds. Biol Plant 44:587–589

del Rio LA, Sandalio LM, Corpas FJ, Palma JM, Barroso JB (2006) Reactive oxygen species and reactive nitrogen species in peroxisomes. Production, scavenging, and role in cell signaling. Plant Physiol 141:330–335

Demiral T, Turkan I (2005) Comparative lipid peroxidation, antioxidant defense systems and proline content in roots of two rice cultivars differing in salt tolerance. Environ Exp Bot 53:247–257

Devoto A, Turner JG (2003) Regulation of jasmonate-mediated plant responses in Arabidopsis. Ann Bot 92:329–337

Doares SH, Narváez-Vásquez J, Conconi A, Ryan CA (1995) Salicylic acid inhibits synthesis of proteinase inhibitors in tomato leaves induced by systemin and jasmonic acid. Plant Physiol 108:1741–1746

Duan J, Li J, Guo S, Kang Y (2008) Exogenous spermidine affects polyamine metabolism in salinity-stressed *Cucumis sativus* roots and enhances short-term salinity tolerance. J Plant Physiol 165:1620–1635

Durner J, Klessig DF (1995) Inhibition of ascorbate peroxidase by salicylic acid and 2,6-dichloroisonicotinic acid, two inducers of plant defense responses. Proc Natl Acad Sci USA 92:11312–11316

Egamberdieva D (2009) Alleviation of salt stress by plant growth regulators and IAA producing bacteria in wheat. Acta Physiol Plant 31:861–864

El-Tayeb MA (2005) Response of barley grains to the interactive effect of salinity and salicylic acid. Plant Growth Regul 45:215–225

Engelberth J, Koch T, Schuler G, Bachmann N, Rechtenbach J, Boland W (2001) Ion channel-forming alamethicin is a potent elicitor of volatile biosynthesis and tendril coiling. Cross talk between jasmonate and salicylate signaling in lima bean. Plant Physiol 125:369–377

Food and Agricultural Organization, FAO (2008). Land and plant nutrition management service. Available online at: http://www.fao.org/ag/agl/agll/spush/. Accessed 25 April

Flexas J, Diaz-Espejo A, Galmes J, Kaldenhoff R, Medrano H, Ribas-Carbo M (2007) Rapid variations of mesophyll conductance in response to changes in CO_2 concentration around leaves. Plant Cell Environ 30:1284–1298

Gemes K, Poor P, Horvath E, Kolbert Z, Szopko D, Szepesi A, Tari I (2011) Cross-talk between salicylic acid and NaCl-generated reactive oxygen species and nitric oxide in tomato during acclimation to high salinity. Physiol Plant 142:179–192

Gfeller A, Liechti R, Farmer EE (2010) Arabidopsis jasmonate signaling pathway. Sci Signal 3:109 cm4

Glazebrook J, Chen W, Estes B, Chang HS, Nawrath C, Metraux JP, Zhu T, Katagiri F (2003) Topology of the network integrating salicylate and jasmonate signal transduction derived from global expression phenotyping. Plant J 34:217–228

Giraud E, Ho LHM, Clifton R, Carroll A, Estavillo G, Tan YF, Howell KA, Ivanova A, Pogson BJ, Millar AH, Whelan J (2008) The absence of alternative oxidase1a in Arabidopsis results in acute sensitivity to combined light and drought stress. Plant Physiol 147:595–610

Greenway H, Munns R (1980) Mechanism of salt tolerance in non halophytes. Annu Rev Plant Physiol 31:149–190

Gul B, Khan MA, Weber DJ (2000) Alleviation salinity and dark enforced dormancy in Allenrolfea occidentalis seeds under various thermoperiods. Aust J Bot 48:745–752

Gunes A, Inal A, Alpaslan M, Eraslan F, Bagci EG, Cicek N (2007) Salicylic acid induced changes on some physiological parameters symptomatic for oxidative stress and mineral nutrition in maize (*Zea mays* L.) grown under salinity. J Plant Physiol 164:728–736

Guo Y, Halfter U, Ishitani M, Zhu JK (2001) Molecular characterization of functional domains in the protein kinase SOS2 that is required for plant salt tolerance. Plant Cell 13:1383–1400

Halfter U, Ishitani M, Zhu JK (2000) The Arabidopsis SOS2 protein kinase physically interacts with and is activated by the calcium-binding protein SOS3. Proc Natl Acad Sci USA 97:3735–3740

Hamayun M, Khan SA, Khan AL, Shin JH, Ahmad B, Shin DH, Lee IJ (2010) Exogenous gibberellic acid reprograms soybean to higher growth and salt stress tolerance. J Agric Food Chem 58:7226–7232

Harms K, Ramirez I, Penacortes H (1998) Inhibition of wound-induced accumulation of allene oxide synthase transcripts in flax leaves by aspirin and salicylic acid. Plant Physiol 118:1057–1065

Hasegawa PM, Bressan RA, Zhu JK, Bohnert HJ (2000) Plant cellular and molecular responses to high salinity. Annu Rev Plant Mol Plant Physiol 51:463–499

Heck S, Grau T, Buchala A, Metraux JP, Nawrath C (2003) Genetic evidence that expression of NahG modifies defence pathways independent of salicylic acid biosynthesis in the Arabidopsis–Pseudomonas syringae pv. tomato interaction. Plant J 36:342–352

Horváth E, Szalai G, Janda T (2007) Induction of abiotic stress tolerance by salicylic acid signaling. J Plant Growth Regul 26:290–300

Hoyos M, Zhang SQ (2000) Calcium-independent activation of salicylic acid-induced protein kinase and a 40-kilodalton protein kinase by hyper osmotic stress. Plant Physiol 122:1355–1363

Hussein MM, Balbaa LK, Gaballah MS (2007) Salicylic Acid and Salinity Effects on Growth of Maize Plants. Res J Agric Biol Sci 3:321–328

Hyun Y, Choi S, Hwang HJ, Yu J, Nam SJ, Ko J, Park JY, Seo YS, Kim EY, Ryu SB, Kim WT, Lee YH, Kang H, Lee, I (2008) Cooperation and functional diversification of two closely related galactolipase genes for jasmonate biosynthesis. Dev Cell 14:183–192

Iqbal M, Ashraf M, Jamil A (2006) Seed enhancement with cytokinins: changes in growth and grain yield in salt stressed wheat plants. Plant Growth Regul 50:29–39

Ishitani M, Liu J, Halfter U, Kim CS, Shi W, Zhu JK (2000) SOS3 function in plant salt tolerance requires N myristoylation and calcium binding. Plant Cell 12:1667–1677

Iuchi S, Yamaguchi-Shinozaki K, Urao T, Shinozaki K (1996) Characterization of two cDNAs for novel drought-inducible genes in the highly drought-tolerant cowpea. J Plant Res 109:415–424

Jackson M (1997) Hormones from roots as signals for the shoots of stressed plants. Elsevier Trends J 2:22–28

Jamil M, Lee KB, Jung KY, Lee DB, Han MS, Rha ES (2007) Salt stress inhibits germination and early seedling growth in cabbage (*Brassica oleracea capitata* L.). Pak J Biol Sci 10:910–914

Jin S, Chen CCS, Plant AL (2000) Regulation by ABA of osmotic-stress-induced changes in protein synthesis in tomato roots. Plant Cell Environ 23:51–60

Jonak C, Okresz L, Bogre L, Hirt H (2002) Complexity, crosstalk and integration of plant MAP kinase signalling. Curr Opin Plant Biol 5:415–424

Joseph B, Jini D, Sujatha S (2010) Insight into the role of exogenous salicylic acid on plants grown under environment. Asian J Crop Sci 2:226–235

Jumali SS, Said IM, Ismail I, Zainal Z (2011) Genes induced by high concentration of salicylic acid in Mtragyna speciosa. Aust J Crop Sci 5:296–303

Jung J, Park C (2011) Auxin modulation of salt stress signaling in Arabidopsis seed germination. Plant Signal Behav 6:1198–1200

Kabar K (1987) Alleviation of salinity stress by plant growth regulators on seed germination. J Plant Physiol 128:179–183

Kang DJ, Seo YJ, Lee JD, Ishii R, Kim KU, Shin DH, Park SK, Jang SW, Lee IJ (2005) Jasmonic acid differentially affects growth, ion uptake and abscisic acid concentration in salt-tolerant and salt-sensitive rice cultivars. J Agron Crop Sci 191:273–282

Kawano T, Muto S (2000) Mechanism of peroxidase actions for salicylic acid induced generation of active oxygen species and an increase in cytosolic calcium in tobacco cell suspension culture. J Exp Bot 51:685–693

Kawano T, Furuichi T, Muto S (2004) Controlled free salicylic acid levels and corresponding signaling mechanisms in plants. Plant Biotechnol 21:319–335

Keskin BC, Sarikaya AT, Yuksel B, Memon AR (2010) Abscisic acid regulated gene expression in bread wheat. Aust J Crop Sci 4:617–625

Keutgen AJ, Pawelzik E (2009) Impact of NaCl stress on plant growth and mineral nutrient assimilation in two cultivars of strawberry. Environ Exp Bot 65:170–176

Khadri M, Tejera NA, Lluch C (2007) Sodium chloride-ABA interaction in two common bean (*Phaseolus vulgaris*) cultivars differing in salinity tolerance. Environ Exp Bot 60:211–218

Khan NA, Nazar R, Anjum NA (2009) Growth, photosynthesis and antioxidant metabolism in mustard (*Brassica juncea* L.) cultivars differing in ATP-sulfurylase activity under salinity stress. Sci Hortic 122:455–60

Khan NA, Syeed S, Masood A, Nazar R, Iqbal N (2010) Application of salicylic acid increases contents of nutrients and antioxidative metabolism in mungbean and alleviates adverse effects of salinity stress. Int J Plant Biol 1:e1

Khayyat M, Rajaee S, Sajjadinia A, Eshghi S, Tafazoli E (2009) Calcium effects on changes in chlorophyll contents, dry weight and micronutrients of strawberry (*Fragaria ananassa* Duch.) plants under salt stress conditions. Fruits 64:1–10

Kim EH, Kim YS, Park SH, Koo YJ, Choi YD, Chung YY (2009) Methyl jasmonate reduces grain yield by mediating stress signals to alter spikelet development in rice. Plant Physiol 149:1751–1760

Klessig DF, Durner J, Noad R, Navarre DA, Wendehenne D, Kumar D, Zhou JM, Shali S, Zhang S, Kachroo P, Trifa Y, Pontier D, Lam E, Silva H (2000) Nitric oxide and salicylic acid signaling in plant defense. Proc Natl Acad Sci USA 97:8849–8855

Kocsy G, von Ballmoos P, Ruegsegger A, Szalai G, Galiba G, Brunold C (2001) Increasing the glutathione content in a chilling-sensitive maize genotype using safeners increased protection against chilling-induced injury. Plant Physiol 127:1147–1156

Kramell R, Atzorn R, Schneider G, Miersch O, Bruckner C, Schmidt J, Sembdner G, Parthier B (1995) Occurrence and identification of jasmonic acid and its amino acid conjugates induced by osmotic stress in barley leaf tissue. J Plant Growth Regul 14:29–36

Kramell R, Miersch O, Atzorn R, Parthier B, Wasternack C (2000) Octadecanoid-derived alteration of gene expression and the oxylipin signature in stressed barley leaves. Implications for different signaling pathways. Plant Physiol 123:177–188

Kumari GJ, Reddy AM, Naik ST, Kumar SG, Prasanthi J, Sriranganayakulu G, Reddy PC, Sudhakar C (2006) Jasmonic acid induced changes in protein pattern, antioxidative enzyme activities and peroxidase isozymes in peanut seedlings. Biol Planta 50:219–226

Laurie S, Feeney KA, Maathuis FJM, Heard PJ, Brown SJ, Leigh RA (2002) A role for HTK1 in sodium uptake by wheat roots. Plant J 32:139–149

Lawlor DW, Cornic G (2002) Photosynthetic carbon assimilation and associated metabolism in relation to water deficits in higher plants. Plant Cell Environ 25:275–294

Lee DS, Nioche P, Hamberg M, Raman CS (2008) Structural insights into the evolutionary paths of oxylipin biosynthetic enzymes. Nature 455:363–68

Leon J, Shulaev V, Yalpani N, Lawton MA, Raskin I (1995) Benzoic acid 2- hydroxylase, a soluble oxygenase from tobacco, catalyzes salicylic acid biosynthesis. Proc Natl Acad Sci USA 92:10413–10417

Li ZR, Wakao S, Fischer BB, Niyogi KK (2009) Sensing and responding to excess light. Annu Rev Plant Biol 60:239–260

Liu J, Zhu JK (1998) A calcium sensor homolog required for plant salt tolerance. Science 280:1943–1945

Liu J, Ishitani M, Halfter U, Kim CS, Zhu JK (2000) The Arabidopsis thaliana SOS2 gene encodes a protein kinase that is required for salt tolerance. Proc Natl Acad Sci USA 97:3730–3734

Lopez-Gomez E, Sanjuán MA, Diaz-Vivancos P, Mataix Beneyto J, García-Legaz MF, Hernández JA (2007) Effect of salinity and rootstocks on antioxidant systems of loquat plants (Eriobotrya japonica Lindl.): response to supplementary boron addition. Environ Exp Bot 160:151–158

Luo JP, Jiang ST, Pan LJ (2001) Enhanced somatic embryogenesis by salicylic acid of Astragalus adsurgens Pall.: relationship with H_2O_2 production and H2O2-metabolising enzyme activities. Plant Science 161:125–132

Mass EV, Grieve CM (1987) Sodium induced calcium deficiency in salt-stressed corn. Plant Cell Environ. 10:559–564

Manchanda G, Garg N (2008) Salinity and its effect on the functional biology of legumes. Acta Physiol Plant 30:595–618

Meneguzzo S, Navarri-Izzo F, Izzo R (1999) Antioxidative responses of shoots and roots of wheat to increasing NaCl concentrations. J Plant Physiol 155:274–280

Menezes-Benavente L, Teixeira FK, Kamei CLV, Pinheiro MM (2004) Salt stress induces altered expression of genes encoding antioxidant enzymes in seedlings of a Brazilian indica rice (*Oryza sativa* L.). Plant Sci 166:323–331

Meyer Y, Siala W, Bashandy T (2008) Glutaredoxins and thioredoxins in plants. Biochim Biophys Acta 1783:589–600

Miller G, Suzuki N, Ciftci-Yilmazi N, Mittler R (2010) Reactive oxygen species homeostasis and signaling during drought and salinity stresses. Plant Cell Environ 33:453–467

Moons A, Prinsen E, Bauw G, Van Montagu M (1997) Antagonistic effects of abscisic acid and jasmonates on salt stress-inducible transcripts in rice roots. Plant Cell 9:2243–2259

Moharekar ST, Lokhande SD, Hara T, Tanaka R, Tanaka A, Chavan PD (2003) Effect of salicylic acid on chlorophyll and carotenoid contents of wheat and moong seedlings. Photosynthetica 41:315–317

Moreno PRH, Van der Heijden R, Verpoorte R (1994) Elicitormediated induction of isochorismate synthase and accumulation of 2,3-dihydroxybenzoic acid in *Catharanthus roseus* cell suspension and shoot cultures. Plant Cell Rep 14:188–191

Mulholland BJ, Taylor IB, Jackson AC, Thompson AJ (2003) Can ABA mediate responses of salinity stressed tomato. Environ Exp Bot 50:17–28

Munns R (2002) Comparative physiology of salt and water stress. Plant Cell Environ 25:239–250

Munns R, James RA, Lauchli A (2006) Approaches to increasing the salt tolerance of wheat and other cereals. J Exp Bot 57:1025–1043

Munns R, Tester M (2008) Mechanisms of salinity tolerance. Annu Rev Plant Biol 59:651–681

Mustafa NR, Kim HK, Choi YH, Erkelens C, Lefeber AWM, Spijksma G, Van der Heijden R, Verpoorte R (2009) Biosynthesis of salicylic acid in fungus elicited *Catharanthus roseus* cells. Phytochemistry 70:532–539

Navrot N, Rouhier E, Gelhaye JJP (2007) Reactive oxygen species generation and antioxidant systems in plant mitochondria. Physiol Plant 129:85–195

Nazar R, Iqbal N, Syeed S, Khan NA (2011) Salicylic acid alleviates decreases in photosynthesis under salt stress by enhancing nitrogen and sulfur assimilation and antioxidant metabolism differentially in two mungbean cultivars. J Plant Physiol 168:807–815.

Ndamukong I, Abdallat AA, Thurow C, Fode B, Zander M, Weigel R, Gatz C (2007) SA-inducible Arabidopsis glutaredoxin interacts with TGA factors and suppresses JA-responsive PDF1.2 transcription. Plant J 50:128–139

Niki T, Mitsuhara I, Seo S, Ohtsubo N, Ohashi YS (1998) Antagonistic effect of salicylic acid and jasmonic acid on the expression of pathogenesis-related (PR) protein genes in wounded mature tobacco leaves. Plant Cell Physiol 39:500–507

Parida AK, Das AB (2005) Salt tolerance and salinity effects on plants: a review. Ecotoxicol Environ Saf 60:324–349

Parra-Lobato MC, Fernandez-Garcia N, Olmos E, Alvarez-Tinauta MC, Gomez-Jimeneza MC (2009) Methyl jasmonate-induced antioxidant defence in root apoplast from sunflower seedlings. Environ Exp Bot 66:9–17

Parvaiz A, Satyawati S (2008) Salt stress and phyto-biochemical responses of plants – a review. Plant Soil Environ 54:89–99

Pedranzani H, Racagni G, Alemano S, Miersch O, Ramirez I, Pena-Cortes H, Taleisnik E, Machado-Domenech E, Abdala G (2003) Salt tolerant tomato plants show increased levels of jasmonic acid. Plant Growth Regul 41:149–158

Pieterse CM, van Wees SC, van Pelt JA, Knoester M, Laan R, Gerrits H, Weisbeek PJ, van Loon LC (1998) A novel signaling pathway controlling induced systemic resistance in Arabidopsis. Plant Cell 10:1571–1580

Popova L, Ananieva E, Hristova V, Christov K, Georgieva K, Alexieva V, Stoinova ZH (2003) Salicylic acid and methyl jasmonate-induced protection on photosynthesis to paraquat oxidative stress. Bulg J Plant Physiol 2003:133–152

Prasad TK, Anderson MD, Martin BA, Stewart CR (1994) Evidence for chilling induced oxidative stress in maize seedlings and a regulatory role of H_2O_2. Plant Cell 6:65–74

Rus A, Yokoi S, Sharkhuu A, Reddy M, Lee BH, Matsumoto TK, Koiwa H, Zhu JK, Bressan RA, Hasegawa PM (2001) AtHKT1 is a salt tolerance determinant that controls Na^+ entry into plant roots. Proc Natl Acad Sci USA 98:14150–14155

Sakhabutdinova AR, Fatkhutdinova DR, Bezrukova MV, Shakirova FM (2003) Salicylic acid prevents the damaging action of stress factors on wheat plants. Bulg J Plant Physiol (Spl Issue):314–319

Sanchez-Casas P, Klessig DF (1994) A salicylic acid-binding activity and a salicylic acid-inhibitable catalase activity are present in a variety of plant species. Plant Physiol 106:1675–1679

Sanders D (2000) Plant biology: the salty tale of Arabidopsis. Curr Biol 10:486–488

Sanders PM, Lee PY, Biesgen C, Boone JD, Beals TP (2000) The Arabidopsis delayed dehiscence1 gene encodes an enzyme in the jasmonic acid synthesis pathway. Plant Cell 12:1041–1062

Sawada H, Shim IS, Usui K (2006) Induction of benzoic acid 2- hydroxylase and salicylic acid biosynthesis Modulation by salt stress in rice seedlings. Plant Sci 171:263–270

Seckin B, Sekmen AH, Turkan I (2009) An enhancing effect of exogenous mannitol on the antioxidant enzyme activities in roots of wheat under salt stress. J Plant Growth Regul 28:12–20

Sembdner G, Parthier B (1993) The biochemistry and physiology and molecular actions of jasmonates. Annu Rev Plant Physiol Plant Mol Biol 44:569–586

Sharma N, Abrams SR, Waterer DR (2005) Uptake, movement, activity, and persistence of an abscisic acid analog (80 acetylene ABA methyl ester) in marigold and tomato. J Plant Growth Regul 24:28–35

Shaterian J, Waterer D, De Jong H, Tanino KK (2005) Differential stress responses to NaCl salt application in early- and late maturing diploid potato (Solanum sp.) clones. Environ Exp Bot 54:202–212

Silva C, Martínez V, Carvajal M (2008) Osmotic versus toxic effects of NaCl on pepper plants. Biol Plant 52:72–79

Spoel SH, Koornneef A, Claessens SMC, Korzelius JP, van Pelt JA, Mueller MJ, Buchala AJ, Metraux J, Brown R, Kazan K (2003) NPR1 modulates cross-talk between salicylate- and jasmonate-dependent defense pathways through a novel function in the cytosol. Plant Cell 15:760–770

Steduto P, Albrizio R, Giorio P, Sorrentino G (2000) Gas exchange response and stomatal and non-stomatal limitations to carbon assimilation of sunflower under salinity. Environ Exp Bot 44:243–55

Stintzi A, Browse J (2000) The Arabidopsis male-sterile mutant, opr3, lacks the 12-oxophytodienoic acid reductase required for jasmonate synthesis. Proc Natl Acad Sci USA 97:10625–10630

Syeed S, Anjum NA, Nazar R, Iqbal N, Masood A, Khan NA (2011) Salicylic acid-mediated changes in photosynthesis, nutrients content and antioxidant metabolism in two mustard (*Brassica juncea* L.) cultivars differing in salt tolerance. Acta Physiol Plant 33:877–886

Tabur S, Demir K (2010) Role of some growth regulators on cytogenetic activity of barley under salt stress. Plant Growth Regul 60:99–104

Thaler JS (1999) Induced resistance in agricultural crops: effects of jasmonic acid on herbivory and yield in tomato plants. Environ Entomol 28:30–37

Thaler JS, Fidantsef AL, Duffey SS, Bostock RM (1999) Trade-offs in plant defense against pathogens and herbivores: a field demonstration of chemical elicitors of induced resistance. J Chem Ecol 25:1597–1609

Tseng MJ, Liu CW, Yiu JC (2007) Enhanced tolerance to sulfur dioxide and salt stress of transgenic Chinese cabbage plants expressing both superoxide dismutase and catalase in chloroplasts. Plant Physiol Biochem 45:822–833

Tsonev TD, Lazova GN, Stoinova ZG, Popova LP (1998) A possible role for jasmonic acid in adaptation of barley seedlings to salinity stress. J Plant Growth Regul 17:153–159

Turkan I, Demiral T (2009) Recent developments in understanding salinity tolerance. Environ Exp Bot 1:2–9
Tuteja N, Sopory SK (2008) Chemical signaling under abiotic stress environment in plants. Plant Signal Behav 3:525–536
Unno H, Maeda Y, Yamamoto S, Okamoto M, Takenaga H (2002) Relationship between salt tolerance and Ca2+ retention among plant species. Jpn J Soil Sci Plant Nutr 73:715–718
Van Breusegem F, Vranova E, Dat JF, Inze D (2001) The role of active oxygen species in plant signal transduction. Plant Sci 161:405–414
Van Loon LC, Rep M, Pieterse CMJ (2006) Significance of inducible defense-related proteins in infected plants. Annu Rev Phytopathol 44:135–162
Velitcukova M, Fedina I (1998) Response of photosynthesis of Pisum sativum to salt stress as affected by methyl jasmonate. Photosynthetica 35:89–97
Verberne MC, Budi Muljono RA, Verpoorte R (1999) Salicylic acid biosynthesis. In: Hooykaas PPJ, Hall MA, Libbenga KR (eds) Biochemistry and molecular biology of plant hormones. Elsevier Science, Amsterdam
Vick BA, Zimmerman DC (1983) The biosynthesis of jasmonic acid: a physiological role for plant lipoxygenase. Biochem Biophys Res Commun 111:470–477
Wahid A, Perveen M, Gelani S, Basra SMA (2007) Pretreatment of seed with H_2O_2 improves salt tolerance of wheat seedlings by alleviation of oxidative damage and expression of stress proteins. J Plant Physiol 164:283–294
Wang X (1999) The role of phospholipase D in signaling cascades. Plant Physiol 120:645–651
Wang Y, Mopper S, Hasentein KH (2001) Effects of salinity on endogenous ABA, IAA, JA, and SA in Iris hexagona. J Chem Ecol 27:327–342
Wasternack C, Hause B (2002) Jasmonates and octadecanoids: signals in plant stress responses and plant development. Prog Nucleic Acid Res Mol Biol 72:165–221
Wasternack C, Parthier B (1997) Jasmonate-signalled plant gene expression. Trends Plant Sci 2:302–307
Weber H, Vick BA, Farmer EE (1997) Dinor-oxo-phytodienoic acid: a new hexadecanoid signal in the jasmonate family. Proc Natl Acad Sci USA 94:10473–10478
Wenxue W, Bilsborrow PE, Hooley P, Fincham DA, Lombi E, Forster BP (2003) Salinity induced difference in growth, ion distribution and partitioning in barley between the cultivar Maythorpe and its derived mutant Golden Promise. Plant Soil 250:183–191
Wong CE, Li Y, Labbe A, Guevara D, Nuin P, Whitty B (2006) Transcriptional profiling implicates novel interactions between abiotic stress and hormonal responses in Thellungiella, a close relative of Arabidopsis. Plant Physiol 140:1437–1450
Wildermuth MC, Dewdney J, Wu G, Ausubel FM (2001) Isochorismate synthase is required to synthesize salicylic acid for plant defence. Nature 414:562–565
Wilen RW, Ewan BE, Gusta LV (1994) Interaction of abscisic acid and jasmonic acid on the inhibition of seed germination and the induction of freezing tolerance. Can J Bot 72:1009–1017
Xiong L, Yang Y (2003) Disease resistance and abiotic stress tolerance in rice are inversely modulated by an abscisic acid-inducible mitogen activated protein kinase. Plant Cell 15:745–759
Yalpani N, Leôn J, Lawton MA, Raskin I (1993) Pathway of salicylic acid biosynthesis in healthy and virus-inoculated tobacco. Plant Physiol 103:315–321
Yokoi S, Quintero FJ, Cubero B, Ruiz MT, Bressan RA, Hasegawa PM, Pardo JM (2002) Differential expression and function of Arabidopsis thaliana NHX Na+/H + antiporters in the salt stress response. Plant J 30:529–539
Yoon JY, Hamayun M, Lee SK, Lee IJ (2009) Methyl jasmonate alleviated salinity stress in soybean. J Crop Sci Biotechnol 12:63–68
Zhang S, Klessig D (1997) Salicylic acid activates a 48-kD MAP kinase in tobacco. Plant Cell 9:809–824

Zhang S, Klessig DF (2001) MAPK cascades in plant defense signaling. Trends Plant Sci 11:520–527

Zholkevich VN, Pustovoytova TN (1993) The role of *Cucumis sativum* L leaves and content of phytohormones under soil drought. Russ J Plant Physiol 40:676–680

Zhu J-K, Liu J, Xiong L (1998) Genetic analysis of salt tolerance in Arabidopsis thaliana: evidence of a critical role for potassium nutrition. The Plant Cell 10:1181–1192

Zhou J, Wang X, Jiao Y, Qin Y, Liu X, He K (2007) Global genome expression analysis of rice in response to drought and high-salinity stresses in shoot, flag leaf, and panicle. Plant Mol Biol 63:591–608

Zhu JK (2000) Genetic analysis of plant salt tolerance using Arabidopsis. Plant Physiol 124:941–948

Zhu JK (2001) Plant salt tolerance. Trends Plant Sci 6:66–71

Ziegler J, Stenzel I, Hause B, Maucher H, Hamberg M (2000) Molecular cloning of allene oxide cyclase. The enzyme establishing the stereochemistry of octadecanoids and jasmonates. J Biol Chem 275:19132–19138

Index

N.A. Khan et al. (eds.), *Phytohormones and Abiotic Stress Tolerance in Plants*,
DOI 10.1007/978-3-642-25829-9, © Springer-Verlag Berlin Heidelberg 2012

O

P

R

S